LONDON MATHEMATICAL SOCIETY STUDENT TEXTS

Managing editor: Professor E.B. Davies, Department of Mathematics, King's College, Strand, London WC2R 2LS

1 Introduction to combinators and λ-calculus, J.R. HINDLEY & J.P. SELDIN
2 Building models by games, WILFRID HODGES
3 Local fields, J.W.S. CASSELS
4 An introduction to twistor theory, S.A. HUGGETT & K.P. TOD
5 Introduction to general relativity, L. HUGHSTON & K.P. TOD
6 Lectures on stochastic analysis: diffusion theory, DANIEL W. STROOCK
7 The theory of evolution and dynamical systems, J. HOFBAUER & K. SIGMUND
8 Summing and nuclear norms in Banach space theory, G.J.O. JAMESON
9 Automorphisms of surfaces after Nielsen and Thurston, A.CASSON & S. BLEILER
10 Nonstandard analysis and its applications, N.CUTLAND (ed)
11 Spacetime and singularities, G. NABER
12 Undergraduate algebraic geometry, MILES REID
13 An introduction to Hankel operators, J.R. PARTINGTON
16 An introduction to noncommutative noetherian rings, K.R.GOODEARL & R.B.WARFIELD,JR.
17 Aspects of quantum field theory in curved space-time, S.A. FULLING

London Mathematical Society Student Texts. 16

An Introduction to Noncommutative Noetherian Rings

K.R.GOODEARL
University of Utah
R.B.WARFIELD,JR.
University of Washington

CAMBRIDGE UNIVERSITY PRESS
Cambridge
New York Port Chester Melbourne Sydney

Published by the Press Syndicate of the University of Cambridge
The Pitt Building, Trumpington Street, Cambridge CB2 1RP
40 West 20th Street, New York, NY 10011, USA
10, Stamford Road, Oakleigh, Melbourne 3166, Australia

First published 1989

Printed in Great Britain at the University Press, Cambridge

Library of Congress cataloguing in publication data: available

British Library cataloguing in publication data: available

ISBN 0 521 36086 2 Hardcover
ISBN 0 521 36925 8 Paperback

CONTENTS

INTRODUCTION

Noncommutative noetherian rings are presently the subject of very active research. Recently the theory has attracted particular interest due to its applications in related areas, especially the representation theories of groups and Lie algebras. We find the subject of noetherian rings an exciting one, for its own sake as well as for its applications, and our primary purpose in writing this volume was to attract more participants into the area.

This book is an introduction to the subject intended for anyone who is potentially interested, but primarily for students who are at the level which in the United States corresponds to having completed one year of graduate study. Since the topics included in an American first year graduate course vary considerably, and since those in analogous courses in other countries (e.g., third year undergraduate or M.Sc. courses in Britain) vary even more, we have attempted to minimize the actual prerequisites in terms of material, by reviewing some topics that many readers may already have in their repertoires. More importantly, we have concentrated on developing the basic tools of the subject, in order to familiarize the student with current methodology. Thus we focus on results which can be proved from a common point of view and steer away from miraculous arguments which can be used only once. In this spirit, our treatment is deliberately not encyclopedic, but is rather aimed at what we see as the major threads and key topics of current interest.

It is our hope that this book can be read by a student without the benefit of a course or an instructor. To encourage this possibility, we have tried to include details when they might have been omitted, and to discuss the motivations for proceeding as we do. Moreover, we have woven an extensive selection of exercises into the text. These exercises are particularly designed to give the novice some experience and familiarity with both the material and the tools being developed.

One of the fundamental differences between the theories of commutative and noncommutative rings is that the former arise naturally as rings of functions, whereas the latter arise naturally as rings of operators. For example, early in the twentieth century, some of the first noncommutative rings that received serious study were certain rings of differential operators. More generally, given any set of linear transformations of a vector space, we can form the algebra of linear transformations generated by this set, and many problems of interest concerning the original transformations become module–theoretic questions, where we view the original vector space as a module over the algebra we have created. In many modern applications, in turn, it is essential to regard noncommutative rings as rings of transformations or operators of various kinds. We are partial to this point of view. This has led us to emphasize the role of modules when studying a ring, for modules are simply ways of representing the ring at issue in terms of endomorphisms of abelian groups. Also, when defining a ring we have tended to present it as a ring of operators of some sort rather than by taking a more formal approach, such as giving generators and relations. For example, when constructing rings of fractions, we have preferred to find them as rings of endomorphisms rather than as sets of equivalence classes of ordered pairs of elements.

Although the noetherian condition is very natural in commutative ring theory, since it holds for the rings of integers in algebraic number fields and for the coordinate rings crucial to algebraic geometry, it was originally less clear that this condition would be useful in the noncommutative setting. For instance, Jacobson's definitive book of 1956 makes only minimal mention of noetherian rings. Similarly, prime ideals, essential in the commutative theory, seemed to have relatively less importance for noncommutative rings; in fact, because of the fundamental role of representation–theoretic ideas in the development of the noncommutative theory, the initial emphasis in the subject was almost exclusively on irreducible representations (i.e., simple modules) and primitive ideals (i.e., annihilators of irreducible representations). In the meantime, however, it has turned out that various important types of noncommutative rings – in particular, certain infinite group rings and the enveloping algebras of finite–dimensional Lie algebras – are in fact noetherian. This has been used to good effect in recent work on the representation theory of the corresponding groups and Lie algebras, just as the theory of finite–dimensional algebras and artinian rings has played a key role in research on the representations of finite groups. Also, as soon as noetherian rings and their modules received serious attention, prime ideals forced themselves into the picture, even in contexts where the original interest had been entirely in primitive ideals. As a consequence, we have made prime ideals a major theme in our text.

The first important result in the theory of noncommutative noetherian rings was proved relatively recently, in 1958. This was Goldie's Theorem, which gives an analog of a field of fractions for factor rings R/P where R is a noetherian ring and P a prime ideal of R. Once this milestone had been reached, noetherian ring theory proceeded apace, partly from its own impetus and partly through feedback from neighboring areas in which noetherian ideas found applications. One of our aims in this book has been to develop those aspects of the theory of noetherian rings which have the strongest connections with the representation–theoretic areas to which we have alluded. However, as these areas have their own extensive theories, it was impossible to treat them in any generality in this volume. Instead, we present a brief discussion in the prologue, giving some representative examples to which the theory in the text can be applied relatively directly, without extensive side trips into technical intricacies.

To give the reader an idea of the historical sources of the theory, we have included some bibliographical notes at the end of each chapter. We have sought to make these notes as accurate as possible, but as with any evolving theory complete precision is difficult to attain, especially since in many research papers sources are not well documented. Some inaccuracies are thus probably inevitable, and we apologize in advance for any that may have occurred.

In an appendix we discuss some open problems in noetherian ring theory; we hope that our readers will be stimulated to solve them.

For helpful comments on various drafts of the book, we would like to thank A. D. Bell, K. A. Brown, D. A. Jordan, T. H. Lenagan, P. Perkins, L. W. Small, and J. T. Stafford. We would also like to thank our competitors J. C. McConnell and J. C. Robson for letting us see early drafts of various chapters from their noetherian rings book [1987].

PROLOGUE

Since much of the current interest in noncommutative noetherian rings stems from applications of the general theory to several specific types, we present here a very sketchy introduction to four major areas of application: polynomial identity rings, group algebras, rings of differential operators and enveloping algebras. Each of these areas has a very extensive theory of its own, far too voluminous to be incorporated into a book of this size. (See for instance Rowen [1980], Passman [1977], and McConnell–Robson [1987].) Instead, we shall concentrate on surrogates – some classes of rings which are either simple prototypes or analogs of the four major types just mentioned, which we can investigate by relatively direct methods while still exhibiting the flavor of the areas they represent. These surrogates are module–finite algebras over commutative rings (for polynomial identity rings), skew–Laurent rings (for group algebras), and formal differential operator rings (for both rings of differential operators and enveloping algebras). They will be introduced below and studied in greater detail in the following chapter.

We will conclude the prologue with a few comments about our notation and terminology.

♦ POLYNOMIAL IDENTITY RINGS

Commutativity in a ring may be phrased in terms of a relation that holds identically, namely $xy - yx = 0$ for all choices of x and y from the ring. More complicated identical relations sometimes also hold in noncommutative rings. For example, if x and y are any 2×2 matrices over a commutative ring S, then the trace of $xy - yx$ is zero, and so it follows from the Cayley–Hamilton Theorem that $(xy - yx)^2$ is a scalar matrix. Consequently, $(xy - yx)^2$ commutes with every 2×2 matrix z, and hence the relation

$$(xy - yx)^2 z - z(xy - yx)^2 = 0$$

holds for all choices of x,y,z from the matrix ring $M_2(S)$. A much deeper result, the Amitsur–Levitzki Theorem, asserts that for all choices of $2n$ matrices $x_1,...,x_{2n}$ from the $n \times n$ matrix ring $M_n(S)$,

$$\sum_{\sigma \in S_{2n}} \operatorname{sgn}(\sigma) x_{\sigma(1)} x_{\sigma(2)} \cdots x_{\sigma(2n)} = 0,$$

where S_{2n} is the symmetric group on $\{1,2,...,2n\}$ and $\operatorname{sgn}(\sigma)$ denotes the sign of a permutation σ (namely $+1$ or -1 depending on whether σ is even or odd).

Such an "identical relation" on a ring may be thought of as saying that a certain polynomial – with noncommuting variables! – vanishes identically on the ring. In this context, the polynomials are usually restricted to having integer coefficients. Thus a *polynomial identity* on a ring R is a polynomial $p(x_1,...,x_n)$ in noncommuting variables $x_1,...,x_n$ with coefficients from $\mathbb{Z}$ such that $p(r_1,...,r_n) = 0$ for all $r_1,...,r_n \in R$. A *polynomial identity ring*, or *P.I. ring* for short, is a ring R which satisfies some *monic* polynomial identity $p(x_1,...,x_n)$ (that is, among the monomials of highest total degree which appear in p, at least one has coefficient 1).

The Amitsur–Levitzki Theorem implies that every matrix ring over a commutative ring is a P.I. ring, and consequently so is every factor ring of a subring of such a matrix ring. For example, the endomorphism ring of a finitely generated module A over a commutative ring S has this form. To see that, identify A with S^n/K for some $n \in \mathbb{N}$ and some submodule K of S^n, and identify the matrix ring $M_n(S)$ with the endomorphism ring of S^n. Then the set

$$T = \{f \in M_n(S) \mid f(K) \leq K\}$$

is a subring of $M_n(S)$, the set $I = \{f \in M_n(S) \mid f(S^n) \leq K\}$ is an ideal of T, and $T/I \cong \operatorname{End}_S(A)$. Therefore $\operatorname{End}_S(A)$ is a P.I. ring.

Certain algebras over commutative rings fit naturally into this context. Recall that an *algebra* over a commutative ring S is just a ring R equipped with a specified ring homomorphism ϕ from S to the center of R. (The map ϕ is not assumed to be injective.) Then ϕ is used to define products of elements of S with elements of R: for $s \in S$ and $r \in R$, we set sr and rs equal to $\phi(s)r$ (or $r\phi(s)$, which is the same because $\phi(s)$ is in the center of R). Using this product, we can view R as an S–module. We say that R is a *module–finite* S–algebra if R is a finitely generated S–module. Note that $R \cong \operatorname{End}_R(R_R) \subseteq \operatorname{End}_S(R)$ as rings, and so any polynomial identity satisfied in $\operatorname{End}_S(R)$ will also be satisfied in R. Taking the preceding paragraph into account, we conclude that any module–finite algebra over a commutative ring is a P.I. ring.

The class of module–finite algebras over commutative noetherian rings provides us with a supply of prototypical examples of noetherian P.I. rings. To illustrate some applications of the noetherian theory to P.I. rings, we shall at times work out consequences of the former for our class of examples. In this setting, we will be able to replace P.I. theory by some much more direct methods from commutative ring theory.

♦ GROUP ALGEBRAS

One of the earliest stimuli to the modern development of noncommutative ring theory came from the study of *group representations*. The key idea was to study a group G by "representing" it in terms of linear transformations on a vector space V, namely by studying a group homomorphism ϕ from G to the group of invertible linear transformations on V. Linear algebra can then by used to study the group $\phi(G)$, and the information gleaned can be pulled back to G via the *representation* ϕ. Using ϕ, there is an "action" of G on V, namely a product $G \times V \to V$ given by the rule $g \cdot v = \phi(g)(v)$, and since ϕ is a homomorphism, $(gh) \cdot v = g \cdot (h \cdot v)$ for all $g,h \in G$ and $v \in V$. This looks a lot like a module multiplication, if we ignore the lack of an addition for elements of G, and in fact V is called a *G–module* in this situation.

To make V into an actual module over a ring, we build G and its multiplication into a ring, along with whichever field k we are using for scalars. Just make up a vector space with a basis which is in one–to–one correspondence with the elements of G, identify each element of G with the corresponding basis element, and then extend the multiplication from G to this vector space linearly:

$$\left(\sum_{g \in G} \alpha_g g\right)\left(\sum_{h \in G} \beta_h h\right) = \sum_{g,h \in G} (\alpha_g \beta_h)(gh).$$

The result is a k–algebra called the *group algebra* of G over k, denoted $k[G]$ or just kG. Except for the obvious changes in terminology, $k[G]$–modules are the same as representations of G on vector spaces over k.

In the case of a finite group G, the group algebra $k[G]$ is finite–dimensional, and the theory of finite–dimensional algebras has much to say about representations of G. A noetherian group algebra is known to occur when G is *polycyclic–by–finite*, that is, when G has a series of subgroups

$$(1) = G_0 \subset G_1 \subset \dots \subset G_n \subseteq G_{n+1} = G$$

such that each G_{i-1} is a normal subgroup of G_i and G_i/G_{i-1} is infinite cyclic for $i = 1,\dots,n$, while G/G_n is finite. (It is an open problem whether $k[G]$ is noetherian only when G is polycyclic–by–finite.) One of the simplest infinite non–abelian examples is the group G with two generators x,y and the sole relation $yxy^{-1} = x^{-1}$. In this case elements of $k[G]$ can all be put in the form $\sum_{i=-n}^{n} p_i(x)y^i$ where each $p_i(x)$ is a Laurent polynomial (i.e., a polynomial in x and x^{-1}). From the relation $yxy^{-1} = x^{-1}$ it follows that $yp(x)y^{-1} = p(x^{-1})$ for all Laurent polynomials $p(x)$. Hence, the Laurent polynomial ring $k[x,x^{-1}]$ is sent into itself by the map $p(x) \mapsto yp(x)y^{-1}$, and this map coincides with the map $p(x) \mapsto p(x^{-1})$, which is an automorphism of $k[x,x^{-1}]$.

The pattern of this example suggests a construction that starts with a ring R and an automorphism α of R, and then builds a ring T whose elements look like Laurent polynomials over R in a new indeterminate y, except that instead of commuting with y, elements $r \in R$ satisfy the relation $yry^{-1} = \alpha(r)$, or $yr = \alpha(r)y$. Since the usual multiplication of polynomials has been "skewed" through α, the ring T is called a *skew–Laurent ring*. Thus the group algebra of the previously discussed group with the relation $yxy^{-1} = x^{-1}$ may be viewed as a skew–Laurent ring with coefficient ring $k[x,x^{-1}]$.

We shall see that any skew–Laurent ring with a noetherian coefficient ring is itself noetherian. This fact actually provides the method used to show that the group algebra of any polycyclic–by–finite group G is noetherian. Namely, if

$$(1) = G_0 \subset G_1 \subset \ldots \subset G_n \subseteq G_{n+1} = G$$

is the series of subgroups of G occurring in the definition of "polycyclic–by–finite", it can be shown that for $i = 1,\ldots,n$ the group algebra $k[G_i]$ is isomorphic to a skew–Laurent ring whose coefficient ring is $k[G_{i-1}]$. Starting at the bottom with $k[G_0] = k$, it follows immediately by induction that $k[G_n]$ is noetherian. It then just remains to observe that $k[G]$ is a finitely generated right or left module over $k[G_n]$ to conclude that $k[G]$ itself is noetherian. In particular, we see from this discussion that (iterated) skew–Laurent rings are a better match for group algebras of polycyclic–by–finite groups than might have been suggested by the very special example given above.

♦ RINGS OF DIFFERENTIAL OPERATORS

Another early stimulus to noncommutative ring theory came from the study of differential equations. Late in the nineteenth century, it was realized that, just as polynomial functions provide a useful means of dealing with algebraic equations, "differential operators" are convenient for handling linear differential equations. For example, a homogeneous linear differential equation

$$a_n(x)y^{(n)} + a_{n-1}(x)y^{(n-1)} + \ldots + a_1(x)y' + a_0(x)y = 0$$

can be rewritten very compactly as $d(y) = 0$, where d denotes the *linear differential operator*

$$a_n(x)\frac{d^n}{dx^n} + a_{n-1}(x)\frac{d^{n-1}}{dx^{n-1}} + \ldots + a_1(x)\frac{d}{dx} + a_0(x).$$

From this viewpoint, d is a linear transformation on some vector space of functions, and the solution space of the original differential equation is just the null space of d.

To be a bit more specific, let us consider the special case in which coefficients and solutions are real–valued rational functions. Then our differential operators are $\mathbb{R}$–linear transformations on the field $\mathbb{R}(x)$. The composition of two differential operators is certainly a linear transformation, but it takes a minute to see that such a composition is actually another differential operator. In order to make the notation more convenient, we use the symbol D to denote the operator d/dx. If we form the operator composition Da, which means "first multiply by the function a(x) and then differentiate", we see that

$$(Da)(y) = (ay)' = ay' + a'y = aD(y) + a'y$$

for any function y, and so $Da = aD + a'$. Iterated use of this identity then allows us to write the composition of any two differential operators in the standard form of a differential operator, i.e., as a sum of terms a_iD^i where $a_i \in \mathbb{R}(x)$. Thus the collection of differential operators on $\mathbb{R}(x)$ forms a ring, which is sometimes denoted $B_1(\mathbb{R})$. We may think of $B_1(\mathbb{R})$ as a polynomial ring $\mathbb{R}(x)[D]$, in which, however, the multiplication is twisted to make a noncommutative ring. This ring attracted particular attention early in the twentieth century, when it was proved that it is a principal ideal domain (that is, all left and right ideals are principal) and that it satisfies a form of unique factorization.

We can of course proceed in the same way using for coefficients other rings of functions which are closed under differentiation. For example, if we start with the ring $\mathbb{C}[x]$ of complex polynomials, the ring of differential operators we obtain looks like a twisted polynomial ring in two variables, $\mathbb{C}[x][D]$. This ring is called the *first complex Weyl algebra,* and is denoted $A_1(\mathbb{C})$. More generally, we may start with a polynomial ring $\mathbb{C}[x_1,...,x_n]$ in several variables and build differential operators using the partial derivatives $\partial/\partial x_i$, abbreviated D_i. This results in a twisted polynomial ring in $2n$ variables, $\mathbb{C}[x_1,...,x_n][D_1,...,D_n]$, which is called the nth *complex Weyl algebra* and is denoted $A_n(\mathbb{C})$.

Examples such as $B_1(\mathbb{R})$ and $A_n(\mathbb{C})$, which will often recur in the text, can be taken as representative of a more general class that has assumed some importance in recent years: rings of differential operators on algebraic varieties. We cannot discuss these in detail, but will content ourselves with indicating how they can be decribed. We recall that a *complex algebraic variety* is a subset V of $\mathbb{C}^n$ which is the set of common zeroes of some collection I of polynomials in $\mathbb{C}[x_1,...,x_n]$. If I contains all the polynomials that vanish on V, then I is an ideal in the polynomial ring, and the factor ring $R = \mathbb{C}[x_1,...,x_n]/I$ is the *coordinate ring of* V. The *ring of differential operators on* V, denoted $\mathscr{D}(V)$, consists of those differential operators on $\mathbb{C}[x_1,...,x_n]$ that induce operators on R, modulo those that induce the zero operator on R. More precisely, the set

$$S = \{s \in A_n(\mathbb{C}) \mid s(I) \subseteq I\}$$

is a subring of $A_n(\mathbb{C})$, the set

$$J = \{s \in A_n(\mathbb{C}) \mid s(\mathbb{C}[x_1,...,x_n]) \subseteq I\}$$

is an ideal of S, and $\mathscr{D}(V) = S/J$. It has been proved that $\mathscr{D}(V)$ is noetherian in case V has no singularities, and in case V is a curve, but it appears that for higher–dimensional varieties with singularities $\mathscr{D}(V)$ is usually not noetherian.

♦ ENVELOPING ALGEBRAS

A *Lie algebra* over a field k is a vector space L over k equipped with a non–

associative product $[\cdot\cdot]$ satisfying the usual distributive laws as well as the rules

$$[xx] = 0 \qquad \text{and} \qquad [x[yz]] + [y[zx]] + [z[xy]] = 0$$

for all $x,y,z \in L$. For example, $\mathbb{R}^3$ equipped with the usual vector cross product is a real Lie algebra. The standard model for the product in a Lie algebra is the *additive commutator* operation $[x,y] = xy - yx$ in an associative ring (more precisely, any associative k–algebra when equipped with the operation $[\cdot,\cdot]$ becomes a Lie algebra over k). Conversely, starting with a Lie algebra L, one can build an associative k–algebra U(L) using the elements of L as generators, together with relations $xy - yx = [xy]$ for all $x,y \in L$. The algebra U(L) is called the *(universal) enveloping algebra* of L, and it is known to be noetherian in case L is finite–dimensional. (Whether it is possible for the enveloping algebra of an infinite–dimensional Lie algebra to be noetherian is an open problem.)

The simplest Lie algebra L with a nonzero product is 2–dimensional, with a basis $\{x,y\}$ such that $[yx] = x$. Elements of the enveloping algebra U(L) can in that case all be put into the form $\sum_{i=0}^{n} p_i(x)y^i$, where each $p_i(x)$ is an ordinary polynomial in the variable x. In U(L), the relation $[yx] = x$ becomes $[y,x] = x$, and from this it follows easily that $[y,p(x)] = x\frac{d}{dx}(p(x))$ for all polynomials $p(x)$. In other words, $[y,-]$ maps the polynomial ring $k[x]$ into itself, and its action on polynomials is given by the operator $x\frac{d}{dx}$. The reader should note that this is very similar to the ring $A_1(\mathbb{C})$ discussed above, the difference being that in $A_1(\mathbb{C})$ we have the relation $[D,p(x)] = \frac{d}{dx}(p(x))$. (In fact, U(L) in our example is isomorphic to the subalgebra of $A_1(k)$ generated by x and xD.)

Abstracting this pattern, we may start with a ring R and a map $\delta : R \to R$ which is a *derivation* (that is, δ is additive and satisfies the usual product rule for derivatives), and then build a larger ring T using an indeterminate y such that $[y,r] = \delta(r)$ for all $r \in R$. The elements of T look like differential operators $\sum r_i\delta^i$ on R, except that it may be possible for $\sum r_i\delta^i$ to be the zero operator without all the coefficients r_i being zero. Thus the elements $\sum r_iy^i$ in T are called *formal differential operators*, and T is called a *(formal) differential operator ring*.

We shall see that all formal differential operator rings with noetherian coefficient rings are themselves noetherian, and we shall view them as representative analogs of enveloping algebras. The analogy is actually a little better than one might think knowing only the single example mentioned above. Namely, if L is a finite–dimensional Lie algebra which can be realized as a Lie algebra of upper triangular matrices over k (using $[\cdot,\cdot]$ for the Lie product), then U(L) can be built as an *iterated differential operator ring* through a series of extensions

$$k = T_0 \subset T_1 \subset \ldots \subset T_m = U(L)$$

where each T_i is isomorphic to a differential operator ring with coefficients from T_{i-1}. (Over $\mathbb{C}$, the finite–dimensional Lie algebras which can be realized as upper triangular matrices are precisely the *solvable* Lie algebras.)

♦ NOTATION AND TERMINOLOGY

The background needed for this book is fairly standard, and may be found in most graduate level texts on algebra, such as Cohn [1974, 1977], Hungerford [1974], or Jacobson [1974, 1980]. The following short lists, giving some reference sources and some notation, are not meant to be exhaustive but to help keep the reader on track. We emphasize one convention – our rings, modules, and ring homomorphisms are assumed to be unital except in a few rare, specified cases. Also, all our homomorphisms and other functions are written on the left of their arguments, i.e., in the form $f(x)$.

The following are some references for frequently used notions in this book. These references contain more information than we actually need, and the reader to whom the italicized terms are all familiar should not feel it necessary to read through these references unless a particular problem arises. *Algebras.* Cohn [1977, § 3.2], Hungerford [1974, § 4.7], Jacobson [1974, § 7.1]. *Direct Sums and Products.* Cohn [1974, § 10.3; 1977, § 4.4], Hungerford [1974, §§ 3.2, 4.1], Jacobson [1974, § 3.5; 1980, § 3.4]. *Domains.* Cohn [1974, § 6.1], Jacobson [1974, § 2.2]. *Epimorphisms, Monomorphisms, and Isomorphisms.* Cohn [1977, § 4.1], Hungerford [1974, § 4.1], Jacobson [1980, § 1.2]. *Free and Projective Modules.* Cohn [1974, § 10.4; 1977, § 4.3], Hungerford [1974, §§ 4.2, 4.3], Jacobson [1974, § 3.4; 1980, §§ 1.7, 3.10]. *Indecomposable Rings and Modules.* Cohn [1977, § 4.4], Jacobson [1980, § 3.4]. *Independent Families of Submodules.* Cohn [1974, § 10.3], Hungerford [1974, § 9.4], Jacobson [1974, § 3.5; 1980, § 3.5]. *Opposite Rings.* Cohn [1974, §10.2], Hungerford [1974, § 7.1], Jacobson [1974, § 2.8]. *Ring Homomorphisms.* Cohn [1974, § 10.1], Hungerford [1974, § 3.1], Jacobson [1974, § 2.7]. *Sums of Submodules.* Cohn [1974, § 10.3], Hungerford [1974, §4.1], Jacobson [1974, § 3.3]. *Tensor Products.* Cohn [1977, §§ 3.1, 3.2], Hungerford [1974, § 4.5], Jacobson [1980, §§ 3.7, 3.9]. *Zorn's Lemma.* Cohn [1977, § 1.3], Hungerford [1974, § 0.7], Jacobson [1980, § 0.1].

$\subset$ and $\supset$	Proper inclusions.
$\mathbb{N}$	The set of natural numbers (i.e., positive integers).
$\mathbb{Z}^+$	The set of nonnegative integers.
$\mathbb{Z}$	The ring of integers.
$\mathbb{Q}$	The field of rational numbers.
$\mathbb{R}$	The field of real numbers.

$\mathbb{C}$	The field of complex numbers.
$\mathbb{H}$	The division ring of real quaternions.
A_R	A right module A over a ring R. (In case A can be considered as a module over several different rings, the notation A_R is used to indicate that A is being viewed as an R–module.)
${}_RA$	A left module A over a ring R.
R_R	A ring R viewed as a right module over itself.
${}_RR$	A ring R viewed as a left module over itself.
$\text{Hom}_R(A,B)$	The abelian group of all R–module homomorphisms from an R–module A to an R–module B.
$\text{End}_R(A)$	The ring of all R–module endomorphisms of an R–module A.
A^n or $\oplus^n A$	The direct sum of n copies of a module A. (For a right or left ideal A in a ring, $\oplus^n A$ is used to avoid confusion with multiplicative powers of A.)
I^n	The nth multiplicative power of a right or left ideal I in a ring (i.e., the set of all sums of n–fold products $i_1i_2 \cdots i_n$ where $i_1,\ldots,i_n \in I$).
IJ	The multiplicative product of right or left ideals I and J in a ring (i.e., the set of all sums of products ij where $i \in I$ and $j \in J$).

1 A FEW NOETHERIAN RINGS

After a review of the definition and basic properties of noetherian modules and rings, we introduce a few classes of examples of noetherian rings, which will serve to illustrate and support the later theory. We concentrate particularly on the "surrogate" examples outlined in the prologue, namely module–finite algebras over commutative rings, differential operator rings, and skew–Laurent rings.

♦ THE NOETHERIAN CONDITION

We begin with several basic equivalent conditions which are abbreviated by the adjective "noetherian", honoring E. Noether, who first demonstrated the importance and usefulness of these conditions. Recall that a collection $\mathcal{A}$ of subsets of a set A satisfies the *ascending chain condition* (or *ACC*) if there does not exist a properly ascending infinite chain $A_1 \subset A_2 \subset \dots$ of subsets from $\mathcal{A}$. Recall also that a subset $B \in \mathcal{A}$ is a *maximal element* of $\mathcal{A}$ if there does not exist a subset in $\mathcal{A}$ which properly contains B. To emphasize the order–theoretic nature of these considerations, we use the notation of inequalities ($\leq$, $<$, $\lneq$, etc.) for inclusions among submodules and/or ideals. In particular, if A is a module the notation $B \leq A$ means that B is a submodule of A, and the notation $B < A$ (or $A > B$) means that B is a proper submodule of A.

PROPOSITION 1.1. *For a module A, the following conditions are equivalent:*

(a) A has ACC on submodules.

(b) Every nonempty family of submodules of A has a maximal element.

(c) Every submodule of A is finitely generated.

Proof. (a) $\Rightarrow$ (b): Suppose that $\mathcal{A}$ is a nonempty family of submodules of A without a maximal element. Choose $A_1 \in \mathcal{A}$. Since A_1 is not maximal, there exists $A_2 \in \mathcal{A}$ such that $A_2 > A_1$. Continuing in this manner, we obtain a properly ascending infinite chain $A_1 < A_2 < A_3 < \dots$ of submodules of A, contradicting the ACC.

(b) $\Rightarrow$ (c): Let B be a submodule of A, and let $\mathcal{B}$ be the family of all finitely generated submodules of B. Note that $\mathscr{B}$ contains 0 and so is nonempty. By (b), there exists a maximal element $C \in \mathscr{B}$. If $C \neq B$, choose an element $x \in C - B$, and let C' be the submodule of B generated by C and x. Then $C' \in \mathscr{B}$ and $C' > C$, contradicting the maximality of C. Thus $C = B$, whence B is finitely generated.

(c) ⇒ (a): Let $B_1 \leq B_2 \leq \ldots$ be an ascending chain of submodules of A. Let B be the union of the B_n. By (c), there exists a finite set X of generators for B. Since X is finite, it is contained in some B_n, whence $B_n = B$. Thus $B_m = B_n$ for all $m \geq n$, establishing the ACC for submodules of A. □

DEFINITION. A module A is *noetherian* if and only if the equivalent conditions of Proposition 1.1 are satisfied. As follows from the proof of (b) ⇒ (c), a further equivalent condition is that A have ACC on *finitely generated* submodules.

For example, any finite–dimensional vector space V over a field k is a noetherian k–module, since a properly ascending chain of submodules (subspaces) of V cannot contain more than $\dim_k(V) + 1$ subspaces.

DEFINITION. A ring R is *right (left) noetherian* if and only if the right module R_R (left module ${}_RR$) is noetherian. If both conditions hold, R is called a *noetherian ring*.

Rephrasing Proposition 1.1 for the ring itself, we see that a ring R is right (left) noetherian if and only if R has ACC on right (left) ideals, if and only if all right (left) ideals of R are finitely generated. For example, $\mathbb{Z}$ is a noetherian ring because all its ideals are principal (singly generated).

EXERCISE 1A. Show that the 2×2 matrices over $\mathbb{Q}$ of the form $\begin{pmatrix} a & b \\ 0 & c \end{pmatrix}$ with $a \in \mathbb{Z}$ and $b,c \in \mathbb{Q}$ make a ring which is right noetherian but not left noetherian. □

A large class of examples of commutative noetherian rings is revealed by the Hilbert Basis Theorem (which states that a polynomial ring over a commutative noetherian ring is again noetherian). We do not prove this here because we shall shortly derive a slightly more general form of it (Theorem 1.12).

PROPOSITION 1.2. *Let R be an algebra over a field k. If R is commutative and finitely generated as a k–algebra, then R is noetherian.*

Proof. Let $x_1,\ldots,x_n$ generate R as a k–algebra, and let $S = k[y_1,\ldots,y_n]$ be a polynomial ring over k in n independent indeterminates. Since R is commutative, there exists a k–algebra map $\phi : S \to R$ such that $\phi(y_i) = x_i$ for each i, and ϕ is surjective because the x_i generate R. Hence, $R \cong S/\ker(\phi)$.

By the Hilbert Basis Theorem, S is a noetherian ring, and so all ideals of S are finitely generated. If I is any ideal of R, then $\phi^{-1}(I)$ is an ideal of S, and $\phi\phi^{-1}(I) = I$ because ϕ is surjective. As $\phi^{-1}(I)$ is finitely generated, so is I. Therefore R is noetherian. □

Noncommutative finitely generated algebras need not be noetherian, as the following example shows. (For readers familiar with free algebras, we mention that the free algebra on two letters over any field k is a finitely generated k–algebra which is neither right nor left noetherian.)

EXERCISE 1B. Let k be a field, and let V be a vector space over k with a countably infinite basis $v_1, v_2, \ldots$. Define $s, t \in \mathrm{End}_k(V)$ so that $s(v_i) = v_{i+1}$ for all i while $t(v_i) = v_{i-1}$ for all $i > 1$ and $t(v_1) = 0$, and let R be the k–subalgebra of $\mathrm{End}_k(V)$ generated by s and t. Show that R is neither right nor left noetherian. [Hint: Define $e_1, e_2, \ldots$ in $\mathrm{End}_k(V)$ so that $e_i(v_i) = v_i$ for all i while $e_i(v_j) = 0$ for all $i \neq j$, and show that each $e_i \in R$. Then show that $\sum e_iR$ and $\sum Re_i$ are not finitely generated.] □

PROPOSITION 1.3. *Let B be a submodule of a module A. Then A is noetherian if and only if B and A/B are both noetherian.*

Proof. First assume that A is noetherian. Since any ascending chain of submodules of B is also an ascending chain of submodules of A, it is immediate that B is noetherian. If $C_1 \leq C_2 \leq \ldots$ is an ascending chain of submodules of A/B, each C_i is of the form A_i/B for some submodule A_i of A that contains B, and $A_1 \leq A_2 \leq \ldots$. Since A is noetherian there is some n such that $A_i = A_n$ for all $i \geq n$, and then $C_i = C_n$ for all $i \geq n$. Thus A/B is noetherian.

Conversely, assume that B and A/B are noetherian, and let $A_1 \leq A_2 \leq \ldots$ be an ascending chain of submodules of A. There are ascending chains of submodules

$$A_1 \cap B \leq A_2 \cap B \leq \ldots$$

$$(A_1 + B)/B \leq (A_2 + B)/B \leq \ldots$$

in B and in A/B. Hence, there is some n such that $A_i \cap B = A_n \cap B$ and $(A_i + B)/B = (A_n + B)/B$ for all $i \geq n$, and the latter equation yields $A_i + B = A_n + B$. For all $i \geq n$, we conclude that

$$A_i = A_i \cap (A_i + B) = A_i \cap (A_n + B) = A_n + (A_i \cap B) = A_n + (A_n \cap B) = A_n$$

(using the modular law for the third equality). Therefore A is noetherian. □

COROLLARY 1.4. *Any finite direct sum of noetherian modules is noetherian.*

Proof. By induction on the number of summands, it suffices to prove that the direct sum of two noetherian modules A_1 and A_2 is noetherian. The module $A = A_1 \oplus A_2$ has a submodule $B = A_1 \oplus 0$ such that $B \cong A_1$ and $A/B \cong A_2$. Then B and A/B are noetherian, whence A is noetherian by Proposition 1.3. □

COROLLARY 1.5. *If R is a right noetherian ring, all finitely generated right R–modules are noetherian.*

Proof. If A is a finitely generated right R–module, then $A \cong F/K$ for some finitely generated free right R–module F and some submodule $K \leq F$. Since F is isomorphic to a finite direct sum of copies of the noetherian module R_R, it is noetherian by Corollary 1.4. Then by Proposition 1.3, A must be noetherian. □

COROLLARY 1.6. *Let S be a subring of a ring R. If S is right noetherian and R is finitely generated as a right S–module, then R is right noetherian.*

Proof. By Corollary 1.5, R is noetherian as a right S–module. Since all right ideals of R are also right S–submodules, R must have ACC on right ideals. □

Using Corollary 1.6 we obtain some easy examples of noncommutative noetherian rings.

PROPOSITION 1.7. *If R is a module–finite algebra over a commutative noetherian ring S, then R is a noetherian ring.*

Proof. The image of S in R is a noetherian subring S' of the center of R such that R is a finitely generated (right or left) S'–module. Apply Corollary 1.6. □

For instance, Proposition 1.7 shows that for any positive integer n, the ring of all $n \times n$ matrices over a commutative noetherian ring is noetherian. This also holds for matrix rings over noncommutative noetherian rings, as follows.

DEFINITION. Given a ring R and a positive integer n, we use $M_n(R)$ to denote the ring of all $n \times n$ matrices over R. The *standard $n \times n$ matrix units* in $M_n(R)$ are the matrices e_{ij} (for $i, j = 1,...,n$) such that e_{ij} has 1 for the i,j–entry and 0 for all other entries.

PROPOSITION 1.8. *Let R be a right noetherian ring, and let S be a subring of a matrix ring $M_n(R)$. If S contains the subring*

$$R' = \{diag(r,r,...,r) \mid r \in R\}$$

of all "scalar matrices", then S is right noetherian. In particular, $M_n(R)$ is a right noetherian ring.

Proof. Clearly $R' \cong R$, whence R' is a right noetherian ring. Observe that $M_n(R)$ is generated as a right R'–module by the standard $n \times n$ matrix units. Hence, Corollary 1.5 implies that $M_n(R)$ is a noetherian right R'–module. As all right ideals of S are also right R'–submodules of $M_n(R)$, we conclude that S is right noetherian. □

♦ FORMAL TRIANGULAR MATRIX RINGS

One way to construct rings to which Corollary 1.6 and Proposition 1.8 apply is to take an upper (or lower) triangular matrix ring over a known ring, or to take a subring of a triangular matrix ring. For instance, if S and T are subrings of a ring B, the set R of all matrices of the form $\begin{pmatrix} s & b \\ 0 & t \end{pmatrix}$ (for $s \in S$, $b \in B$, $t \in T$) is a subring of $M_2(B)$. (If S and T are right noetherian, and B_T is finitely generated, it follows easily from Corollary 1.6 that R is right noetherian.) Note that B need not be a ring itself in order for R to be a ring – rather B must be closed under addition, left multiplication by elements of S, and right multiplication by elements of T. More formally, symbols $\begin{pmatrix} s & b \\ 0 & t \end{pmatrix}$ will form a ring under matrix addition and multiplication provided only that B is simultaneously a left S–module and a right T–module satisfying an associative law connecting its left and right module structures. We focus on this ring construction because it provides a convenient source for any number of interesting examples. Later, we shall see such left/right modules as B appearing for their own sake in noetherian ring theory.

DEFINITION. Let S and T be rings. An *(S,T)–bimodule* is an abelian group B equipped with a left S–module structure and a right T–module structure (both utilizing the given addition) such that $s(bt) = (sb)t$ for all $s \in S$, $b \in B$, $t \in T$. The symbol ${}_SB_T$ is used to denote this situation. An *(S,T)–sub–bimodule* of B (or just a *sub–bimodule*, if S and T are clear from the context) is any subgroup of B which is both a left S–submodule and a right T–submodule. Note that if C is a sub–bimodule of B, the factor group B/C is a bimodule in the obvious manner.

For instance, if S is a ring and T is a subring, then S itself (or an ideal of S) can be regarded as an (S,T)–bimodule (or as a (T,S)–bimodule). For another example, if B is a right module over a ring T and S is a subring of $\mathrm{End}_T(B)$, then B is an (S,T)–bimodule. Perhaps most importantly, if $I \leq J$ are ideals in a ring S, then J/I is an (S,S)–bimodule. The next exercise shows that in a sense every bimodule appears this way, as an ideal of a formal triangular matrix ring.

EXERCISE 1C. Let ${}_SB_T$ be a bimodule, and write $\begin{pmatrix} S & B \\ 0 & T \end{pmatrix}$ for the abelian group $S \oplus B \oplus T$, where triples (s,b,t) from $S \oplus B \oplus T$ are written as formal 2×2 matrices $\begin{pmatrix} s & b \\ 0 & t \end{pmatrix}$.

(a) Show that formal matrix addition and multiplication make sense in $\begin{pmatrix} S & B \\ 0 & T \end{pmatrix}$, and that using those operations $\begin{pmatrix} S & B \\ 0 & T \end{pmatrix}$ becomes a ring.

(b) Show that there is also a ring $\begin{pmatrix} T & 0 \\ B & S \end{pmatrix}$ of formal lower triangular matrices, and that $\begin{pmatrix} T & 0 \\ B & S \end{pmatrix} \cong \begin{pmatrix} S & B \\ 0 & T \end{pmatrix}$.

(c) Observe that the set $\begin{pmatrix} 0 & B \\ 0 & 0 \end{pmatrix}$ of matrices $\begin{pmatrix} 0 & b \\ 0 & 0 \end{pmatrix}$ is an ideal of $\begin{pmatrix} S & B \\ 0 & T \end{pmatrix}$, and that under the obvious abelian group isomorphism of B onto $\begin{pmatrix} 0 & B \\ 0 & 0 \end{pmatrix}$, left S–submodules (right T–submodules, (S,T)–sub–bimodules) of B correspond precisely to left ideals (right ideals, two–sided ideals) of $\begin{pmatrix} S & B \\ 0 & T \end{pmatrix}$ contained in $\begin{pmatrix} 0 & B \\ 0 & 0 \end{pmatrix}$. □

DEFINITION. A *formal triangular matrix ring* is any ring of the form $\begin{pmatrix} S & B \\ 0 & T \end{pmatrix}$ or $\begin{pmatrix} T & 0 \\ B & S \end{pmatrix}$ as described in Exercise 1C. By way of abbreviation, we write "let $\begin{pmatrix} S & B \\ 0 & T \end{pmatrix}$ be a formal triangular matrix ring" in place of "let S and T be rings, let B be an (S,T)–bimodule, and let $\begin{pmatrix} S & B \\ 0 & T \end{pmatrix}$ be the corresponding formal triangular matrix ring".

Observe that if S and T are subrings of a ring U, and B is an (S,T)–sub–bimodule of U, the formal triangular matrix ring $\begin{pmatrix} S & B \\ 0 & T \end{pmatrix}$ is isomorphic to the subring of $M_2(U)$ consisting of all honest matrices of the form $\begin{pmatrix} s & b \\ 0 & t \end{pmatrix}$ with $s \in S$, $b \in B$, $t \in T$.

PROPOSITION 1.9. *Let* $R = \begin{pmatrix} S & B \\ 0 & T \end{pmatrix}$ *be a formal triangular matrix ring. Then* R *is right noetherian if and only if* S *and* T *are right noetherian and* B_T *is finitely generated. Similarly,* R *is left noetherian if and only if* S *and* T *are left noetherian and* ${}_SB$ *is finitely generated.*

Proof. Assume first that S and T are right noetherian and B_T is finitely generated. Observe that the diagonal subring $\begin{pmatrix} S & 0 \\ 0 & T \end{pmatrix}$ is isomorphic to $S \times T$ and so is right noetherian. Observe also that if elements $b_1,\dots,b_n$ generate B as a right T–module, then the matrices

$$\begin{pmatrix} 1 & 0 \\ 0 & 1 \end{pmatrix}, \begin{pmatrix} 0 & b_1 \\ 0 & 0 \end{pmatrix}, \begin{pmatrix} 0 & b_2 \\ 0 & 0 \end{pmatrix}, \dots, \begin{pmatrix} 0 & b_n \\ 0 & 0 \end{pmatrix}$$

generate R as a right $\begin{pmatrix} S & 0 \\ 0 & T \end{pmatrix}$–module. Consequently, Corollary 1.6 shows that R is right noetherian.

Conversely, assume that R is right noetherian. Observing that the projection maps $\begin{pmatrix} s & b \\ 0 & t \end{pmatrix} \mapsto s$ and $\begin{pmatrix} s & b \\ 0 & t \end{pmatrix} \mapsto t$ are ring homomorphisms of R onto S and R onto T, we see that S and T must be right noetherian. Moreover, $\begin{pmatrix} 0 & B \\ 0 & 0 \end{pmatrix}$ is a right ideal of R and must have a finite list of generators

$$\begin{pmatrix} 0 & b_1 \\ 0 & 0 \end{pmatrix}, \begin{pmatrix} 0 & b_2 \\ 0 & 0 \end{pmatrix}, \ldots, \begin{pmatrix} 0 & b_n \\ 0 & 0 \end{pmatrix},$$

from which we infer that the elements $b_1,\ldots,b_n$ generate B_T.

The left noetherian analog is proved in the same fashion. □

For example, it is immediate from Proposition 1.9 that the ring $\begin{pmatrix} \mathbb{Z} & \mathbb{Q} \\ 0 & \mathbb{Q} \end{pmatrix}$ is right noetherian but not left noetherian (Exercise 1A).

EXERCISE 1D. Let $R = \begin{pmatrix} S & B \\ 0 & T \end{pmatrix}$ be a formal triangular matrix ring. The purpose of this exercise is to give a description of all R–modules in terms of S–modules and T–modules.

(a) Let A be a right S–module, C a right T–module, and $f \in \mathrm{Hom}_T(A \otimes_S B, C)$. For $(a,c) \in A \oplus C$ and $\begin{pmatrix} s & b \\ 0 & t \end{pmatrix} \in R$, define

$$(a,c)\begin{pmatrix} s & b \\ 0 & t \end{pmatrix} = (as, f(a \otimes b) + ct).$$

Show that using this multiplication rule, $A \oplus C$ is a right R–module.

(b) Show that the R–module $A \oplus C$ in (a) is finitely generated if and only if A is a finitely generated S–module and $C/f(A \otimes_S B)$ is a finitely generated T–module.

(c) Show that every right R–module is isomorphic to one of the type $A \oplus C$ constructed in (a). □

EXERCISE 1E. Let ${}_SB_T$ be a bimodule, and form the ring $R = S^{op} \otimes_{\mathbb{Z}} T$, where S^{op} denotes the *opposite ring* of S. (That is, S^{op} is the same abelian group as S, but with the opposite multiplication: the product of s_1 and s_2 in S^{op} is s_2s_1.) Show that B can be made into a right R–module where $b(s \otimes t) = sbt$ for all $s \in S$, $t \in T$, $b \in B$, and that the right R–submodules of B are precisely its (S,T)–sub–bimodules. Conversely, show that every right R–module can be made into an (S,T)–bimodule. □

♦ CONSTRUCTION OF SKEW POLYNOMIAL RINGS

In the prologue we have seen two examples of rings which look like polynomial rings in one indeterminate but in which the indeterminate does not commute with everything. We now want to consider the general question of polynomials in a non–central indeterminate, over some coefficient ring R. Let us use θ for this non–central indeterminate. If we want our polynomials all to have left–hand coefficients, our polynomial ring should be a free left R–module with basis $1,\theta,\theta^2,\ldots$. If multiplication is to respect degree in the usual way, $\theta^n r$ should have degree at most n, for any $n \in \mathbb{N}$ and $r \in R$. In particular, for any $r \in R$ the product θr should be linear, so that $\theta r = \alpha(r)\theta + \delta(r)$ for some $\alpha(r), \delta(r) \in R$. Now α and δ are maps from R to R, and we may ask what properties α and δ must satisfy for a polynomial ring to be constructed on this pattern.

The distributive law says that $\theta(r + s) = \theta r + \theta s$ (for $r, s \in R$), and it follows that α and δ must be additive maps. From $\theta 1 = \theta$ we obtain $\alpha(1) = 1$ and $\delta(1) = 0$. The associative law says that $\theta(rs) = (\theta r)s$ (for $r, s \in R$), and a computation shows that $\alpha(rs) = \alpha(r)\alpha(s)$ and $\delta(rs) = \alpha(r)\delta(s) + \delta(r)s$. Thus α must be a ring endomorphism of R. In typical examples, α will usually be an automorphism.

DEFINITION. Let α be an endomorphism of a ring R. An *α–derivation* of R is any additive map $\delta : R \to R$ such that $\delta(rs) = \alpha(r)\delta(s) + \delta(r)s$ for all $r, s \in R$. (Strictly speaking, we have defined a *left α–derivation*, but we shall not need a *right α–derivation*, which is any additive map $\delta : R \to R$ satisfying the rule $\delta(rs) = \delta(r)\alpha(s) + r\delta(s)$.) In case α is the identity map on R, an α–derivation of R is just called a *derivation*. Thus a derivation of R is any additive map $\delta : R \to R$ that satisfies the product rule for derivatives: $\delta(rs) = \delta(r)s + r\delta(s)$ for all $r, s \in R$.

We do not include the condition $\delta(1) = 0$ in the definition of an α–derivation, since it is automatically satisfied. Namely,

$$\delta(1) = \delta(1\cdot 1) = \alpha(1)\delta(1) + \delta(1)1 = 2\delta(1),$$

whence $\delta(1) = 0$.

For example, in any polynomial ring $K[x]$ the usual formal derivative d/dx, given by the rule

$$\frac{d}{dx}(a_0 + a_1x + \ldots + a_nx^n) = a_1 + 2a_2x + \ldots + na_nx^{n-1}$$

(for $a_0,a_1,\ldots,a_n \in K$), is a derivation. For another example, given any endomorphism α of a ring R, and any element $d \in R$, the rule $\delta(r) = dr - \alpha(r)d$ defines an α–derivation of R.

EXERCISE 1F. Let α be an endomorphism of a ring R, let δ be any map from R to R, and define a map $\phi : R \to M_2(R)$ by the rule $\phi(r) = \begin{pmatrix} \alpha(r) & \delta(r) \\ 0 & r \end{pmatrix}$. Show that δ is an α–derivation of R if and only if ϕ is a ring homomorphism. □

EXERCISE 1G. Let $R[x]$ be a polynomial ring over a ring R, let α be an endomorphism of R, and let δ be an α–derivation of R. Extend α to an endomorphism α' of $R[x]$ such that $\alpha'(x) = x$. Suppose that there is an element p of $R[x]$ satisfying $pa = \alpha(a)p$ for all $a \in R$. (E.g., if α is the identity map on R, this just requires that p be central in $R[x]$.) Show that δ extends uniquely to an α'–derivation δ' on $R[x]$ such that $\delta'(x) = p$. □

Given a ring R, an endomorphism α of R, and an α–derivation δ of R, we would like to construct a skew polynomial ring S following the pattern suggested above. We could form a free left R–module with basis $1,\theta,\theta^2,\ldots$ and start generating a multiplication from the rules $\theta^i\theta^j = \theta^{i+j}$ and $\theta r = \alpha(r)\theta + \delta(r)$. The skew polynomial ring can be constructed in this way, but getting the multiplication to be fully defined and checking the ring axioms is a rather messy business. With a slight change of perspective, we can obtain our skew polynomial ring as a subring of the endomorphism ring of a suitable abelian group, which will give us the ring axioms for free.

We start with an analog of the fact that over an ordinary polynomial ring $R[x]$, the ring R can be made into a left $R[x]$–module isomorphic to $R[x]/R[x]x$. In the skew polynomial case, notice that if an S of the desired form exists, then the left S–module $S/S\theta$ has the same left R–module structure as R, and θ acts on it like δ (since $\theta(r + S\theta) = \delta(r) + S\theta$ for all $r \in R$). Thus the left S–module structure on R, which amounts to a ring homomorphism from S to $\mathrm{End}_{\mathbb{Z}}(R)$, would give us a homomorphic image of S in $\mathrm{End}_{\mathbb{Z}}(R)$, namely the subring generated by δ and all left multiplications by elements of R. In case the powers of δ are left linearly independent over R this subring would be isomorphic to S. (This holds for example in $A_1(\mathbb{C})$, as discussed in the prologue, where $R = \mathbb{C}[x]$ and δ is the usual derivative d/dx.) Under these circumstances we could obtain S just by constructing this subring. Since that need not be the case (for instance, δ might be zero), we build S as a subring of something larger.

Notice that if S exists, then in S we would have

$$\theta\left(\sum_{i=0}^{n} r_i\theta^i\right) = \sum_{i=0}^{n} (\alpha(r_i)\theta^{i+1} + \delta(r_i)\theta^i)$$

for all $r_0,\ldots,r_n \in R$. Hence, we may model this by an additive endomorphism of the polynomial ring $R[x]$, namely the map

$$\sum_{i=0}^{n} r_i x^i \mapsto \sum_{i=0}^{n} (\alpha(r_i)x^{i+1} + \delta(r_i)x^i).$$

The subring of $\mathrm{End}_{\mathbb{Z}}(R[x])$ generated by this map together with all left multiplications by elements of R will then provide our skew polynomial ring S.

PROPOSITION 1.10. *Let R be a ring, let α be an endomorphism of R, and let δ be an α–derivation of R. Then there exists a ring S, containing R as a subring, such that S is a free left R–module with a basis of the form $1,\theta,\theta^2,\ldots$ and $\theta r = \alpha(r)\theta + \delta(r)$ for all $r \in R$.*

Proof. Let $E = \mathrm{End}_{\mathbb{Z}}(R[x])$ where x is an indeterminate. There is an injective ring map $R \to E$ under which any element $r \in R$ corresponds to left multiplication by r on $R[x]$. We identify R with its image under this map, so that R is now a subring of E.

We next extend α and δ to maps in E where $\alpha(rx^i) = \alpha(r)x^i$ and $\delta(rx^i) = \delta(r)x^i$ for all $r \in R$ and $i = 0,1,\dots$. Then α becomes a ring endomorphism of $R[x]$ and δ becomes an α–derivation on $R[x]$. Now define $\theta \in E$ according to the rule $\theta(f) = \alpha(f)x + \delta(f)$. For $r \in R$ and $f \in R[x]$ we compute that

$$(\theta r)(f) = \alpha(rf)x + \delta(rf) = \alpha(r)\alpha(f)x + \alpha(r)\delta(f) + \delta(r)f = \alpha(r)\theta(f) + \delta(r)f.$$

Thus $\theta r = \alpha(r)\theta + \delta(r)$ for all $r \in R$. In particular, $\theta R \subseteq R\theta + R$.

From the relation $\theta R \subseteq R\theta + R$, it follows by induction that

$$\theta^i R \subseteq R\theta^i + R\theta^{i-1} + \dots + R\theta + R$$

for all $i = 0,1,2,\dots$. Consequently,

$$(R\theta^i)(R\theta^j) \subseteq R\theta^{i+j} + R\theta^{i+j-1} + \dots + R\theta^{j+1} + R\theta^j$$

for all $i, j = 0,1,2,\dots$, whence the set $\sum R\theta^i$ is closed under multiplication. Hence, $\sum R\theta^i$ is a subring of E, which we now label S. Thus S is generated as a left R–module by $1,\theta,\theta^2,\dots$.

It remains to show that $1,\theta,\theta^2,\dots$ are left linearly independent over R. Observe that $\theta(x^j) = x^{j+1}$ for all $j = 0,1,\dots$, whence $\theta^i(x^0) = x^i$ for all $i = 0,1,\dots$. Given an element $s = r_0 + r_1\theta + \dots + r_n\theta^n$ in S (with $r_0,\dots,r_n \in R$), it follows that $s(x^0) = r_0 + r_1x + \dots + r_nx^n$, and hence $s = 0$ only if $r_0 = \dots = r_n = 0$. Thus the θ^i are indeed left linearly independent over R, and therefore they form a basis for S as a free left R–module. □

The ring S constructed in Proposition 1.10 is essentially unique, in the following sense.

EXERCISE 1H. Let R be a ring, let α be an endomorphism of R, and let δ be an α–derivation of R. Let S_1 and S_2 be ring extensions of R such that each S_i is a free left R–module with a basis of the form $1,\theta_i,\theta_i^2,\dots$ and $\theta_i r = \alpha(r)\theta_i + \delta(r)$ for all $r \in R$.

(a) If $\phi : R \to T$ is a ring homomorphism and there is an element $\xi \in T$ such that $\xi\phi(r) = \phi\alpha(r)\xi + \phi\delta(r)$ for all $r \in R$, show that ϕ extends uniquely to a ring homomorphism $\Phi : S_1 \to T$ such that $\Phi(\theta_1) = \xi$. [Hint: First define a map Φ, and then show that $\Phi(\theta_1^n r) = \xi^n\phi(r)$ for $n \in \mathbb{N}$ and $r \in R$.]

(b) Show that the identity map on R extends to a ring isomorphism $\Phi : S_1 \to S_2$ such that $\Phi(\theta_1) = \theta_2$. □

DEFINITION. Let R be a ring, let α be an endomorphism of R, and let δ be an α–derivation of R. The ring S constructed in Proposition 1.10 is denoted $R[\theta;\alpha,\delta]$ and is called a *skew polynomial ring* or an *Ore extension* of R. In case α is the identity map on R, we abbreviate $R[\theta;\alpha,\delta]$ to $R[\theta;\delta]$ and call this ring a *(formal) differential operator ring*. In case $\delta = 0$, we abbreviate $R[\theta;\alpha,\delta]$ to $R[\theta;\alpha]$.

The reader should be warned that some authors prefer their skew polynomial rings to have *right–hand* coefficients. To achieve this, one starts with a ring R, an endomorphism α of R, and a *right* α–derivation δ on R. The corresponding skew polynomial ring is a free *right* R–module with a basis $1,\theta,\theta^2,\ldots$ where $r\theta = \theta\alpha(r) + \delta(r)$ for all $r \in R$.

EXERCISE 1I. Let δ be a derivation on a ring R. Show that R can be made into a left $R[\theta;\delta]$–module with a module multiplication $*$ such that $(\sum r_i\theta^i) * r = \sum r_i\delta^i(r)$ for all $r_i,r \in R$. (Thus the elements of $R[\theta;\delta]$ act on R as "linear differential operators".) □

DEFINITION. Let $R[\theta;\alpha,\delta]$ be a skew polynomial ring. Any nonzero element p in $R[\theta;\alpha,\delta]$ may be uniquely expressed in the form

$$p = r_n\theta^n + r_{n-1}\theta^{n-1} + \ldots + r_1\theta + r_0$$

for some nonnegative integer n and some elements $r_i \in R$ with $r_n \neq 0$. The integer n is called the *degree* of p, abbreviated deg(p), and the element r_n is called the *leading coefficient* of p. (In the differential operator ring case, namely $R[\theta;\delta]$, it is common to call n the *order* of p rather than the degree.) The zero element of $R[\theta;\alpha,\delta]$ is defined to have degree $-\infty$ and leading coefficient 0.

In case R is a domain and α is injective, $\deg(pq) = \deg(p) + \deg(q)$ for all p,q in $R[\theta;\alpha,\delta]$. In particular, $R[\theta;\alpha,\delta]$ is a domain in this case.

EXERCISE 1J. Let $R[\theta;\alpha,\delta]$ be a skew polynomial ring. For any $r \in R$ and any nonnegative integer n, show that $\theta^n r = \alpha^n(r)\theta^n$ + [terms of degree less than n]. Hence, if α is injective and r is nonzero, $\theta^n r$ has degree n and leading coefficient $\alpha^n(r)$. □

EXERCISE 1K. Let $R[\theta;\delta]$ be a differential operator ring. For any $r \in R$ and any nonnegative integer n, show that $\theta^n r = \sum_{i=0}^{n} \binom{n}{i}\delta^{n-i}(r)\theta^i$. □

EXERCISE 1L. Let $R[\theta;\alpha,\delta]$ be a skew polynomial ring, and assume that α is an automorphism of R. Show that α^{-1} is an automorphism of the opposite ring R^{op}, that $-\delta\alpha^{-1}$ is an α^{-1}–derivation of R^{op}, and that there is a ring isomorphism

$$\phi : R[\theta;\alpha,\delta]^{op} \to R^{op}[\theta;\alpha^{-1},-\delta\alpha^{-1}]$$

such that $\phi(r) = r$ for all $r \in R$ and $\phi(\theta) = \theta$. Conclude that if R is commutative and α is the identity map, then $R[\theta;\delta]^{op} \cong R[\theta;\delta]$. □

In the case of a skew polynomial ring $R[\theta;\alpha,\delta]$ with α an automorphism, Exercise 1L allows us to transfer easily between right–handed and left–handed results. For instance, it follows that $R[\theta;\alpha,\delta]$ is a free *right* R–module with basis $1,\theta,\theta^2,\ldots$. We shall use this principle several times, starting with an analog of the result that the polynomial ring in one variable over a field is a P.I.D.

♦ IDEALS IN SKEW POLYNOMIAL RINGS

DEFINITION. A *principal right ideal domain* is a domain in which all right ideals are principal. *Principal left ideal domains* are defined analogously.

THEOREM 1.11. *Let R be a division ring, let α be an endomorphism of R, and let δ be an α–derivation of R. Then the skew polynomial ring $S = R[\theta;\alpha,\delta]$ is a principal left ideal domain. If α is an automorphism of R, then S is also a principal right ideal domain.*

Proof. Note that as R is a division ring, α must be injective, whence S is a domain.

Given a nonzero left ideal J of S, let m be the minimum degree for nonzero elements of J, and choose $p \in J$ with degree m. If r is the leading coefficient of p, then p may be replaced by $r^{-1}p$, and so there is no loss of generality in assuming that p has leading coefficient 1.

We claim that $J = Sp$. Obviously $Sp \leq J$, and we prove the reverse inclusion by induction on degree. The only element of J with degree less than m is 0, and certainly $0 \in Sp$. Now assume, for some integer $k \geq m$, that all elements of J with degree less than k lie in Sp. Let q be any element of J with degree k, and let s be the leading coefficient of q. Now $s\theta^{k-m}p$ has degree k and leading coefficient s, whence $q - s\theta^{k-m}p$ is an element of J with degree less than k. By the induction hypothesis, $q - s\theta^{k-m}p$ lies in Sp, and so $q \in Sp$. This completes the induction step, proving that $J = Sp$.

Therefore S is a principal left ideal domain.

Now assume that α is an automorphism of R. By Exercise 1L, $R[\theta;\alpha,\delta]^{op}$ is isomorphic to $R^{op}[\theta;\alpha^{-1},-\delta\alpha^{-1}]$. Since R^{op} is a division ring, our results above show that $R[\theta;\alpha,\delta]^{op}$ is a principal left ideal domain, and therefore $R[\theta;\alpha,\delta]$ is a principal right ideal domain. □

For example, let $R = k(x)$ be a rational function field over a field k, and let δ be the formal derivative d/dx on R. Then the differential operator ring $R[\theta;\delta]$ is a principal right and left ideal domain by Theorem 1.11, and this ring is clearly not commutative. This particular differential operator ring $R[\theta;\delta]$ is sometimes denoted $B_1(k)$. (A special case is discussed in the prologue.)

EXERCISE 1M. Let $R = k(x)$ be a rational function field over a field k, and let α be the k–algebra endomorphism of R given by the rule $\alpha(f) = f(x^2)$. Show that $R[\theta;\alpha]$ is not a principal right ideal domain. [Hint: Look at the right ideal generated by θ and $x\theta$.] (Actually $R[\theta;\alpha]$ is not even right noetherian, as in the analogous example in Exercise 1N.) □

We now derive a Hilbert Basis Theorem for skew polynomial rings. Note the hypothesis that α be an automorphism of R.

THEOREM 1.12. *Let R be a ring, let α be an automorphism of R, and let δ be an α–derivation of R. If R is right (left) noetherian, then the skew polynomial ring $S = R[\theta;\alpha,\delta]$ is right (left) noetherian.*

Proof. First assume that R is right noetherian, and let J be any nonzero right ideal of S. Let L be the set of leading coefficients of elements of J. We claim that L is a right ideal of R.

Obviously $0 \in L$. Given any nonzero $r, s \in L$, there exist $p,q \in J$ such that p has leading coefficient r and q has leading coefficient s. If $r + s = 0$ then $r + s \in L$, so assume that $r + s \neq 0$. Let m and n be the degrees of p and q. If $m \leq n$, then $p\theta^{n-m} + q$ is an element of J with leading coefficient $r + s$, while if $m \geq n$, then $p + q\theta^{m-n}$ is an element of J with leading coefficient $r + s$. In either case, $r + s \in J$. Moreover, if $t \in R$ and $rt \neq 0$ then we see that $rt \in L$ since $p\alpha^{-m}(t)$ is an element of J with leading coefficient rt (Exercise 1J). Thus L is a right ideal of R, as claimed.

Since R is right noetherian, we may choose nonzero generators $r_1,\ldots,r_k$ for L, and then for $i = 1,\ldots,k$ we may choose $p_i \in J$ with leading coefficient r_i and degree $n(i)$. If $n = \max\{n(1),\ldots,n(k)\}$, then each p_i may be replaced by $p_i\theta^{n-n(i)}$. Hence, there is no loss of generality in assuming that $p_1,\ldots,p_k$ all have the same degree n.

Set $N = \{p \in S \mid \deg(p) < n\}$ and observe that N is the right R–submodule of S generated by $1,\theta,\ldots,\theta^{n-1}$. By Corollary 1.5, N is a noetherian right R–module. As a result, $J \cap N$ is a finitely generated right R–module, say generated by $q_1,\ldots,q_t$.

Now set $J_0 = p_1S + \ldots + p_kS + q_1S + \ldots + q_tS$. We will show that $J_0 = J$. Since $q_1,\ldots,q_t$ generate $J \cap N$ as a right R–module, $J \cap N \subseteq J_0$. Thus J_0 contains all elements of J with degree less than n. Now assume, for some integer $m \geq n$, that J_0 contains all elements of J with degree less than m. Let p be any element of J with degree m, and let r be the leading coefficient of p. Then $r = r_1s_1 + \ldots + r_ks_k$ for some $s_i \in R$. Set

$$q = (p_1\alpha^{-n}(s_1) + \ldots + p_k\alpha^{-n}(s_k))\theta^{m-n},$$

and observe that q is an element of J_0 with degree m and leading coefficient r. Now $p - q$ is an element of J with degree less than m, whence $p - q \in J_0$ and so $p \in J_0$.

Hence, by induction, we conclude that $J = J_0$. Thus J is a finitely generated right ideal of S.

Therefore S is right noetherian.

The left noetherian case follows via Exercise 1L. □

EXERCISE 1N. Let $R = k[x]$ be a polynomial ring over a field k, and let α be the k–algebra endomorphism of R given by the rule $\alpha(f) = f(x^2)$. Show that $R[\theta;\alpha]$ is neither right nor left noetherian. [Hint: Look at the right ideal generated by $x\theta, \theta x\theta, \theta^2 x\theta, \ldots$, and at the left ideal generated by $x, x\theta, x\theta^2, \ldots$.] □

Of course, Theorem 1.12 may be applied to *iterated skew polynomial rings,* i.e., rings of the form

$$R[\theta_1;\alpha_1,\delta_1][\theta_2;\alpha_2,\delta_2]\cdots[\theta_n;\alpha_n,\delta_n],$$

where α_1 is an automorphism of R and δ_1 is an α_1–derivation of R, while α_2 is an automorphism of $R[\theta_1;\alpha_1,\delta_1]$ and δ_2 is an α_2–derivation of $R[\theta_1;\alpha_1,\delta_1]$, etc. (Rings of this sort can arise as enveloping algebras of Lie algebras.)

A case of particular interest is an iterated differential operator ring built from a ring R and a finite list $\delta_1,\ldots,\delta_n$ of commuting derivations on R, in the following manner. First construct $R[\theta_1;\delta_1]$. Then define a map ∂_2 on $R[\theta_1;\delta_1]$ according to the rule $\partial_2(\Sigma\, r_i\theta_1^i) = \Sigma\, \partial_2(r_i)\theta_1^i$. Since δ_2 commutes with δ_1, an easy computation shows that ∂_2 is a derivation on $R[\theta_1;\delta_1]$, and we form the ring $R[\theta_1;\delta_1][\theta_2;\partial_2]$. Similarly, δ_3 extends to a derivation ∂_3 on $R[\theta_1;\delta_1][\theta_2;\partial_2]$, and we form the ring $R[\theta_1;\delta_1][\theta_2;\partial_2][\theta_3;\partial_3]$. The final differential operator ring, namely

$$R[\theta_1;\delta_1][\theta_2;\partial_2]\cdots[\theta_n;\partial_n],$$

is denoted $R[\theta_1,\ldots,\theta_n;\, \delta_1,\ldots,\delta_n]$. Note that up to isomorphism, $R[\theta_1,\ldots,\theta_n;\, \delta_1,\ldots,\delta_n]$ is independent of the order in which the δ_i are listed.

COROLLARY 1.13. *Let R be a right (left) noetherian ring. If $\delta_1,\ldots,\delta_n$ are commuting derivations of R, then the iterated differential operator ring $R[\theta_1,\ldots,\theta_n;\, \delta_1,\ldots,\delta_n]$ is a right (left) noetherian ring.* □

A similar construction can be performed given a finite list of commuting automorphisms of a ring R.

♦ WEYL ALGEBRAS

DEFINITION. Let $K[x_1,\ldots,x_n]$ be a polynomial ring in n independent indeterminates over a ring K. The formal partial derivatives $\partial/\partial x_1,\ldots,\partial/\partial x_n$ are commuting derivations on $K[x_1,\ldots,x_n]$, and so we may form the iterated differential operator ring

$$K[x_1,\dots,x_n][\theta_1,\dots,\theta_n; \frac{\partial}{\partial x_1},\dots,\frac{\partial}{\partial x_n}].$$

This ring is called the nth *Weyl algebra over* K, and is denoted $A_n(K)$. (Unless K is commutative, the term "algebra" is out of place, but is commonly used.)

By Corollary 1.13, all the Weyl algebras over a right (left) noetherian ring are right (left) noetherian. In many instances, Weyl algebras have only a very meager supply of ideals. We make this more precise, and proceed by induction, using the following concepts.

DEFINITION. A *simple ring* is any nonzero ring R such that the only ideals of R are 0 and R.

DEFINITION. Let R be a ring, and let $y \in R$. The rule $\delta_y(r) = yr - ry$ defines a derivation δ_y on R, called the *inner derivation induced by* y. Any derivation on R which is not an inner derivation is called an *outer derivation.*

DEFINITION. Let δ be a derivation on a ring R. A *δ-ideal* of R is any ideal I of R such that $\delta(I) \subseteq I$. The ring R is called *δ-simple* if R is nonzero and the only δ-ideals of R are 0 and R.

PROPOSITION 1.14. *Let R be a $\mathbb{Q}$-algebra, and let δ be a derivation on R. If R is δ-simple and δ is outer, then $R[\theta;\delta]$ is a simple ring.*

Proof. Let $S = R[\theta;\delta]$, let I be any nonzero ideal of S, and let n be the minimum degree for nonzero elements of I. Let L be the subset of R consisting of 0 together with the leading coefficients of those elements of I which have degree n, and observe that L is a nonzero ideal of R.

We claim that L is a δ-ideal. Any $r \in L$ is the leading coefficient of some $p \in I$ having degree n. Observe that $\theta p - p\theta \in I$, and that

$$\theta p - p\theta = \delta(r)\theta^n + [\text{terms of degree less than } n].$$

Hence, either $\delta(r) = 0$ or $\delta(r)$ is the leading coefficient of an element of I of degree n, and in either case $\delta(r) \in L$. Thus L is a δ-ideal, as claimed.

Since R is δ-simple, $L = R$. Hence, I contains an element q with degree n and leading coefficient 1. If $n = 0$, then $q = 1$ and $I = S$. We will show that the assumption that $n > 0$ leads to a contradiction.

Write $q = \theta^n + x\theta^{n-1} + [\text{terms of degree less than } n-1]$ for some $x \in R$. For any $s \in R$, observe that $sq - qs \in I$, and (using Exercise 1K) that

$$sq - qs = (sx - n\delta(s) - xs)\theta^{n-1} + [\text{terms of degree less than } n-1].$$

By the minimality of n, we must have $sq - qs = 0$, and hence $sx - n\delta(s) - xs = 0$. Since $n > 0$ and R is a $\mathbb{Q}$-algebra, we obtain $\delta(s) = (-x/n)s - s(-x/n)$ for all $s \in R$, contradicting the assumption that δ is outer.

Thus $n = 0$ and $I = S$. Therefore S is a simple ring. □

COROLLARY 1.15. *If R is a simple $\mathbb{Q}$–algebra, then all the Weyl algebras $A_n(R)$ are simple rings.*

Proof. Observe that $A_n(R) \cong A_1(A_{n-1}(R))$ for all $n = 2,3,\ldots$. Hence, it suffices to prove that A_1 of any simple $\mathbb{Q}$–algebra is simple.

Thus consider $A_1(R) = R[x][\theta;\delta]$ where x is an indeterminate and $\delta = d/dx$. As $\delta(x) = 1$ and x is central in $R[x]$, we see that δ cannot be an inner derivation of $R[x]$. We next show that $R[x]$ is δ–simple.

Let I be any nonzero δ–ideal of $R[x]$, let n be the minimal degree for nonzero elements of I, and choose $p \in I$ with degree n. If p has leading coefficient r, then

$$\delta(p) = nrx^{n-1} + [\text{terms of degree less than } n-1].$$

Since $\delta(p) \in I$, the minimality of n forces $\delta(p) = 0$, and so $nr = 0$. As $r \neq 0$ and R is a $\mathbb{Q}$–algebra, $n = 0$, whence p is a nonzero element of R. Now $RpR = R$ (because R is simple), and hence $I = R[x]$. Thus $R[x]$ is δ–simple.

By Proposition 1.14, $A_1(R)$ is a simple ring. □

In characteristic $p > 0$, the Weyl algebras are not simple rings. For example, if R is a nonzero ring of characteristic p, then $\frac{d}{dx}(x^p) = 0$ in the polynomial ring $R[x]$ (see Exercise 1U), whence x^p is in the center of the Weyl algebra $A_1(R)$. Consequently, $x^pA_1(R)$ is a nontrivial ideal of $A_1(R)$.

♦ SKEW–LAURENT RINGS

One way to view a differential operator ring $R[\theta;\delta]$ is that it is a ring extension of R in which δ becomes an inner derivation (since $\delta(r) = \theta r - r\theta$ for all $r \in R$). If we start with an automorphism α of R and seek a ring extension in which α becomes an inner automorphism, we need a ring extension containing a unit θ such that $\alpha(r) = \theta r\theta^{-1}$ for all $r \in R$, that is, $\theta r = \alpha(r)\theta$ for all $r \in R$. This suggests constructing a ring extension of the skew polynomial ring $R[\theta;\alpha]$ in which θ has an inverse. We will give a direct construction of such a ring by analogy with Proposition 1.10, where this time we will work with additive endomorphisms of the Laurent polynomial ring $R[x,x^{-1}]$.

PROPOSITION 1.16. *Let R be a ring and let α be an automorphism of R. Then there exists a ring S, containing R as a subring, with a unit $\theta \in S$ such that S is a free left R–module with a basis of the form $1,\theta,\theta^{-1},\theta^2,\theta^{-2},\ldots$ and $\theta r = \alpha(r)\theta$ for all $r \in R$.*

Proof. Let $E = \mathrm{End}_{\mathbb{Z}}(R[x,x^{-1}])$ where x is an indeterminate, and embed R in E (as a subring) via left multiplications. Extend α to an automorphism of $R[x,x^{-1}]$ where

$\alpha(rx^i) = \alpha(r)x^i$ for all $r \in R$ and $i \in \mathbb{Z}$. Then define $\theta \in E$ according to the rule $\theta(f) = \alpha(f)x$, and observe that $\theta r = \alpha(r)\theta$ for all $r \in R$. Moreover, θ is invertible in E, and $\theta^{-1}(f) = \alpha^{-1}(f)x^{-1}$ for all $f \in R[x,x^{-1}]$. As in the proof of Proposition 1.10, the set $S = \sum_{i \in \mathbb{Z}} R\theta^i$ is a subring of E, and the powers of θ are left linearly independent over R. □

EXERCISE 1O. Let α be an automorphism of a ring R. Let S_1 and S_2 be ring extensions of R with units $\theta_i \in S_i$ such that S_i is a free left R–module with basis $1,\theta_i,\theta_i^{-1},\theta_i^2,\theta_i^{-2},\ldots$ and $\theta_i r = \alpha(r)\theta_i$ for all $r \in R$.

(a) If $\phi : R \to T$ is a ring homomorphism and there is a unit $\xi \in T$ such that $\xi\phi(r) = \phi\alpha(r)\xi$ for all $r \in R$, show that ϕ extends uniquely to a ring homomorphism $\Phi : S_1 \to T$ such that $\Phi(\theta_1) = \xi$.

(b) Show that the identity map on R extends to a ring isomorphism $\Phi : S_1 \to S_2$ such that $\Phi(\theta_1) = \theta_2$. □

DEFINITION. Let R be a ring and α an automorphism of R. The ring S constructed in Proposition 1.16 is denoted $R[\theta,\theta^{-1};\alpha]$ and is called a *skew–Laurent ring* or a *skew–Laurent extension* of R. Note that if α is the identity map on R, then $R[\theta,\theta^{-1};\alpha]$ is the ring of ordinary Laurent polynomials over R. Note also that the subring of $R[\theta,\theta^{-1};\alpha]$ generated by R and θ may be identified with the skew polynomial ring $R[\theta;\alpha]$.

EXERCISE 1P. If α is an automorphism of a ring R, show that $R[\theta,\theta^{-1};\alpha]^{op}$ is isomorphic to $R^{op}[\theta,\theta^{-1};\alpha^{-1}]$. □

THEOREM 1.17. *Let R be a ring and α an automorphism of R. If R is right (left) noetherian, then the skew–Laurent ring $S = R[\theta,\theta^{-1};\alpha]$ is right (left) noetherian.*

Proof. Let $T = R[\theta;\alpha] \subseteq S$. To prove the right noetherian part of the theorem, it suffices to show that $I = (I \cap T)S$ for all right ideals I of S. For if R is right noetherian, then so is T (by Theorem 1.12), whence $I \cap T$ is a finitely generated right ideal of T, and consequently $I = (I \cap T)S$ is a finitely generated right ideal of S. The left noetherian case follows similarly from showing that $J = S(J \cap T)$ for all left ideals J of S, or by using Exercise 1P.

Thus consider a right ideal I of S. Obviously $(I \cap T)S \leq I$. Any element $x \in I$ may be written as $x = \sum_{i=-n}^{n} a_i\theta^i$ for some $n \in \mathbb{N}$ and some $a_i \in R$. Then $x\theta^n \in I \cap T$, and so $x = x\theta^n\theta^{-n} \in (I \cap T)S$. Therefore $I = (I \cap T)S$, as desired. □

EXERCISE 1Q. Let α be an automorphism of a nonzero ring R. If 0 and R are the only ideals of R invariant under α, and if α^n is not inner for any positive integer n, show that $R[\theta,\theta^{-1};\alpha]$ is a simple ring. (See also Exercise 1Z.) □

♦ ADDITIONAL EXERCISES

1R. Prove Proposition 1.3 using the condition "all submodules finitely generated" in place of the ACC on submodules. □

1S. Let $R[\theta;\alpha,\delta]$ be a skew polynomial ring. If there exists $d \in R$ such that $\delta(r) = dr - \alpha(r)d$ for all $r \in R$, then δ is called an *inner α–derivation* of R. In this case, show that $R[\theta;\alpha,\delta]$ is isomorphic to a skew polynomial ring $R[\theta';\alpha]$, where $\theta' = \theta - d$. In particular, if α is the identity map and δ is an inner derivation, then $R[\theta;\delta]$ is isomorphic to an ordinary polynomial ring over R. □

1T. Let $R[\theta;\alpha,\delta]$ be a skew polynomial ring. Assume that α is an inner automorphism of R; say there exists a unit $a \in R$ with $\alpha(r) = a^{-1}ra$ for all $r \in R$. Show that $a\delta$ is a derivation on R, and that $R[\theta;\alpha,\delta]$ is isomorphic to a differential operator ring $R[\theta';a\delta]$, where $\theta' = a\theta$. □

1U. Let δ be a derivation on a ring R. For $r, s \in R$ and positive integers n, prove *Leibniz's Rule:* $\delta^n(rs) = \sum_{i=0}^{n} \binom{n}{i}\delta^{n-i}(r)\delta^i(s)$. If $\delta(r)$ commutes with r, show that $\delta(r^n) = nr^{n-1}\delta(r)$. □

1V. Let R be a commutative domain, and let K be the quotient field of R. Show that any derivation on R induces a derivation on K via the usual quotient rule for derivatives. [Hint: Exercise 1F.] □

1W. If $F \subseteq K$ are fields of characteristic zero, show that any derivation δ on F extends to a derivation on K. [Hint: By Zorn's Lemma, we may assume that δ cannot be extended to any larger subfield of K. If $x \in K$ is transcendental over F, extend δ to $F[x]$ and then to $F(x)$ using Exercises 1G, 1V. If $x \in K$ is algebraic over F, there is only one possible way to extend δ to $F(x)$.] □

1X. Let $S = R[\theta;\delta]$ be a differential operator ring. If I is any δ–ideal of R, show that $IS = SI$ and hence that IS is an ideal of S. □

1Y. Let R be a $\mathbb{Q}$–algebra and let δ be a derivation on R. If $R[\theta;\delta]$ is a simple ring, show that R is δ–simple and that δ is outer. □

1Z. Let α be an automorphism of a ring R, and set $S = R[\theta;\alpha]$. If 0 and R are the only ideals of R invariant under α, and if α^n is not inner for any positive integer n, show that $0, S, \theta S, \theta^2 S, \ldots$ are the only ideals of S. (See also Exercise 1Q.) □

1ZA. Let R be a ring and α an endomorphism of R.

(a) Construct a *skew power series ring* $R[[\theta;\alpha]]$ consisting of power series of the form $\sum_{i=0}^{\infty} r_i\theta^i$ with coefficients $r_i \in R$, where $\theta r = \alpha(r)\theta$ for all $r \in R$. [Hint: Let R and θ act on an ordinary power series ring $R[[x]]$, and show that infinite sums $\sum_{i=0}^{\infty} r_i\theta^i$ with coefficients $r_i \in R$ give well–defined additive endomorphisms of $R[[x]]$.]

(b) If α is an automorphism, construct a *skew Laurent series ring* $R((\theta;\alpha))$ consisting of Laurent series $\sum_{i=n}^{\infty} r_i\theta^i$ with $n \in \mathbb{Z}$ and coefficients $r_i \in R$, where $\theta r = \alpha(r)\theta$ for all $r \in R$.

(c) If α is an automorphism and R is right (left) noetherian, show that $R[[\theta;\alpha]]$ and $R((\theta;\alpha))$ are right (left) noetherian. □

1ZB. Let R be a ring, α an automorphism of R, and δ an α–derivation of R.

(a) Construct a *skew inverse Laurent series ring* $R((\theta^{-1};\alpha,\delta))$ consisting of inverse Laurent series $\sum_{i=-\infty}^{n} r_i\theta^i$ with $n \in \mathbb{Z}$ and coefficients $r_i \in R$, where $\theta r = \alpha(r)\theta + \delta(r)$ for all $r \in R$. (In case α is the identity map, $R((\theta^{-1};\delta))$ is called a *formal pseudo–differential operator ring*.) Observe that the subset $R[[\theta^{-1};\alpha,\delta]]$ of $R((\theta^{-1};\alpha,\delta))$ consisting of inverse Laurent series of the form $\sum_{i=-\infty}^{0} r_i\theta^i$ is a subring of $R((\theta^{-1};\alpha,\delta))$.

(b) If R is right (left) noetherian, show that $R((\theta^{-1};\alpha,\delta))$ and $R[[\theta^{-1};\alpha,\delta]]$ are right (left) noetherian. □

1ZC. Let R be a ring and δ a *locally nilpotent* derivation of R, meaning that for each $r \in R$ there is some $n \in \mathbb{N}$ such that $\delta^n(r) = 0$.

(a) Construct a *skew power series ring* $R[[\theta;\delta]]$ consisting of power series of the form $\sum_{i=0}^{\infty} r_i\theta^i$ with coefficients $r_i \in R$, where $\theta r = r\theta + \delta(r)$ for all $r \in R$.

(b) Suppose that $R = k[t]$ and $\delta = d/dt$, where k is a field of characteristic zero and t is an indeterminate. Show that in this case $R[[\theta;\delta]] \cong \mathrm{End}_k(R)$, and conclude that $R[[\theta;\delta]]$ is not noetherian on either side. □

♦ NOTES

Noetherian Rings. The introduction of commutative rings with ACC, and the source of their subsequent influence, was a paper of Noether [1921].

ACC Versus Finite Generation. Noether proved that a commutative ring has the ACC on ideals if and only if all ideals are finitely generated [1921, Satz I and comment following].

Skew Polynomial Rings. Skew polynomial rings in several variables with coefficients from a field K were introduced by Noether and Schmeidler [1920]; they were particularly interested in the cases $K[\theta_1,\dots,\theta_n;\, \delta_1,\dots,\delta_n]$ and $K[\theta_1,\dots,\theta_n;\, \alpha_1,\dots,\alpha_n]$ where K consists of (C^∞) functions in variables $x_1,\dots,x_n$ and each $\delta_i = \partial/\partial x_i$ while α_i is the automorphism of K sending x_i to $x_i + 1$ and fixing the other x_j. Later, Ore produced a systematic investigation of skew polynomial rings in one variable over a division ring [1933]; he in particular observed that in the relation $\theta r = \alpha(r)\theta + \delta(r)$, the map α must be a ring endomorphism and the map δ must be an α–derivation.

Skew Polynomial PIDs. The key ingredients in proving that a skew polynomial ring over a division ring is a PID, namely the left–hand division algorithm (in general) and the right–hand analog (in case α is an automorphism), are due to Ore [1933, pp. 483–484 and Theorem 6].

Noetherian Skew Polynomial Rings. Hilbert's original basis theorems, for homogeneous ideals in polynomial rings over fields and over $\mathbb{Z}$, are [1890, Theorems I, II]. Finite generation of left ideals for skew polynomial rings in several variables over a field was proved by Noether and Schmeidler [1920, Satz III].

Weyl Algebras. Manipulations with relations of the form $pq - qp = 1$ arising from quantum mechanics occurred in work of Dirac [1926] and Weyl [1928, §§ 10, 18, 44]. The first investigation of an algebra with generators satisfying such a relation was carried out by Littlewood [1933], who proved that $A_1(\mathbb{R})$ and $A_1(\mathbb{C})$ are nonzero simple domains satisfying the left common multiple condition and possessing division algebras of left fractions [1933, Theorems VII, X, XII, XIX, XXI]. The appelation "Weyl algebra" was introduced into ring theory by Dixmier [1968, Introduction], following an analogous usage of "infinitesimal Weyl algebras" in differential geometry by Segal [1968, § 2].

Simplicity of Weyl Algebras. Simplicity of $A_n(k)$ for k a field of characteristic zero was proved by Hirsch [1937, Theorem].

Skew–Laurent Rings. The skew–Laurent ring $\mathbb{Q}((x))[\theta,\theta^{-1};\alpha]$, where α is the automorphism of $\mathbb{Q}((x))$ sending x to $2x$, was constructed by Hilbert to show the existence of a noncommutative ordered division ring [1903, Theorem 39].

2 PRIME IDEALS

In trying to understand the ideal theory of a commutative ring, one quickly sees that it is important to first understand the prime ideals. We recall that a proper ideal P in a commutative ring R is prime if whenever we have two elements a and b of R such that $ab \in P$, it follows that $a \in P$ or $b \in P$; equivalently, P is a prime ideal if and only if the factor ring R/P is a domain. (The terminology comes from algebraic number theory, where, for instance, one replaces the prime *numbers* in $\mathbb{Z}$ by the prime *ideals* in a Dedekind domain in order to preserve the unique factorization property.) The importance of prime ideals is perhaps clearest in the setting of algebraic geometry, for if R is the coordinate ring of an algebraic variety, the prime ideals of R correspond to irreducible subvarieties.

In the noncommutative setting, we define an integral domain just as we do in the commutative case (as a nonzero ring in which the product of any two nonzero elements is nonzero), but it turns out not to be a good idea to concentrate our attention on ideals P such that R/P is a domain. In fact, many noncommutative rings have no factor rings which are domains. (Consider a matrix ring over a field.) Thus a more relaxed definition for the concept of a prime ideal in the noncommutative case is desirable. The key is to change the commutative definition by replacing products of elements with products of ideals, which was first proposed in 1928 by Krull.

In the commutative case, there is a close connection between prime ideals and nilpotent elements. In particular, the intersection of all prime ideals is the set of nilpotent elements. The noncommutative analogue of this theory is presented in the opening sections of this chapter. We then see how prime ideals arise as annihilators, which is responsible for much of their significance. The most important class of prime ideals which arises in this way is the class of primitive ideals. We close the chapter by giving an analysis of the prime ideals in certain differential operator rings.

♦ PRIME IDEALS

We remind the reader of our convention that inclusions among ideals or submodules will be denoted by $\leq, <$, etc.

DEFINITION. A *prime ideal* in a ring R is any proper ideal P of R such that whenever I and J are ideals of R with $IJ \leq P$, either $I \leq P$ or $J \leq P$. A *prime ring* is a ring in which 0 is a prime ideal. (Note that a prime ring must be nonzero.)

That this definition coincides with the usual one for the commutative case follows from part (e) of the next proposition.

PROPOSITION 2.1. *For a proper ideal P in a ring R, the following conditions are equivalent:*

(a) P is a prime ideal.

(b) R/P is a prime ring.

(c) If I and J are any right ideals of R such that $IJ \leq P$, either $I \leq P$ or $J \leq P$.

(d) If I and J are any left ideals of R such that $IJ \leq P$, either $I \leq P$ or $J \leq P$.

(e) If $x,y \in R$ with $xRy \subseteq P$, either $x \in P$ or $y \in P$.

Proof. (a) $\Rightarrow$ (b): Given ideals I and J in R/P, there exist ideals $I' \geq P$ and $J' \geq P$ in R such that $I'/P = I$ and $J'/P = J$. If $IJ = 0$, then $I'J' \leq P$, whence either $I' \leq P$ or $J' \leq P$, and so either $I = 0$ or $J = 0$.

(b) $\Rightarrow$ (a): If I and J are ideals of R satisfying $IJ \leq P$, then $(I+P)/P$ and $(J+P)/P$ are ideals of R/P whose product is zero. Then either $(I + P)/P = 0$ or $(J + P)/P = 0$, whence either $I \leq P$ or $J \leq P$.

(a) $\Rightarrow$ (c): Since I is a right ideal, $(RI)(RJ) = RIJ \leq P$. Thus either $RI \leq P$ or $RJ \leq P$.

(c) $\Rightarrow$ (e): Since $(xR)(yR) \leq P$, either $xR \leq P$ or $yR \leq P$.

(e) $\Rightarrow$ (a): Given ideals $I \nleq P$ and $J \nleq P$, choose elements $x \in I - P$ and $y \in J - P$. Then $xRy \nsubseteq P$, whence $IJ \nleq P$.

(a) $\Leftrightarrow$ (d) by symmetry. □

It follows immediately (by induction) from Proposition 2.1 that if P is a prime ideal in a ring R and $J_1,\dots,J_n$ are right ideals of R such that $J_1J_2\cdots J_n \leq P$, then some $J_i \leq P$.

Recall that by a *maximal ideal* in a ring is meant a maximal *proper* ideal, i.e., an ideal which is a maximal element in the collection of proper ideals.

PROPOSITION 2.2. *Every maximal ideal M of a ring R is a prime ideal.*

Proof. If I and J are ideals of R not contained in M, then $I + M = R$ and $J + M = R$. Now

$$R = (I + M)(J + M) = IJ + IM + MJ + M^2 \leq IJ + M,$$

and hence $IJ \nleq M$. □

Proposition 2.2 together with Zorn's Lemma guarantees that every nonzero ring has at least one prime ideal.

EXERCISE 2A. Let $R = \mathbb{Z} + \mathbb{Z}i + \mathbb{Z}j + \mathbb{Z}k$ be the ring of quaternions with integer coefficients. (In algebraic number theory, one meets a *ring of integer quaternions* which is slightly larger than the ring R.) Since R is contained in the division ring $\mathbb{H}$, it is a domain, and so 0 is a prime ideal of R.

(a) For any odd prime integer p, show that pR is a maximal ideal of R. [Hint: If I is a proper ideal containing pR, show that the cosets $1+I, i+I, j+I, k+I$ are linearly independent over $\mathbb{Z}/p\mathbb{Z}$.]

(b) For any odd prime integer p, show that R/pR is not a domain. [Hint: Find integers a,b in $\{0,1,\ldots,(p-1)/2\}$ such that $a^2 \equiv -b^2 - 1 \pmod{p}$, and look at $1+ai+bj$ and $1-ai-bj$.] (Readers who are familiar with the Wedderburn–Artin Theorem can now conclude that $R/pR \cong M_2(\mathbb{Z}/p\mathbb{Z})$.)

(c) Show that 2R is not a prime ideal of R, and that $2R + (1+i)R + (1+j)R$ is a maximal ideal of R.

(d) Show that the only prime ideals of R are 0 and $2R + (1+i)R + (1+j)R$ together with pR for all odd prime integers p. [Hint: If P is a prime ideal of R not containing any prime integer, show that R/P is a free abelian group with basis $\{1+P, i+P, j+P, k+P\}$.] □

DEFINITION. A *minimal prime ideal* in a ring R is any prime ideal of R which does not properly contain any other prime ideals.

For instance, if R is a prime ring, then 0 is a minimal prime ideal of R, and is the only one.

EXERCISE 2B. Given an integer $n > 1$, show that the minimal prime ideals of $\mathbb{Z}/n\mathbb{Z}$ are exactly the ideals $p\mathbb{Z}/n\mathbb{Z}$ where p is any prime divisor of n. □

PROPOSITION 2.3. *Any prime ideal P in a ring R contains a minimal prime ideal.*

Proof. Let $\mathscr{X}$ be the set of those prime ideals of R which are contained in P. We may use Zorn's Lemma going downward in $\mathscr{X}$ provided we show that any chain $\mathscr{Y} \subseteq \mathscr{X}$ has a lower bound in $\mathscr{X}$.

Since $\mathscr{Y}$ is a chain, the set $Q = \bigcap \mathscr{Y}$ is an ideal of R, and it is clear that $Q \leq P$. We claim that Q is a prime ideal.

Thus consider any $x,y \in R$ such that $xRy \subseteq Q$ but $x \notin Q$. Then $x \notin P'$ for some $P' \in \mathcal{Y}$. For any $P'' \in \mathcal{Y}$ such that $P'' \leq P'$, we have $x \notin P''$ and $xRy \subseteq Q \subseteq P''$, whence $y \in P''$. In particular, $y \in P'$. If $P'' \in \mathcal{Y}$ and $P'' \nleq P'$, then $P' \leq P''$, and so $y \in P''$. Hence, $y \in P''$ for all elements P'' of $\mathcal{Y}$, and so $y \in Q$, which proves that Q is a prime ideal.

Now $Q \in \mathcal{X}$, and Q is a lower bound for $\mathcal{Y}$.

Thus Zorn's Lemma applies, giving us a prime ideal $P^* \in \mathcal{X}$ which is minimal among the ideals in $\mathcal{X}$. Since any prime ideal contained in P^* is in $\mathcal{X}$, we conclude that P^* is a minimal prime ideal of R. □

Given an ideal I in a ring R and a prime ideal P containing I, we may apply Proposition 2.3 in the ring R/I to see that the prime ideal P/I contains a minimal prime Q/I of R/I. Then Q is a prime ideal of R which contains I and is minimal among the primes containing I. By way of abbreviation, we say that Q is a *prime minimal over* I.

THEOREM 2.4. *In a right noetherian ring R, there exist only finitely many minimal prime ideals, and there is a finite product of minimal prime ideals (repetitions allowed) that equals zero.*

Proof. It suffices to prove that there exist prime ideals $P_1,\dots,P_n$ in R such that $P_1P_2\cdots P_n = 0$. To see this, note that after replacing each P_i by a minimal prime ideal contained in it, we may assume that each P_i is minimal. Since any minimal prime P contains $P_1P_2\cdots P_n$, it must contain some P_j, whence $P = P_j$ by minimality. Thus the minimal prime ideals of R are contained in the finite set $\{P_1,\dots,P_n\}$.

Suppose that no finite product of prime ideals in R is zero. Let X be the set of those ideals K in R that do not contain a finite product of prime ideals. Since X contains 0, it is nonempty. By the noetherian hypothesis (not Zorn's Lemma!), there exists a maximal element $K \in X$.

As R/K is a counterexample to the theorem, we may replace R by R/K. Thus we may assume, without loss of generality, that no finite product of prime ideals in R is zero, while all nonzero ideals of R contain finite products of prime ideals.

In particular, 0 cannot be a prime ideal. Hence, there exist nonzero ideals I,J in R such that $IJ = 0$. Then there exist prime ideals $P_1,\dots,P_m,Q_1,\dots,Q_n$ in R with $P_1P_2\cdots P_m \leq I$ and $Q_1Q_2\cdots Q_n \leq J$. But then $P_1P_2\cdots P_mQ_1Q_2\cdots Q_n = 0$, contradicting our supposition.

Therefore some finite product of prime ideals in R is zero. □

The use of the noetherian condition in the proof of Theorem 2.4 to pass from R to R/K is known as *noetherian induction*. Since R/K is as small as possible among factor rings of R violating the theorem, it is known as a *minimal criminal*. (For this terminology we are indebted to Reinhold Baer, who remarked that as in the larger world, it is the minimal criminal who is apprehended.)

In general, a ring may have infinitely many minimal prime ideals, as the following example shows.

EXERCISE 2C. Let X be an infinite set, let k be a field, and let R be the ring of all functions from X to k. For $x \in X$, let P_x be the set of those functions in R which vanish at x. Show that each P_x is a minimal prime ideal of R, and also a maximal ideal. □

♦ SEMIPRIME IDEALS AND NILPOTENCE

DEFINITION. A *semiprime ideal* in a ring R is any ideal of R which is an intersection of prime ideals. (By convention, the intersection of the empty family of prime ideals of R is R, so that R is a semiprime ideal of itself.) A *semiprime ring* is any ring in which 0 is a semiprime ideal. Note that an ideal P in a ring R is semprime if and only if R/P is a semiprime ring.

In $\mathbb{Z}$, the intersection of any infinite family of prime ideals is 0. The intersection of a finite list $p_1\mathbb{Z},\dots,p_k\mathbb{Z}$ of prime ideals, where $p_1,\dots,p_k$ are distinct prime integers, is the ideal $p_1p_2\cdots p_k\mathbb{Z}$. Hence, the nonzero semiprime ideals of $\mathbb{Z}$ consist of $\mathbb{Z}$ together with the ideals $n\mathbb{Z}$ where n is any square–free integer.

The development of ideas in this section is made clearer if we first review the commutative case, even though it is well known.

LEMMA 2.5. *Let R be a ring and let X be a subset of R such that $0 \notin X$ and such that X is closed under multiplication. Let P be an ideal of R chosen maximal with respect to the property that P and X are disjoint. Then P is a prime ideal.*

Proof. We must show that R/P is a prime ring – that is, that the product of two nonzero ideals of R/P is nonzero. Equivalently, it suffices to show that if I and J are ideals such that $I > P$ and $J > P$, then $IJ \nsubseteq P$. If $I > P$, then by the maximality of P, there is an $x \in X$ such that $x \in I$, and, similarly, if $J > P$, there is a $y \in X$ such that $y \in J$. Then $xy \in IJ$, and $xy \in X$ (since X is multiplicatively closed). Since P and X are disjoint, it follows that $IJ \nsubseteq P$ as desired. □

PROPOSITION 2.6. *If R is a commutative ring, then*

(a) The intersection of all prime ideals of R is precisely the set of nilpotent elements of R.

(b) For every ideal I, the intersection of all of the prime ideals of R containing I is the set of elements r in R such that $r^n \in I$ for some positive integer n.

(c) R is semiprime if and only if it contains no nonzero nilpotent elements.

Proof. (a) If r is a nilpotent element of R then r must be contained in every prime ideal, since if P is a prime ideal then R/P has no nonzero nilpotent elements. Hence all nilpotent elements are in the intersection of the prime ideals. Conversely, if r is not nilpotent, then letting $X = \{r^n \mid n \in \mathbb{N}\}$, we can apply Lemma 2.5 to obtain a prime ideal P such that $r \notin P$, and so r is not in the intersection of the prime ideals.

Clearly, (b) follows from (a) by passing to the factor ring R/I, and (c) is a special case of (a). □

It follows from Proposition 2.6 that an ideal I in a commutative ring R is semprime if and only if whenever $x \in R$ and $x^2 \in I$, it follows that $x \in I$. The example of a matrix ring over a field shows that this criterion fails in the noncommutative case. However, there is an analogous criterion, as we will see in the next theorem. One should think of it as parallel to condition (e) in Proposition 2.1 describing a prime ideal. The method of the proof is to generalize the idea of Lemma 2.5 to a set which is not quite multiplicatively closed. (This idea is further developed in Exercise 2ZB.)

THEOREM 2.7. [Levitzki, Nagata] *An ideal I in a ring R is semiprime if and only if*

(*) Whenever $x \in R$ with $xRx \subseteq I$, then $x \in I$.

Proof. First suppose that I equals the intersection of some family $\{P_j \mid j \in J\}$ of prime ideals. Given $x \in R$ with $xRx \subseteq I$, we have $xRx \subseteq P_j$ for each $j \in J$. Then x lies in P_j for each $j \in J$, whence $x \in I$.

Conversely, assume that (*) holds. We shall prove that I equals the intersection of all those prime ideals of R which contain I. Hence, given any $x \in R - I$, we need a prime ideal $P \geq I$ such that $x \notin P$.

Set $x_0 = x$. By (*), $x_0Rx_0 \nsubseteq I$, and so we may choose an element x_1 in $x_0Rx_0 - I$. Applying (*) to x_1, we may choose an element x_2 in $x_1Rx_1 - I$. Continuing in this manner, we obtain elements $x_0, x_1, x_2, \ldots$ in $R - I$ such that $x_{i+1} \in x_iRx_i$ for all i. Note (by induction) that if J is any ideal, and some $x_j \in J$, then $x_n \in J$ for all $n \geq j$.

Now $x_i \notin I$ for all i. By Zorn's Lemma, there is an ideal $P \geq I$ maximal with respect to the property that $x_i \notin P$ for all i. In particular, $x = x_0 \notin P$, and P is a proper ideal. We claim that P is a prime ideal.

The argument is now the same as that in the proof of Lemma 2.5. If P is not prime, then R/P is not a prime ring, and so R/P has two nonzero ideals whose product is zero. Hence, R has ideals J and K, properly containing P, such that $JK \leq P$. By maximality of P, some $x_j \in J$ and some $x_k \in K$. If m is the maximum of j and k, then $x_m \in J \cap K$, and so

$$x_{m+1} \in x_m R x_m \subseteq JK \subseteq P,$$

contradicting our choice of P.

Thus P is a prime ideal, as claimed. Therefore I does equal the intersection of all prime ideals of R containing I, whence I is semiprime. □

COROLLARY 2.8. *For an ideal I in a ring R, the following conditions are equivalent:*

(a) I is a semiprime ideal.

(b) If J is any ideal of R such that $J^2 \leq I$, then $J \leq I$.

(c) If J is any right ideal of R such that $J^2 \leq I$, then $J \leq I$.

(d) If J is any left ideal of R such that $J^2 \leq I$, then $J \leq I$.

Proof. (a) $\Rightarrow$ (c): For any $x \in J$, we have $xRx \subseteq J^2 \subseteq I$, whence $x \in I$ by Theorem 2.7. Thus $J \leq I$.

(c) $\Rightarrow$ (b) *a priori*.

(b) $\Rightarrow$ (a): Given any $x \in R$ such that $xRx \subseteq I$, we have $(RxR)^2 = RxRxR \leq I$ and so $RxR \leq I$, whence $x \in I$. By Theorem 2.7, I is semiprime.

(a) $\Leftrightarrow$ (d) by symmetry. □

COROLLARY 2.9. *Let I be a semiprime ideal in a ring R. If J is a right or left ideal of R such that $J^n \leq I$ for some positive integer n, then $J \leq I$.*

Proof. In case $n = 1$, there is nothing to prove. Now let $n > 1$, and assume the corollary holds for lower powers. Since $n \geq 2$, we have $2n - 2 \geq n$, whence

$$(J^{n-1})^2 = J^{2n-2} \leq J^n \leq I.$$

Then $J^{n-1} \leq I$ by Corollary 2.8, and so $J \leq I$ by the induction hypothesis. This completes the induction step. □

DEFINITION. A right or left ideal J in a ring R is *nilpotent* provided $J^n = 0$ for some positive integer n. More generally, J is *nil* provided every element of J is nilpotent. Exercise 2G provides an example of a nil ideal which is not nilpotent. On the other hand, in noetherian rings all nil one–sided ideals are nilpotent (Corollary 5.19; Exercise 5K).

Corollary 2.9 shows that in a semiprime ring, the only nilpotent right or left ideal is 0. Conversely, if a ring R has no nonzero nilpotent ideals, then R is a semiprime ring by Corollary 2.8.

DEFINITION. The *prime radical* of a ring R is the intersection of all the prime ideals of R.

If R is the zero ring, it has no prime ideals, and the prime radical equals R. If R is nonzero, it has at least one maximal ideal, which is prime by Proposition 2.2. Thus the prime radical of a nonzero ring is a proper ideal.

Note that a ring R is semiprime if and only if its prime radical is zero. In any case, the prime radical of R is the smallest semiprime ideal of R, and because the prime radical is semiprime, it contains all nilpotent one–sided ideals of R (see Corollary 2.9).

EXERCISE 2D. Given an integer $n > 1$, find the prime radical of $\mathbb{Z}/n\mathbb{Z}$. □

EXERCISE 2E. Show that the prime radical of any ring is nil. □

PROPOSITION 2.10. *In any ring R, the prime radical equals the intersection of the minimal prime ideals of R.*

Proof. This is immediate from the fact that every prime ideal of R contains a minimal prime (Proposition 2.3). □

The next theorem completes our development of a noncommutative analogue to Proposition 2.6. (Note that we have only done this in the noetherian case.)

THEOREM 2.11. *Let R be a right noetherian ring, and let N be the prime radical of R. Then N is a nilpotent ideal of R containing all the nilpotent right or left ideals of R.*

Proof. Since N is a semiprime ideal, it contains all the nilpotent one–sided ideals of R by Corollary 2.9. By Theorem 2.4, there exist (minimal) prime ideals $P_1,\ldots,P_k$ in R such that $P_1P_2\cdots P_k = 0$. Since N is contained in each P_i, we conclude that $N^k = 0$. □

EXERCISE 2F. Prove Theorem 2.11 without using Theorem 2.4, by considering a maximal nilpotent ideal of R. □

EXERCISE 2G. Show that the prime radical of the ring $R = \prod_{n=1}^{\infty} \mathbb{Z}/2^n\mathbb{Z}$ is nil but not nilpotent. [Hint: Show that the prime radical contains elements $x_1, x_2, \ldots$ such that $x_n{}^n \neq 0$.] □

EXERCISE 2H. If R is a prime (semiprime) ring, show that each matrix ring $M_n(R)$ is prime (semiprime). □

♦ ANNIHILATORS AND ASSOCIATED PRIME IDEALS

A basic principle of algebraic geometry is to study algebraic varieties via rings of functions on them. A key part of the theory is a correspondence between certain ideals and subvarieties that arises from "annihilation". (Such terminology is commonly used in reference to any process that results in a zero. E.g., if $f(x) = 0$ or $cx = 0$, one says that x has been "annihilated" or even "killed" by f or c.) If R is a ring of functions on a set X, we can build a correspondence that takes a subset Y of X to an ideal

$$\{r \in R \mid r(y) = 0 \text{ for all } y \in Y\}$$

in R, and that takes an ideal I of R to a "zero–set"

$$\{x \in X \mid r(x) = 0 \text{ for all } r \in I\}$$

in X. In case X is an algebraic variety and R is its coordinate ring, then under this correspondence subvarieties of X correspond precisely to semiprime ideals of R and irreducible subvarieties of X correspond precisely to prime ideals of R.

In the theory of modules, there is a similar correspondence between certain ideals and certain submodules (using annihilation via multiplication in place of annihilation via functions), which we will study in this and the following sections. Here again it is important to study the submodules which correspond to prime ideals, and to see how other modules are made up from these.

DEFINITION. Let A be a right module over a ring R. Given any subset $X \subseteq A$, the *annihilator* of X is the set

$$\text{ann}(X) = \{r \in R \mid xr = 0 \text{ for all } x \in X\},$$

which is a right ideal of R. In case the ring R must be made explicit, we write $\text{ann}_R(X)$. Similarly, to emphasize that we are taking an annihilator on the right side of X (because A is a right module), we can write $\text{r.ann}(X)$ for $\text{ann}(X)$. When X consists of a single element x, we abbreviate $\text{ann}(\{x\})$ to $\text{ann}(x)$. We have already noted that $\text{ann}(X)$ is a right ideal of R; moreover, if X is a submodule of A then $\text{ann}(X)$ is an ideal of R. Annihilators of subsets of left R–modules are defined analogously, and are left ideals of R. In case $A = R$, we must specify whether an annihilator is taken with respect to the right module R_R or the left module ${}_RR$. Thus for $X \subseteq R$ we have a *right annihilator*

$$\text{r.ann}(X) = \{r \in R \mid xr = 0 \text{ for all } x \in X\}$$

as well as a *left annihilator*

$$\text{l.ann}(X) = \{r \in R \mid rx = 0 \text{ for all } x \in X\}.$$

Finally, there are annihilators in modules to define. If A is a right R–module and Y is a subset of R, the *annihilator of* Y *in* A is the set

$$\text{ann}_A(Y) = \{a \in A \mid ay = 0 \text{ for all } y \in Y\},$$

which is an additive subgroup of A. To emphasize that this annihilator is on the left side of Y, we may write $\text{l.ann}_A(Y)$ for $\text{ann}_A(Y)$. The annihilator of Y in a left R–module is defined analogously.

Note that the annihilator of a left ideal of R in a right R–module A is a submodule of A, and similarly the annihilator of a right ideal in a left module is a submodule.

DEFINITION. A module A over a ring R is a *faithful* R–module if $\text{ann}_R(A) = 0$.

Notice that a faithful module over a nonzero ring must be nonzero. Note also that the annihilator of an R–module A is an ideal of R, and that A is a faithful module over $R/\text{ann}(A)$.

An important example to observe is that in a prime ring, every nonzero right or left ideal is faithful.

DEFINITION. A module A over a ring R is *fully faithful* provided A and all nonzero submodules of A are faithful R–modules. If A is an R–module which is fully faithful as a module over $R/\text{ann}_R(A)$ then A is sometimes called a *prime module*. (Observe that the annihilator of a nonzero prime module must be a prime ideal.)

The following proposition provides the key to finding fully faithful submodules over factor rings, at least in modules over noetherian rings.

PROPOSITION 2.12. *Let* A *be a nonzero right module over a ring* R. *Suppose that there exists an ideal* P *maximal among the annihilators of nonzero submodules of* A. *Then* P *is a prime ideal of* R, *and* $\text{ann}_A(P)$ *is a fully faithful right* (R/P)*–module.*

Proof. There is a nonzero submodule B in A such that $P = \text{ann}(B)$. Suppose that I and J are ideals of R, properly containing P, such that $IJ \leq P$. Then $BI \neq 0$ and $\text{ann}(BI) \geq J > P$, contradicting the maximality of P. Thus P is prime.

Now set $C = \text{ann}_A(P)$ and note that C is a submodule of A with $P \leq \text{ann}(C)$. Then $P = \text{ann}(C)$, because $B \leq C$. Thus C is a faithful right (R/P)–module. Given any nonzero submodule $D \leq C$, we have $P = \text{ann}(C) \leq \text{ann}(D)$, whence $P = \text{ann}(D)$ by maximality of P. Therefore C is fully faithful as a right (R/P)–module. □

In general, given an R–module A the family of annihilators of nonzero submodules of A need not have any maximal elements. However, if R is either right or left noetherian, the existence of maximal annihilators is automatic.

DEFINITION. An *annihilator prime* for a module A over a ring R is any prime ideal P of R which equals the annihilator of some nonzero submodule of A. In this case $\text{ann}_A(P)$ is clearly nonzero and is a faithful (R/P)–module. An *associated prime* of A is any annihilator prime P which equals the annihilator of some nonzero submodule B of A such that B is a *fully* faithful (R/P)–module; in other words, not only must P equal $\text{ann}_R(B)$, it must also equal the annihilator of each nonzero submodule of B. (We do not require, however, that $\text{ann}_A(P)$ be a fully faithful (R/P)–module.) The set of all associated primes of A is denoted $\text{Ass}(A)$.

Proposition 2.12 shows that any ideal maximal among the annihilators of nonzero submodules of a module A is an associated prime of A. In particular, it follows that every nonzero module over a right or left noetherian ring has at least one associated prime. Not every associated prime arises as a maximal annihilator, however. For instance, the $\mathbb{Z}$–module $\mathbb{Z} \oplus (\mathbb{Z}/2\mathbb{Z})$ has two associated primes, 0 and $2\mathbb{Z}$, and 0 is certainly not maximal among annihilators of nonzero submodules of this module.

Over the ring $\mathbb{Z}$, the module A which is the direct sum of the cyclic modules $\mathbb{Z}/p\mathbb{Z}$ for all primes p is an example of a module whose annihilator is the prime ideal 0 but for which 0 is not an associated prime. However, this cannot occur for finitely generated modules over a commutative noetherian ring (cf. Exercise 4ZB in which it is shown that annihilator primes and associated primes are the same for finitely generated modules over a commutative noetherian ring.) For finitely generated modules over a noncommutative noetherian ring Exercise 2ZE shows that an annihilator prime need not be associated, but these two classes of primes do coincide for important classes of rings, cf. Exercise 11K.

EXERCISE 2I. If B is a submodule of a module A, show that $\text{Ass}(B) \subseteq \text{Ass}(A)$ and that $\text{Ass}(A) \subseteq \text{Ass}(B) \cup \text{Ass}(A/B)$. If every nonzero submodule of A has nonzero intersection with B, show that $\text{Ass}(A) = \text{Ass}(B)$. □

EXERCISE 2J. Let $P_1,\dots,P_n$ be distinct associated primes of a module A, and for $i = 1,\dots,n$ let B_i be a nonzero submodule of A such that B_i is a fully faithful (R/P_i)–module. Show that $B_1,\dots,B_n$ are independent (that is, their sum is a direct sum). [Hint: Assume by induction that any $n - 1$ of the B_i are independent. Re–index everything so that P_1 is minimal among the P_i and then show that $B_1 \cap (B_2 + \dots + B_n) = 0$.] Conclude that a noetherian module has only finitely many associated primes. □

♦ AFFILIATED PRIME IDEALS

The idea in this section is to take a finitely generated module over a noetherian ring and to break it into a finite number of simpler pieces which will be fully faithful modules over prime factor rings of R. We thus introduce a larger family of primes, the affiliated primes, which contain more information than the associated primes.

DEFINITION. Let A be a nonzero module over a ring R. An *affiliated submodule* of A is any submodule of the form $\mathrm{ann}_A(P)$ where P is an ideal of R maximal among the annihilators of nonzero submodules of A. (We note that such an ideal P is an associated prime of A by Proposition 2.12, and that P equals the annihilator in R of the affiliated submodule $\mathrm{ann}_A(P)$.) An *affiliated series* for A is a series of submodules of the form

$$0 = A_0 < A_1 < \ldots < A_n = A$$

where for each $i = 1,\ldots,n$ the module A_i/A_{i-1} is an affiliated submodule of A/A_{i-1}. If $P_i = \mathrm{ann}_R(A_i/A_{i-1})$ then the series $P_1,\ldots,P_n$ is the series of *affiliated primes* of A corresponding to the given affiliated series. In general, an *affiliated prime* of A is a prime ideal of R which appears in the series of affiliated primes corresponding to some affiliated series of A.

A module A may have many affiliated series, and the corresponding sets of affiliated primes do not necessarily coincide, even disregarding the order in which each series of affiliated primes is listed (Exercise 2O).

EXERCISE 2K. Show that a noetherian module has only finitely many affiliated series, and hence only finitely many affiliated primes. [Hint: If not, choose a minimal criminal and apply Exercise 2J.] □

PROPOSITION 2.13. *If A is a nonzero finitely generated right module over a right noetherian ring R, then A has an affiliated series. If $0 = A_0 < A_1 < \ldots < A_n = A$ is such an affiliated series, and $P_1,\ldots,P_n$ are the corresponding affiliated primes, then each A_i/A_{i-1} is a fully faithful right (R/P_i)–module.*

Proof. It is clear from Proposition 2.12 that any affiliated submodule B in any R–module is fully faithful as a module over $R/\mathrm{ann}_R(B)$. The final statement of the proposition follows, and it only remains to prove the existence of an affiliated series for A.

Set $A_0 = 0$. Since R is right noetherian, there exists an ideal P_1 maximal among the annihilators of nonzero submodules of A, and then $A_1 = \mathrm{ann}_A(P_1)$ is an affiliated submodule of A. Note that $A_1 \neq 0$. Similarly, if $A_1 \neq A$ there exists a (nonzero) affiliated submodule A_2/A_1 in A/A_1. Since A is a noetherian module, the properly ascending chain $A_0 < A_1 < A_2 < \ldots$ of submodules of A must terminate at some stage A_n, whence $A_n = A$, and we have obtained an affiliated series for A. □

EXERCISE 2L. Let A be a right module over a commutative ring R, and assume that A has an affiliated series $0 = A_0 < A_1 < \ldots < A_n = A$, with corresponding affiliated primes $P_1,\ldots,P_n$. Show that $P_i = \mathrm{ann}_R(A_iP_{i-1}P_{i-2}\cdots P_1)$ for each $i = 1,\ldots,n$. [Hint:

Show first that $A_i = \mathrm{ann}_A(P_iP_{i-1}\cdots P_1)$ for all i.] Thus every affiliated prime of A is also an annihilator prime. □

EXERCISE 2M. Let A be a module which has an affiliated series, with corresponding affiliated primes $P_1,\ldots,P_n$.

(a) If B is a nonzero submodule of A, show that B has an affiliated series such that the corresponding affiliated primes are a subsequence of $P_1,\ldots,P_n$.

(b) If P is any annihilator prime for A, show that P is one of $P_1,\ldots,P_n$. Conclude that $\mathrm{Ass}(A) \subseteq \{P_1,\ldots,P_n\}$. Thus annihilator primes and associated primes of A are affiliated in the strong sense that they appear among the affiliated primes for every affiliated series for A. □

PROPOSITION 2.14. *Let R be a right noetherian ring and A a finitely generated right R–module, and let $P_1,\ldots,P_n$ be the affiliated primes corresponding to some affiliated series for A. Then if P is a prime minimal over $\mathrm{ann}_R(A)$, there is an index i such that $P = P_i$.*

Proof. Let $0 = A_0 < A_1 < \ldots < A_n = A$ be an affiliated series for A with corresponding affiliated primes $P_1,\ldots,P_n$. Since $AP_nP_{n-1}\cdots P_1 = 0$, it follows that $P \geq \mathrm{ann}_R(A) \geq P_nP_{n-1}\cdots P_1$. Hence, $P \geq P_i$ for some index i. On the other hand, since $P_i = \mathrm{ann}_R(A_i/A_{i-1})$ it is clear that $P_i \geq \mathrm{ann}_R(A)$, and since P is minimal over $\mathrm{ann}_R(A)$ we conclude that $P = P_i$. □

EXERCISE 2N. Let $T = k[x,y]$ be a polynomial ring over a field k, and set $R = T/(x^2T + xyT)$. Find an affiliated series for the module R_R, and show that one of the corresponding affiliated primes is not a minimal prime ideal of R. □

EXERCISE 2O. Set $R = \begin{pmatrix} \mathbb{Z} & \mathbb{Z} \\ 0 & \mathbb{Z} \end{pmatrix}$, view the row $(\mathbb{Z} \ \ \mathbb{Z})$ as a right R–module, and set $A = (\mathbb{Z} \ \ \mathbb{Z})/(0 \ \ 2\mathbb{Z})$. Find all affiliated series for the right R–module A (there are two). Show that the corresponding sets of affiliated primes are not the same, and that one of these affiliated primes is neither an associated prime nor an annihilator prime. □

Associated primes are always annihilator primes (by definition), and all annihilator primes are affiliated primes (Exercise 2M). However, even for a finitely generated module, annihilator primes need not be associated (Exercise 2ZE), and affiliated primes need not be annihilators (Exercise 2O). We will see in Exercise 4ZB that for a finitely generated module over a commutative noetherian ring, the affiliated primes are precisely the associated primes. In fact, the affiliated primes were introduced partly in order to find a family of primes which would contain the same kind of information which is contained in the associated primes in the commutative case.

♦ PRIMITIVE AND SEMIPRIMITIVE IDEALS

An almost trivial case of Proposition 2.12 is the case of a simple module A. Here A itself is the only nonzero submodule of A, and hence $\mathrm{ann}(A)$ is automatically maximal among annihilators of nonzero submodules of A. Such an annihilator is said to be "primitive", as follows.

DEFINITION. An ideal P in a ring R is *right (left) primitive* provided $P = \mathrm{ann}_R(A)$ for some simple right (left) R–module A. A *right (left) primitive ring* is any ring in which 0 is a right (left) primitive ideal, i.e., any ring which has a faithful simple right (left) module.

Not all right primitive rings are left primitive, but the known examples are too involved to reproduce here. It is not known whether all right primitive noetherian rings are left primitive.

PROPOSITION 2.15. *Every right or left primitive ideal in a ring R is a prime ideal. Every maximal ideal of R is a right and left primitive ideal.*

Proof. The primeness of primitive ideals follows from Proposition 2.12.

Given a maximal ideal M in R, choose a maximal right ideal K containing M. Then R/K is a simple right R–module and $\mathrm{ann}_R(R/K)$ is a right primitive ideal of R. Since $RM = M \leq K$, we have $(R/K)M = 0$, so that $M \leq \mathrm{ann}(R/K)$. Then $M = \mathrm{ann}(R/K)$ by maximality of M, whence M is right primitive. By symmetry, M is also left primitive. □

Over a commutative ring R, any simple module is isomorphic to R/M for some maximal ideal M, and $\mathrm{ann}(R/M) = M$. Thus all primitive ideals of R are maximal; equivalently, all commutative primitive rings are simple rings. This does not hold in all noncommutative rings, as the following examples show.

EXERCISE 2P. Let V be an infinite–dimensional vector space over a field k, and let $R = \mathrm{End}_k(V)$.

(a) Show that the set $\{f \in R \mid \dim_k(f(V)) < \infty\}$ is a proper nonzero ideal of R. Thus R is not a simple ring.

(b) Note that every nonzero element of V generates V as a left R–module. Thus V is a faithful simple left R–module and R is left primitive.

(c) Choose a nonzero vector $v \in V$, and choose a map $e \in R$ such that $e(v) = v$ and $e(V) = kv$. Show that eR is a faithful simple right R–module and Re is a faithful simple left R–module. Thus R is a right and left primitive ring. □

EXERCISE 2Q. Let R be the polynomial ring $k[x]$ where k is a field of characteristic zero, let $\delta = x\frac{d}{dx}$, and let $S = R[\theta; \delta]$. By Theorem 1.12, S is a noetherian ring. The point of this exercise is to show that S is primitive but not simple. (A treatment which uses more machinery but fewer computations is in Exercises 2Y and 2Z.)

(a) Observe that xS is a proper nonzero ideal of S, and so S is not a simple ring.

(b) Show that $(x-1)S$ is a maximal right ideal of S. [Hint: If $s \in S$ and $s \notin (x-1)S$, observe that $s = (x-1)t + \alpha_0 + \alpha_1\theta + \ldots + \alpha_n\theta^n$ for some $t \in S$ and some $\alpha_i \in k$ with $\alpha_n \neq 0$. Then show that $s(x-1)^n = (x-1)u + n!\alpha_n x^n$ for some $u \in S$.]

(c) Show that $S/(x-1)S$ is a faithful simple right S–module. [Hint: If $0 \neq s \in S$, observe that $s = (x-1)^n t$ for some $n \in \mathbb{Z}^+$ and some $t \in S$ with $t \notin (x-1)S$. Then show that $\theta^n s = (x-1)u + n!x^n t$ for some $u \in S$, and conclude that $\theta^n s \notin (x-1)S$.] Thus S is a right primitive ring. (Cf. Exercise 2Y.)

(d) Show that S is also a left primitive ring. □

PROPOSITION 2.16. *In any ring R, the following sets coincide:*

(a) The intersection of all maximal right ideals of R.

(b) The intersection of all maximal left ideals of R.

(c) The intersection of all right primitive ideals of R.

(d) The intersection of all left primitive ideals of R.

Proof. Let J_a, J_b, J_c, J_d denote the four intersections.

Given any maximal right ideal M of R, the annihilator of the simple right module R/M is a right primitive ideal P. Since $(R/M)P = 0$, we have $P \leq M$, whence $J_c \leq P \leq M$. Thus $J_c \leq J_a$.

We next show that J_a is an ideal of R. Consider any $x \in J_a$ and any $r \in R$. Given any maximal right ideal M of R, either $r \in M$ or $M + rR = R$. If $r \in M$, then obviously $rx \in M$. If $r \notin M$, then left multiplication by r induces an isomorphism of R/L onto R/M, where $L = \{y \in R \mid ry \in M\}$. Hence, R/L is a simple right R–module and L is a maximal right ideal of R, from which we obtain $x \in J_a \leq L$ and so $rx \in M$. Now rx lies in all the maximal right ideals of R, whence $rx \in J_a$. Thus J_a is an ideal, as claimed.

Given any right primitive ideal P in R, there is a maximal right ideal M in R such that $\text{ann}(R/M) = P$. Since J_a is an ideal, $RJ_a = J_a \leq M$ and so $(R/M)J_a = 0$, whence $J_a \leq P$. Thus $J_a \leq J_c$.

Therefore $J_a = J_c$. By symmetry, $J_b = J_d$.

We claim that whenever $x \in J_a$, then $1 - x$ has a right inverse in R. If not, $(1-x)R \neq R$ and so $(1-x)R$ is contained in some maximal right ideal M. As $x \in J_a \leq M$, this is impossible. Thus $1 - x$ must have a right inverse, as claimed.

Next we claim that whenever $x \in J_a$, then $1 - x$ is invertible in R. By the previous claim, $1 - x$ has a right inverse $y \in R$. Then $(1 - x)y = 1$ and so $y = 1 + xy$. Since $-xy \in J_a$, a second application of the previous claim shows that y has a right inverse $z \in R$. Then $z = (1 - x)yz = 1 - x$, whence $y(1 - x) = 1$. Thus $1 - x$ is invertible, as claimed.

We can now show that $J_a \leq J_b$. Consider any $x \in J_a$ and any maximal left ideal M of R. If $x \notin M$, then $rx + m = 1$ for some $r \in R$ and some $m \in M$. Since J_a is an ideal, $rx \in J_a$. By the previous claim, the element $m = 1 - rx$ is invertible in R, which is impossible. Hence, $x \in M$. Thus $J_a \leq J_b$.

By symmetry, $J_b \leq J_a$, and therefore $J_a = J_b$. □

DEFINITION. In any ring R, the ideal defined by the intersections given in Proposition 2.16 is called the *Jacobson radical* of R, denoted J(R).

Since all right and left primitive ideals of R are prime (Proposition 2.15), the prime radical of R is contained in J(R). For example, if $R = k[[x]]$ is a power series ring over a field k, then 0 is a prime ideal of R and so the prime radical of R is 0. On the other hand, xR is the unique maximal ideal of R, whence $J(R) = xR$.

DEFINITION. A ring R is *semiprimitive* (or *Jacobson semisimple*) if and only if $J(R) = 0$. A *semiprimitive ideal* (or a *J–ideal*) in a ring R is any ideal I such that $J(R/I) = 0$.

By Proposition 2.16, an ideal I of a ring R is semiprimitive if and only if I is an intersection of right primitive ideals, if and only if I is an intersection of left primitive ideals. Hence the semiprimitive ideals of R stand in the same relation to the primitive ideals as the semiprime ideals do to the prime ideals.

Note that all semiprimitive rings are semiprime. The example k[[x]] of a power series ring over a field shows that not all semiprime (or even prime) rings are semiprimitive.

EXERCISE 2R. Find the Jacobson radical of a formal triangular matrix ring $\begin{pmatrix} S & B \\ 0 & T \end{pmatrix}$. □

EXERCISE 2S. Show that $J(M_n(R)) = M_n(J(R))$ for any ring R and any $n \in \mathbb{N}$. □

EXERCISE 2T. Show that a one–sided ideal I in a ring R is contained in J(R) if and only if $1 - x$ is invertible for each $x \in I$. □

EXERCISE 2U. Show that the Jacobson radical of a ring R contains all nil one–sided ideals of R. □

EXERCISE 2V. If x is an element of a ring R and $x + J(R)$ is invertible in R/J(R), show that x is invertible in R. □

EXERCISE 2W. If R is a commutative semiprime ring, show that the polynomial ring $R[x]$ is semiprimitive. [Hints: First reduce to the case that R is prime. If $p \in J(R[x])$ note that $1 + px$ is invertible in $R[x]$.] □

THEOREM 2.17. [Jacobson, Azumaya] *If A is a finitely generated right module over a ring R and $AJ(R) = A$, then $A = 0$.*

Proof. If A is cyclic, say $A = aR$, then $A = AJ(R) = aJ(R)$, and so $a = ax$ for some $x \in J(R)$. Since $1 - x$ is invertible (Exercise 2T), we obtain $a = 0$ and thus $A = 0$.

Now suppose that $A = a_1R + ... + a_nR$, where $n > 1$, and that the theorem holds for modules with $n - 1$ generators. Since A/a_1R has $n - 1$ generators, and since

$$(A/a_1R)J(R) = A/a_1R,$$

it follows that $A/a_1R = 0$, that is, $A = a_1R$. Therefore, by the cyclic case, $A = 0$. □

Theorem 2.17 is often called *Nakayama's Lemma.* It implies in particular that if A is a finitely generated module and $a_1,...,a_n$ are elements of A whose images generate $A/AJ(R)$, then $a_1,...,a_n$ generate A. (To see this, consider the module $A/(a_1R + ... + a_nR)$.)

♦ PRIME IDEALS IN DIFFERENTIAL OPERATOR RINGS

In this section we will in some sense write down all of the prime ideals of a special class of noetherian rings – the differential operator rings $R[\theta;\delta]$, where R is assumed to be a commutative noetherian $\mathbb{Q}$–algebra. This will illustrate some of the phenomena which occur in more general settings, and will also give us more examples of primitive noetherian rings.

LEMMA 2.18. *Let R be a ring, δ a derivation of R, and $S = R[\theta;\delta]$ the corresponding differential operator ring. Then*

(a) If I is a right ideal of R then IS is a right ideal of S, and $IS \cap R = I$.

(b) If I is a δ–ideal of R then IS is an ideal of S, and $IS = SI$.

(c) If J is an ideal of S, then $J \cap R$ is a δ–ideal of R.

Proof. Most of this is an easy computation. In (a), $IS \cap R = I$ because R is a direct summand of S as a left R–module. For (b), note that if I is a δ–ideal, and $x \in I$, then because $\theta x = x\theta + \delta(x)$ and $\delta(x) \in I$, we have $\theta x \in IS$. For (c), note that if $x \in J \cap R$ then $\delta(x) = \theta x - x\theta \in J \cap R$. □

LEMMA 2.19. *Let R be a commutative integral domain of characteristic zero with a nonzero derivation δ, and let $S = R[\theta;\delta]$. If I is a nonzero ideal of S, then $I \cap R \neq 0$.*

Proof. Pick a nonzero element $s = s_n\theta^n + \ldots$ from I with degree n and leading coefficient s_n, and assume that $n \geq 1$. Choose $r \in R$ such that $\delta(r) \neq 0$, and look at the element $sr - rs$. An immediate calculation shows that

$$sr - rs = ns_n\delta(r)\theta^{n-1} + [\text{terms of degree less than } n-1].$$

Since under our hypotheses $ns_n\delta(r) \neq 0$, we see that I contains a nonzero element of degree $n-1$. Hence, iterating this argument, we conclude that I contains a nonzero element of degree 0. □

LEMMA 2.20. *If R is a ring, δ a derivation of R, and P a minimal prime ideal of R such that R/P has characteristic zero, then P is a δ–ideal.*

Proof. Let $Q = \{r \in R \mid \delta^n(r) \in P \text{ for all } n \geq 0\}$. Using Leibniz's Rule (Exercise 1U), it is clear that Q is an ideal of R and is contained in P. We show that Q is prime as follows. Consider any $x,y \in R - Q$. Choose nonnegative integers r and s as small as possible so that $\delta^r(x)$ and $\delta^s(y)$ are not in P, and then choose $z \in R$ such that $\delta^r(x)z\delta^s(y) \notin P$. Now use Leibniz's Rule to expand $\delta^{r+s}(xzy)$, as follows:

$$\delta^{r+s}(xzy) = \sum_{i=0}^{r+s} \binom{r+s}{i} \delta^{r+s-i}(x)\delta^i(zy) = \sum_{i=0}^{r+s} \sum_{j=0}^{i} \binom{r+s}{i}\binom{i}{j} \delta^{r+s-i}(x)\delta^{i-j}(z)\delta^j(y).$$

Since $\delta^{r+s-i}(x) \in P$ whenever $i > s$ and $\delta^j(y) \in P$ whenever $j < s$, all of the terms in the last summation are in P except for $\binom{r+s}{s}\binom{s}{s}\delta^r(x)z\delta^s(y)$, which is not in P because $\delta^r(x)z\delta^s(y)$ is not and R/P has characteristic zero. Thus $\delta^{r+s}(xzy) \notin P$ and so $xzy \notin Q$, which shows that Q is prime. Since P is a minimal prime, we must have $P = Q$, and then since Q is clearly a δ–ideal, the result follows. □

EXERCISE 2X. Let R be a ring, δ a derivation on R, and $S = R[\theta;\delta]$. Let I be a δ–ideal of R, let δ' denote the derivation on R/I induced by δ, and let $S' = (R/I)[\theta';\delta']$ be the corresponding differential operator ring. Show that there is a ring isomorphism $\phi : S/IS \to S'$ such that $\phi(r + IS) = r + I$ for all $r \in R$, and $\phi(\theta + IS) = \theta'$. [Hint: Exercise 1H.] □

LEMMA 2.21. *Let R be a noetherian ring with a derivation δ, such that R is an algebra over $\mathbb{Q}$. Let $S = R[\theta;\delta]$, and let P be a prime ideal of S. Then $P \cap R$ is a prime ideal of R.*

Proof. Since $P \cap R$ is a δ–ideal of R (Lemma 2.18), we can use Exercise 2X to reduce to a differential operator ring over $R/(P \cap R)$. Hence, we may assume that

$P \cap R = 0$. If Q is any minimal prime of R then R/Q has characteristic zero (since $R \supseteq \mathbb{Q}$), and so by Lemma 2.20, Q is a δ–ideal. According to Theorem 2.4 there are minimal primes $Q_1,\ldots,Q_m$ in R such that $Q_1Q_2\cdots Q_m = 0$. From Lemma 2.18, we infer that each Q_iS is an ideal of S, and that

$$(Q_1S)(Q_2S)\cdots(Q_mS) = Q_1Q_2\cdots Q_mS = 0.$$

Since P is prime, we have $Q_iS \leq P$ for some index i. Hence, $Q_i \leq P \cap R = 0$, and so $P \cap R$ is a prime ideal, as claimed. □

THEOREM 2.22. *Let R be a commutative noetherian ring which is an algebra over $\mathbb{Q}$, let δ be a derivation on R, and let $S = R[\theta;\delta]$ be the corresponding differential operator ring.*

(a) If P is any prime ideal of S, then $P \cap R$ is a prime δ–ideal of R.

(b) If Q is a prime δ–ideal of R, then QS is a prime ideal of S such that $QS \cap R = Q$. Furthermore, if P is any prime ideal of S such that $P \cap R = Q$, then either $P = QS$ or $\delta(R) \subseteq Q$, and in the latter case S/QS and S/P are commutative rings.

(c) All prime factor rings of S are domains.

Proof. (a) This is contained in Lemmas 2.18 and 2.21.

(b) By Lemma 2.18, QS is an ideal of S such that $QS \cap R = Q$. From Exercise 2X we have that $S/QS \cong (R/Q)[\theta';\delta']$ where δ' is the derivation on R/Q induced by δ. Since R/Q is a domain, it is clear that S/QS is a domain (see the remarks following the definition of degree in the previous chapter), and hence QS is a prime ideal of S.

If P is a prime ideal of S such that $P \cap R = Q$ but $P \neq QS$, then the image of P/QS in $(R/Q)[\theta';\delta']$ is a nonzero ideal I such that $I \cap (R/Q) = 0$. It follows from Lemma 2.19 that $\delta' = 0$, whence $\delta(R) \subseteq Q$. Moreover, $(R/Q)[\theta';\delta']$ is then an ordinary polynomial ring over the commutative ring R/Q. Thus in this case S/QS is commutative, as is S/P (since $P \geq QS$).

(c) In the notation of part (b), if $P = QS$ we have already seen that S/P is a domain. Otherwise, S/P is a commutative prime ring, and again it is a domain. □

One way to summarize Theorem 2.22 is that the prime ideals of S are parametrized by the prime δ–ideals of R. If Q is a prime δ–ideal of R and $\delta(R) \not\subseteq Q$, there is a unique prime ideal of S that contracts to Q (that is, whose intersection with R equals Q), namely QS. If Q is a prime δ–ideal of R and $\delta(R) \subseteq Q$, then S/QS is a commutative ring isomorphic to an ordinary polynomial ring $(R/Q)[x]$. In this case the primes of S that contract to Q correspond to the primes of $(R/Q)[x]$ that contract to zero in R/Q; these in turn correspond precisely to the primes in $K[x]$ where K is the quotient field of R/Q.

Exercise 2ZF below shows that Lemmas 2.20 and 2.21 and Theorem 2.22 are all false in characteristic p.

EXERCISE 2Y. Let R be a polynomial ring k[x] where k is an algebraically closed field of characteristic zero, let $\delta = x\frac{d}{dx}$, and let $S = R[\theta;\delta]$. Show that the only δ–ideals of R are 0 and the ideals x^nR (for n = 0,1,...). Show that the only prime ideals of S are 0 and xS together with $xS + (\theta - \alpha)S$ for all $\alpha \in k$. Then show (without the computations used in Exercise 2Q) that S is right primitive. [Hint: If $\alpha \in k$ is nonzero, then $(x - \alpha)S + xS = S$. Hence, no proper right ideal containing $(x - \alpha)S$ can contain a nonzero prime ideal.] □

EXERCISE 2Z. Let R be a polynomial ring k[x] where k is a field of characteristic zero, and let δ be any nonzero k–linear derivation on R. Show that there is a nonzero polynomial $g \in R$ such that $\delta = g\frac{d}{dx}$. If $S = R[\theta;\delta]$, show that S is right and left primitive. □

EXERCISE 2ZA. (This exercise does the entire theory of this section over again for a skew–Laurent extension.) Let R be a ring and α an automorphism of R. An ideal I of R is an *α–ideal* if $\alpha(I) = I$. Let $S = R[\theta,\theta^{-1};\alpha]$. Show that if J is an ideal of S, then $J \cap R$ is an α–ideal of R, and that if I is an α–ideal of R then IS is an ideal of S. Show that the prime radical of R is an α–ideal. Assume now that R is a commutative noetherian ring. Show that an ideal I of R is of the form $P \cap R$ for some prime ideal P of S if and only if there is a prime ideal Q of R and a positive integer m such that $I = Q \cap \alpha(Q) \cap \ldots \cap \alpha^{m-1}(Q)$ and $\alpha^m(Q) = Q$. If I is such an ideal, show that IS is the only prime ideal of S whose intersection with R is I if and only if no nonzero power of the automorphism α induces the identity map on R/I. □

♦ ADDITIONAL EXERCISES

2ZB. (This puts the proof of Theorem 2.7 in a general setting, and is McCoy's noncommutative analogue of Lemma 2.5.) An *m–system* in a ring R is a subset $X \subseteq R - \{0\}$ such that for any $x,y \in X$ there exists $r \in R$ with $xry \in X$. Show that any ideal of R which is maximal with respect to being disjoint from X is a prime ideal. □

2ZC. Let $S = A_1(k) = k[x][\theta; d/dx]$, where k is a field of characteristic zero, and set $R = k + \theta S$, which is a subring of S.

(a) Observe that θS is a proper nonzero ideal of R, so that R is not a simple ring. Show that θS is the only proper nonzero ideal of R. [Hint: If I is a nonzero ideal of R, then $SI\theta S = S$.]

(b) Show that θS is a maximal right ideal of S. [Hint: Given a right ideal $I \geq \theta S$, observe that $\theta p - p\theta \in I$ for all $p \in I$.] Observe also that θS is a maximal right ideal of R.

(c) Show that R is a maximal right R–submodule of S. [Hint: Given $p \in S - R$, observe that $p\theta \notin \theta S$, whence $\theta S + p\theta S = S$.]

(d) Show that $(S/R)_R$ is a faithful simple right R–module. Thus R is a right primitive ring.

(e) Show that R is a right noetherian ring. [Hint: Given a right ideal J of R, show that there exist $a_1,\ldots,a_n \in J$ such that $JS = a_1S + \ldots + a_nS$. Then use the simplicity of $(S/R)_R$ to show that the right R–module $JS/(a_1R + \ldots + a_nR)$ is noetherian.]

The next exercise is to show that R is also left primitive and left noetherian. □

2ZD. (a) In the situation of Exercise 2ZC, show that there is a ring isomorphism $\phi : R \to k + S\theta$ such that $\theta\phi(r) = r\theta$ for all $r \in R$.

(b) Show that there is an anti–automorphism $\psi : S \to S$ such that ψ is the identity map on $k[x]$ while $\psi(\theta) = -\theta$. Observe that $\psi\phi$ gives an anti–automorphism of R. Hence, R is left primitive and left noetherian. □

2ZE. In the situation of Exercise 2ZC, let $A = S/\theta S$, viewed as a right R–module.

(a) Show that A is finitely generated.

(b) Show that the only proper nonzero submodule of A is $R/\theta S$. [Hint: If $B/\theta S$ is a proper submodule of A, observe that $B\theta S + \theta S$ is a proper right ideal of S.]

(c) Show that 0 is an annihilator prime of A but not an associated prime. □

2ZF. Let k be a field of characteristic $p > 0$.

(a) Let z be an indeterminate, and let E be the ring of all $k[z^p]$–module endomorphisms of $k[z]$. (Since $k[z]$ is a free $k[z^p]$–module with basis $1,z,\ldots,z^{p-1}$, there is an isomorphism of E onto $M_p(k[z^p])$.) Let $m_z \in E$ be multiplication by z. Show that E is generated as a k–algebra by m_z and d/dz. [Hint: To get endomorphisms corresponding to thc standard $p \times p$ matrix units with respect to the basis $1,z,\ldots,z^{p-1}$, use maps of the form

$$\sum_{j=0}^{p-1} \alpha_j m_z^{j+i}(d/dz)^j \qquad \text{and} \qquad \sum_{j=0}^{p-1} \beta_j m_z^{j}(d/dz)^{j+i} \text{].}$$

(b) Let $R = k[x]/x^pk[x]$ for an indeterminate x, and let u be the coset $x + x^pk[x]$ in R. Show that there is a unique well–defined k–linear derivation δ on R such that $\delta(u) = 1$.

(c) Show that there is a k–algebra isomorphism of $R[\theta;\delta]$ onto E sending u to $-d/dz$ and θ to m_z. Hence, $R[\theta;\delta] \cong M_p(k[z^p])$. □

2ZG. Let $R = A_1(\mathbb{Z}) = \mathbb{Z}[x][\theta; d/dx]$, and let p be a prime integer.

(a) Show that pR is a prime ideal of R and that R/pR is a domain.

(b) Show that $pR + x^pR$ is a prime ideal of R and that

$$R/(pR + x^pR) \cong M_p((\mathbb{Z}/p\mathbb{Z})[y])$$

for an indeterminate y.

(c) Show that $pR + x^pR + \theta^pR$ is a maximal ideal of R and that $R/(pR + x^pR + \theta^pR)$ is isomorphic to $M_p(\mathbb{Z}/p\mathbb{Z})$. □

2ZH. If all prime ideals in a commutative ring R are finitely generated, show that R is noetherian. [Hint: If not, find an ideal maximal among non–finitely–generated ideals.] □

2ZI. If an ideal I in a commutative ring R is contained in the union of finitely many prime ideals $P_1,\dots,P_n$, show that $I \le P_j$ for some j. [Hint: Without loss of generality, $P_i \nleq P_j$ for $i \neq j$. If $I \nleq P_j$, choose $x_j \in I\cdot \prod_{i \neq j} P_i$ such that $x_j \notin P_j$.] □

2ZJ. Show that if R is a ring which satisfies the ascending chain condition on J–ideals, then every J–ideal is a finite intersection of prime J–ideals. [The most important non–noetherian rings satisfying this condition are the affine P.I.–algebras – that is, algebras over a field which are finitely generated as algebras and which satisfy a polynomial identity.] □

♦ NOTES

Prime Ideals. The definition via products of ideals was introduced by Krull in both the commutative and noncommutative cases [1928a, p. 5; 1928b, Definition 3, p. 486].

Existence of Minimal Primes. That every prime contains a minimal prime was first proved in the commutative case by Krull [1929, Satz 5].

Finiteness of the Set of Minimal Primes. This was proved for rings with ACC on semiprime ideals by Nagata [1951, Corollary to Proposition 34].

Semiprime Ideals. These were introduced in the commutative case by Krull [1929, p. 735] and in the noncommutative case by Nagata [1951, Definition 1].

Ideals Disjoint from Multiplicative Sets. That an ideal in a commutative ring maximal with respect to disjointness from a multiplicative set must be prime was observed by Krull in the proof of [1929, Lemma, p. 732].

Semiprime Commutative Rings. Krull proved that a commutative ring is semiprime if and only if it has no nonzero nilpotent elements [1929, Satz 4].

Criterion for Semiprime Ideals. That an ideal is semiprime precisely when the condition

(*) of Theorem 2.7 holds is equivalent to a result proved independently by Levitzki [1951, Theorem 2] and Nagata [1951, Proposition 8 and Corollary], namely that the "McCoy radical" of a ring R coincides with the "Baer lower radical". The McCoy radical is the set of those elements of R not contained in any m–system (see Exercise 2ZB), and McCoy proved that this equals the intersection of the prime ideals of R [1949, Theorem 2]; the Baer lower radical of R is the smallest ideal N such that R/N has no nonzero nilpotent ideals.

Affiliated Primes. Affiliated series and affiliated primes were first introduced for bimodules by Stafford [1979, pp. 265–266].

Primitive Rings. These were introduced by Jacobson [1945a, Definition 3].

Jacobson Radical. Jacobson defined the radical of an arbitrary ring R to be the sum of those right ideals I of R such that $1 + z$ is right invertible for all $z \in I$ [1945a, Definition 2], proved that this definition is left–right symmetric [1945a, Theorems 1,2], and showed that the radical equals the intersection of the maximal right ideals of R [1945a, Corollary 2 to Theorem 18]. Earlier, Perlis had characterized the radical of a finite–dimensional algebra R as the set of those $z \in R$ such that $u + z$ is a unit for all units u [1942, Theorem 1].

Nakayama's Lemma. This was first proved by Jacobson in the case that A is a right ideal contained in the radical [1945a, Theorem 10]; Azumaya then carried over Jacobson's proof to the module case [1951, Theorem 1] (see also Nagata [1950, Footnote 3, p. 67]). An alternate proof derived from a generalized result was presented by Nakayama [1951, (II)].

Invariance of Minimal Primes under Derivations. This was proved for associated primes as well as minimal primes in commutative noetherian $\mathbb{Q}$–algebras by Seidenberg [1967, Theorem 1]. For minimal completely prime ideals in noncommutative $\mathbb{Q}$–algebras it follows from a result of Dixmier [1966, Lemme 6.1], and his argument was extended to arbitrary minimal primes in $\mathbb{Q}$–algebras by Gabriel [1971, Lemme 3.4].

Contraction of Primes in Differential Operator Rings. Gabriel proved that in a differential operator ring $R[\theta;\delta]$ over a right noetherian $\mathbb{Q}$–algebra R, every prime of $R[\theta;\delta]$ contracts to a prime of R [1971, Proposition 3.3(b)].

3 SEMISIMPLE MODULES, ARTINIAN MODULES, AND NONSINGULAR MODULES

In this chapter we discuss several special types of modules that will make frequent appearances later. These include semisimple modules, artinian modules, and nonsingular modules, which have prototypes in vector spaces, finite abelian groups, and torsionfree abelian groups respectively. That special types of modules have useful roles in the study of arbitrary modules may be seen already in the case of abelian groups (i.e., $\mathbb{Z}$–modules). In studying an arbitrary abelian group, an almost reflexive first step is to look at its torsion part (i.e., the torsion subgroup) and its torsionfree part (i.e., the factor group modulo the torsion subgroup), since entirely different techniques are available (and needed) for dealing with torsion groups and torsionfree groups. On the torsionfree side, vector spaces make an appearance due to the fact that the torsionfree divisible abelian groups are exactly the vector spaces over $\mathbb{Q}$, and that every torsionfree abelian group can be embedded in a divisible one. On the torsion side, many questions can be reduced to the case of finite abelian groups, since every torsion abelian group is a directed union of finite subgroups. In studying torsion abelian groups, one also reduces to the case of p–groups for various primes p. Vector spaces make another appearance here, since in a p–group the set of elements of order p (together with 0) forms a vector space over $\mathbb{Z}/p\mathbb{Z}$.

Viewed module–theoretically, vector spaces are distinguished by many nice decomposition properties. For instance, every vector space is a direct sum of one–dimensional subspaces, and every subspace of a vector space is a direct summand. We view simple modules as analogues to one–dimensional vector spaces, and the corresponding analogues to higher–dimensional vector spaces are the *semisimple* modules: modules which are (direct) sums of simple submodules. Other decomposition properties follow; in particular, we shall find that a module is semisimple if and only if every submodule is a direct summand.

The finite abelian groups are perhaps more appropriately characterized as the finitely generated torsion abelian groups. Their "closest relatives" would then be the finitely generated torsion modules over a polynomial ring $k[x]$ with k a field; these are precisely the finite–dimensional $k[x]$–modules. From this perspective, the essential features are the ascending and descending chain conditions on submodules. It has proved advantageous to study these conditions separately – in the form of *noetherian* modules and *artinian* modules – as well as together.

While it is clear enough what we will mean by a torsionfree module over a commutative domain, it is not at all obvious what the appropriate notion is for modules over an arbitrary ring. (In fact, there exists a large body of research on "torsion theories" which sidesteps the question of which notion to choose by axiomatizing the family of possible choices.) We shall discuss one particular concept – that of a *nonsingular* module – which behaves reasonably well for our purposes. We leave the precise definition until later in the chapter, mentioning here only that over a prime noetherian ring the analogy with torsionfree modules over commutative domains becomes quite direct. Namely, we shall see in Chapter 6 that a module over a prime noetherian ring is nonsingular precisely if no nonzero element of the module is annihilated by a non–zero–divisor from the ring.

♦ SEMISIMPLE MODULES

DEFINITION. The *socle* of a module A is the sum of all simple submodules of A, and is denoted soc(A). (By convention, the sum of the empty family of submodules is the zero submodule. Hence, $\operatorname{soc}(A) = 0$ if and only if A has no simple submodules.) A *semisimple* (or *completely reducible*) module is any module A such that $\operatorname{soc}(A) = A$.

For example, any vector space A over a field k is a semisimple k–module, since all the one–dimensional subspaces of A are simple k–submodules. For another example, the socle of an abelian group A consists of 0 together with all elements of A of finite square–free order.

In any ring R, observe that $\operatorname{soc}(R_R)$ is an ideal of R. (If $r \in R$, recall that the function taking x to rx is a homomorphism of the right module R_R into itself. Hence if A is a simple right ideal of R, then either $rA = 0$ or rA is simple, and so $rA \leq \operatorname{soc}(R_R)$ in either case.) Similarly, $\operatorname{soc}({}_RR)$ is an ideal of R, but these two socles need not coincide, as the following example shows.

EXERCISE 3A. If R is the ring of all upper triangular 2×2 matrices over a field k, show that $\operatorname{soc}(R_R) \neq \operatorname{soc}({}_RR)$. □

EXERCISE 3B. Let R be a semiprime ring.

(a) Show that any simple right or left ideal of R is generated by an idempotent. [Hint: If I is a simple right ideal of R, show that there exists $x \in I$ for which $xI \neq 0$, and observe that $I \cap \operatorname{r.ann}(x) = 0$.]

(b) Given an idempotent $e \in R$, show that eR is a simple right ideal if and only if Re is a simple left ideal, if and only if eRe is a division ring.

(c) Show that $\operatorname{soc}(R_R) = \operatorname{soc}({}_RR)$. □

PROPOSITION 3.1. *The socle of any module A is a direct sum of simple submodules of A.*

Proof. Let $\mathscr{B}$ be a maximal independent family of simple submodules of A, and set $B = \oplus\mathscr{B}$. Then $B \leq \text{soc}(A)$, and if $B \neq \text{soc}(A)$ there is a simple submodule $S \leq A$ such that $S \nleq B$. Now $S \cap B \neq S$. As S is simple, $S \cap B = 0$. But then $\mathscr{B} \cup \{S\}$ is independent, contradicting the maximality of $\mathscr{B}$. Therefore $B = \text{soc}(A)$. □

PROPOSITION 3.2. *A module A is semisimple if and only if every submodule of A is a direct summand of A.*

Proof. Assume first that A is semisimple, and let B be a submodule of A. By Zorn's Lemma, there is a submodule C of A maximal with respect to the property $B \cap C = 0$. If $B \oplus C < A$, there is a simple submodule $S \leq A$ such that $S \nleq B \oplus C$. As in the previous proof, $S \cap (B \oplus C) = 0$, and so $\{B, C, S\}$ is independent. But then $B \cap (C \oplus S) = 0$, contradicting the maximality of C. Therefore $B \oplus C = A$.

Conversely, assume that every submodule of A is a direct summand. In particular, $A = \text{soc}(A) \oplus B$ for some submodule B. If $B \neq 0$, let C be a nonzero cyclic submodule of B. If c is a generator for C, then by Zorn's Lemma C has a submodule M which is maximal with respect to the property $c \notin M$. Then M is a maximal proper submodule of C. Now $A = M \oplus N$ for some submodule N, and $C = M \oplus (C \cap N)$. Note that $C \cap N \cong C/M$, whence $C \cap N$ is a simple submodule of A, and so $C \cap N \leq \text{soc}(A)$. As $C \cap N \leq B$, this is impossible. Thus $B = 0$, and therefore $A = \text{soc}(A)$. □

COROLLARY 3.3. *Any submodule of a semisimple module is semisimple.*

Proof. If B is a submodule of a semisimple module A and $C \leq B$, then C is a submodule of A, whence $A = C \oplus D$ for some submodule D. Therefore $B = C \oplus (B \cap D)$, and so C is a direct summand of B. □

EXERCISE 3C. Let $A = \oplus S_i$ where $\{S_i \mid i \in I\}$ is a collection of simple modules, and let $B \leq A$. Show that there is a subset $J \subseteq I$ such that $A = B \oplus (\bigoplus_{j \in J} S_j)$, and conclude that $B \cong \bigoplus_{i \in I-J} S_i$. □

EXERCISE 3D. (a) A semisimple module A is said to be *homogeneous* if A is a direct sum of pairwise isomorphic simple submodules. Show that A is homogeneous if and only if all simple submodules of A are isomorphic to each other.

(b) Show that an arbitrary semisimple module A is a direct sum of homogeneous submodules A_i such that $\text{Hom}(A_i, A_j) = 0$ when $i \neq j$. □

EXERCISE 3E. Show that the endomorphism ring of any simple module is a division ring. (This result is known as *Schur's Lemma.*) If A is a nonzero finitely generated semisimple module, show that its endomorphism ring is isomorphic to a finite direct product of matrix rings over division rings. □

♦ SEMISIMPLE RINGS

Very few rings share the property with fields that all their modules are semisimple. The rings that do are those that appear in the Wedderburn–Artin Theorem. Since this is a standard part of many developments of the Wedderburn–Artin Theorem (which the reader is likely to have seen elsewhere), we will be somewhat sketchy in our treatment. We begin by recalling key properties of matrix rings over a division ring.

EXERCISE 3F. If $R = M_n(D)$ for some $n \in \mathbb{N}$ and some division ring D, show that R_R is a direct sum of n pairwise isomorphic simple right modules, and that ${}_RR$ is a direct sum of n pairwise isomorphic simple left modules. [Hint: In case $n = 2$, look at the right ideals $I_1 = \begin{pmatrix} D & D \\ 0 & 0 \end{pmatrix}$ and $I_2 = \begin{pmatrix} 0 & 0 \\ D & D \end{pmatrix}$, and observe that $I_2 = \begin{pmatrix} 0 & 0 \\ 1 & 0 \end{pmatrix} I_1$.] Show also that R is a simple ring. □

THEOREM 3.4. [Noether] *For any ring R, the following conditions are equivalent:*

(a) All right R–modules are semisimple.

(b) All left R–modules are semisimple.

(c) R_R is semisimple.

(d) ${}_RR$ is semisimple.

(e) Either R is the zero ring or $R \cong M_{n(1)}(D_1) \times \ldots \times M_{n(k)}(D_k)$ for some positive integers $n(i)$ and some division rings D_i.

Proof. Obviously (a) ⇒ (c). Conversely, if R_R is semisimple then all cyclic right R–modules are semisimple, from which it is immediate that any right R–module is semisimple (since any module is the sum of its cyclic submodules). Thus (a) ⇔ (c), and similarly (b) ⇔ (d).

That (e) ⇒ (c) follows from Exercise 3F. On the other hand, if R_R is semisimple, then by Exercise 3E the endomorphism ring of R_R has the desired form, and since $R \cong \text{End}_R(R_R)$ we have proved that (c) ⇒ (e). Finally, since (e) is clearly a symmetric condition, it follows that all of the given conditions are equivalent. □

DEFINITION. A ring satisfying the conditions of Theorem 3.4 is called a *semisimple ring*.

The reader should be warned that in the older literature, "semisimple" is often used to mean "Jacobson semisimple", i.e., "semiprimitive".

EXERCISE 3G. (This is for readers who are familiar with projective and injective modules.) Show that a ring R is semisimple if and only if all R–modules are projective, if and only if all R–modules are injective. □

♦ ARTINIAN MODULES

Classically, the Jacobson and prime radicals were introduced in the context of artinian rings, (where, as we will see, the notions coincide). To refresh the reader's memory, we recall some of the theory of artinian modules and rings.

DEFINITION. A module A is *artinian* provided A satisfies the *descending chain condition* (or *DCC*) on submodules, i.e., there does not exist a properly descending infinite chain $A_1 > A_2 > \ldots$ of submodules of A. Equivalently, A is artinian if and only if every nonempty family of submodules of A has a minimal element. A ring R is *right (left) artinian* if and only if the right module R_R (left module ${}_RR$) is artinian. If both conditions hold, R is called an *artinian ring*.

For example, any finite–dimensional algebra over a field is an artinian ring.

Our first few results concerning artinian modules are completely analogous to the corresponding results for noetherian modules (see Proposition 1.3 and Corollaries 1.4 and 1.5). The proofs of the artinian results may be obtained by imitating the proofs in the noetherian case, reversing inclusions when necessary.

PROPOSITION 3.5. *Let B be a submodule of a module A. Then A is artinian if and only if B and A/B are both artinian.* □

COROLLARY 3.6. *Any finite direct sum of artinian modules is artinian.* □

COROLLARY 3.7. *If R is a right artinian ring, all finitely generated right R–modules are artinian.* □

DEFINITION. A *composition series* for a module A is a chain of submodules

$$0 = A_0 < A_1 < \ldots < A_n = A$$

such that each of the factors A_i/A_{i-1} is a simple module. The number of gaps (namely n) is called the *length* of the composition series, and the factors A_i/A_{i-1} are called the *composition factors* of A corresponding to this composition series. By convention, the zero module is considered to have a composition series of length zero, with no composition factors. A *module of finite length* is any module which has a composition series.

PROPOSITION 3.8. *A module A has finite length if and only if A is both noetherian and artinian.*

Proof. If A has finite length, then (since simple modules are clearly noetherian and artinian) it follows from Propositions 1.3 and 3.5 that A must be noetherian and artinian. Conversely, assume that A satisfies both chain conditions, and set $A_0 = 0$. If $A \neq 0$, then by the DCC, A contains a minimal nonzero submodule A_1, that is, A_1 is simple. Similarly, if $A_1 < A$ then A/A_1 contains a simple submodule A_2/A_1, and we continue in this manner. By the ACC, the chain $A_0 < A_1 < A_2 < \ldots$ must terminate at some stage n. Then $A_n = A$, and the chain $A_0 < A_1 < \ldots < A_n$ is a composition series for A. □

For instance, if R is an algebra over a field k, then any R–module which is finite–dimensional over k has finite length.

EXERCISE 3H. Show that the $\mathbb{Z}$–modules $\mathbb{Z}/4\mathbb{Z}$ and $\mathbb{Z}/6\mathbb{Z}$ have finite length, and that the first has just one composition series, while the second has two. Show that the $\mathbb{Q}$–module $\mathbb{Q} \oplus \mathbb{Q}$ has finite length, and that it has infinitely many composition series. □

If a module A has finite length, it will turn out that all composition series for A have the same length. This result can be obtained as a consequence of a general refinement theorem concerning arbitrary chains of submodules, and since we will need the refinement theorem later, we prove it next and then return to composition series.

DEFINITION. A *submodule series* (or *normal series*) for a module A is any finite chain of submodules of the form

$$0 = A_0 \leq A_1 \leq \ldots \leq A_n = A.$$

A *refinement* of this series is any submodule series that contains all the A_i, that is, a submodule series

$$0 = B_0 \leq B_1 \leq \ldots \leq B_t = A$$

such that each A_i occurs in the list $B_0,\ldots,B_t$. Two submodule series

$$0 = A_0 \leq A_1 \leq \ldots \leq A_n = A \qquad \text{and} \qquad 0 = B_0 \leq B_1 \leq \ldots \leq B_t = A$$

are said to be *isomorphic* (or *equivalent*) provided $n = t$ and there exists a permutation π of $\{1,2,\ldots,n\}$ such that $A_i/A_{i-1} \cong B_{\pi(i)}/B_{\pi(i)-1}$ for all $i = 1,\ldots,n$.

LEMMA 3.9. [Zassenhaus] *Let* A', A'', B', B'' *be submodules of a module* A *such that* $A' \leq A''$ *and* $B' \leq B''$. *Then*

$$[(A' + B'') \cap A'']/[(A' + B') \cap A''] \cong [(B' + A'') \cap B'']/[(B' + A') \cap B''].$$

Proof. Observe that

$$\begin{aligned}[(A' + B'') \cap A'']/[(A' + B') \cap A''] &= [A' + (B'' \cap A'')]/[A' + (B' \cap A'')] \\ &\cong (B'' \cap A'')/([B'' \cap A''] \cap [A' + (B' \cap A'')]) \\ &\cong (B'' \cap A'')/[(B'' \cap A') + (B' \cap A'')].\end{aligned}$$

By symmetry, we also have

$$[(B' + A'') \cap B'']/[(B' + A') \cap B''] \cong (A'' \cap B'')/[(A'' \cap B') + (A' \cap B'')]. \square$$

THEOREM 3.10. [Schreier] *Any two submodule series for a module* A *have isomorphic refinements.*

Proof. Given submodule series

$$(s_1) \qquad 0 = A_0 \leq A_1 \leq \ldots \leq A_n = A$$

$$(s_2) \qquad 0 = B_0 \leq B_1 \leq \ldots \leq B_t = A,$$

we must find a refinement of each series such that the two refinements are isomorphic. Set

$$A_{ij} = (A_i + B_j) \cap A_{i+1} \qquad (\text{for } i = 0,1,\ldots,n-1 \text{ and } j = 0,1,\ldots,t)$$

$$B_{ji} = (B_j + A_i) \cap B_{j+1} \qquad (\text{for } j = 0,1,\ldots,t-1 \text{ and } i = 0,1,\ldots,n).$$

Then we obtain refinements for (s_1) and (s_2) as follows:

$$(r_1) \qquad 0 = A_{00} \leq A_{01} \leq \ldots \leq A_{0t} = A_1 = A_{10} \leq A_{11} \leq \ldots$$
$$\leq A_{1t} = A_2 = A_{20} \leq \ldots \leq A_{n-1,t} = A_n = A$$

$$(r_2) \qquad 0 = B_{00} \leq B_{01} \leq \ldots \leq B_{0n} = B_1 = B_{10} \leq B_{11} \leq \ldots$$
$$\leq B_{1n} = B_2 = B_{20} \leq \ldots \leq B_{t-1,n} = B_t = A.$$

We leave to the reader the bookkeeping chore of relabelling (r_1) and (r_2) using single indices. Lemma 3.9 shows that $A_{i,j+1}/A_{ij} \cong B_{j,i+1}/B_{ji}$ for $i = 0,1,\ldots,n-1$ and $j = 0,1,\ldots,t-1$, from which we conclude that (r_1) and (r_2) are isomorphic. $\square$

THEOREM 3.11. [Jordan, Hölder] *If a module* A *has finite length, then any two composition series for* A *are isomorphic. In particular, all composition series for* A *have the same length.*

Proof. Consider two composition series

$$0 = A_0 < A_1 < \ldots < A_n = A \qquad \text{and} \qquad 0 = B_0 < B_1 < \ldots < B_t = A.$$

By Theorem 3.10, these two submodule series have isomorphic refinements, say

$$0 = C_0 \leq C_1 \leq \ldots \leq C_m = A \qquad \text{and} \qquad 0 = D_0 \leq D_1 \leq \ldots \leq D_m = A.$$

There is a permutation σ of $\{1,2,\ldots,m\}$ such that $C_k/C_{k-1} \cong D_{\sigma(k)}/D_{\sigma(k)-1}$ for all

$k = 1,\ldots,m$.

Since each of the factors A_i/A_{i-1} is simple, there are no submodules lying strictly between A_{i-1} and A_i. Consequently, the refined series $C_0 \le C_1 \le \ldots \le C_m$ consists of the submodules $A_0, A_1, \ldots, A_n$ in order but with possible repetitions. Hence, among the factors C_k/C_{k-1}, each factor A_i/A_{i-1} occurs exactly once, and the remaining factors are all zero. Similarly, among the factors D_k/D_{k-1}, each factor B_j/B_{j-1} occurs exactly once, and the remaining factors are all zero.

Since $C_k/C_{k-1} \neq 0$ if and only if $D_{\sigma(k)}/D_{\sigma(k)-1} \neq 0$, we conclude that $n = t$ and that there exists a permutation π of $\{1,2,\ldots,n\}$ such that whenever $C_k/C_{k-1} = A_i/A_{i-1}$, then $D_{\sigma(k)}/D_{\sigma(k)-1} = B_{\pi(i)}/B_{\pi(i)-1}$. Therefore $A_i/A_{i-1} \cong B_{\pi(i)}/B_{\pi(i)-1}$ for $i = 1,\ldots,n$, which proves that the two given composition series are isomorphic. □

DEFINITION. If A is a module of finite length, the common length of all composition series for A is called the *length* (or the *composition length*) of A, and we shall denote it by $\text{length}(A)$.

For instance, the only module of length 0 is the zero module, and the modules of length 1 are precisely the simple modules. Note that a finitely generated semisimple module A has finite length, and if A is a direct sum of n simple submodules, then $\text{length}(A) = n$.

PROPOSITION 3.12. *Let A be a module of finite length. If B is any submodule of A, then*

$$\text{length}(A) = \text{length}(B) + \text{length}(A/B).$$

Proof. Since this is clear if either $B = 0$ or $B = A$, we may assume that $0 < B < A$. In this case, choose composition series

$$0 = B_0 < B_1 < \ldots < B_m = B \quad \text{and} \quad 0 = C_0/B < C_1/B < \ldots < C_n/B = A/B$$

for B and A/B. Since the chain

$$0 = B_0 < B_1 < \ldots < B_m < C_1 < \ldots < C_n = A$$

is a composition series for A, the result follows. □

In particular, if $A_1,\ldots,A_n$ are modules of finite length, then

$$\text{length}(A_1 \oplus \ldots \oplus A_n) = \text{length}(A_1) + \ldots + \text{length}(A_n).$$

EXERCISE 3I. If A and B are submodules of a module of finite length, show that $\text{length}(A) + \text{length}(B) = \text{length}(A + B) + \text{length}(A \cap B)$. □

♦ ARTINIAN RINGS

THEOREM 3.13. [Wedderburn, Artin] *For a ring R, the following conditions are equivalent:*

(a) R is right artinian and $J(R) = 0$.

(b) R is left artinian and $J(R) = 0$.

(c) R is semisimple.

Proof. (a) ⇒ (c): Let $\mathscr{B}$ be the set of those right ideals I of R such that R/I is a semisimple module, and note that $\mathscr{B}$ is nonempty (e.g., $R \in \mathscr{B}$). Since R is right artinian, we may choose a right ideal K minimal in $\mathscr{B}$. If $K \neq 0$, then since $J(R) = 0$, there is a maximal right ideal M in R such that $K \nleq M$. Since M is maximal, $K + M = R$, and hence

$$R/(K \cap M) \cong (R/K) \oplus (R/M).$$

But then $R/(K \cap M)$ is semisimple, and since $K \cap M < K$ this contradicts the minimality of K. Therefore $K = 0$, and so R_R is semisimple.

(c) ⇒ (a): Corollary 3.6 shows that R is right artinian (since simple modules are clearly artinian). Now write $R_R = S_1 \oplus ... \oplus S_n$ where each S_i is a simple right R–module. Each of the annihilators $r.ann_R(S_i)$ is a right primitive ideal of R and so contains J(R). Thus $S_iJ(R) = 0$ for each i, and consequently $J(R) = 0$.

(b) ⇔ (c) by symmetry. □

With a little study of socles over artinian rings, we can prove that every artinian ring is also noetherian, and also give a classical characterization of the Jacobson radical of an artinian ring.

DEFINITION. The *socle series* of a module A is the ascending chain

$$soc^0(A) \leq soc^1(A) \leq soc^2(A) \leq ...$$

of submodules of A defined inductively by setting $soc^0(A) = 0$ and

$$soc^{n+1}(A)/soc^n(A) = soc(A/soc^n(A))$$

for all nonnegative integers n.

For example, if $A = \mathbb{Z}/p^k\mathbb{Z}$ for some prime integer p and some positive integer k, then $soc^n(A) = p^{k-n}\mathbb{Z}/p^k\mathbb{Z}$ for $n = 0,1,...,k$ and $soc^n(A) = A$ for all $n \geq k$.

EXERCISE 3J. Describe the socle series of the $\mathbb{Z}$–module $\mathbb{Q}/\mathbb{Z}$. □

PROPOSITION 3.14. *Let R be a ring such that R/J(R) is semisimple, and let A be a right R–module. Then $soc^n(A) = ann_A(J(R)^n)$ for all nonnegative integers n.*

Proof. Since $\mathrm{soc}^0(A) = 0$ and $J(R)^0 = R$, the proposition is clear in case $n = 0$.

Now assume that the proposition holds for some nonnegative integer n, and set $B = \mathrm{soc}^n(A)$. Note from the induction hypothesis that $BJ(R)^n = 0$. Given any simple submodule $S \leq A/B$, note that $\mathrm{ann}_R(S)$ is a right primitive ideal of R. By definition, $J(R) \leq \mathrm{ann}_R(S)$, and so $SJ(R) = 0$. Thus $(\mathrm{soc}(A/B))J(R) = 0$, whence $\mathrm{soc}^{n+1}(A)J(R)^{n+1} = 0$.

Set $C = \mathrm{ann}_A(J(R)^{n+1})$. As $B = \mathrm{ann}_A(J(R)^n)$ (by the induction hypothesis), we see that $B \leq C$ and $CJ(R) \leq B$. Hence, $(C/B)J(R) = 0$, so that C/B is a right module over $R/J(R)$. By Theorem 3.4, C/B must be a semisimple module (over $R/J(R)$, and hence also over R), whence $C/B \leq \mathrm{soc}(A/B)$. Thus $C \leq \mathrm{soc}^{n+1}(A)$. Therefore $\mathrm{soc}^{n+1}(A) = C$, completing the induction step. □

In particular, Proposition 3.14 applies to any right artinian ring R, since $R/J(R)$ is semisimple according to Theorem 3.13.

THEOREM 3.15. [Hopkins, Levitzki] *If R is a right artinian ring, then R is also right noetherian, and $J(R)$ is nilpotent.*

Proof. Since the powers of $J(R)$ form a descending chain of ideals, there must exist a positive integer n such that $J(R)^{n+1} = J(R)^n$. In view of Proposition 3.14, it follows that $\mathrm{soc}^{n+1}(R_R) = \mathrm{soc}^n(R_R)$. Hence if $I = \mathrm{soc}^n(R_R)$, then $\mathrm{soc}(R/I) = 0$.

If $I \neq R$, then R/I has a minimal nonzero submodule M. But then M is a simple submodule of R/I, contradicting the fact that $\mathrm{soc}(R/I) = 0$. Thus $I = R$. Hence, by Proposition 3.14, $\mathrm{r.ann}_R(J(R)^n) = \mathrm{soc}^n(R_R) = R$, and so $J(R)^n = 0$. Therefore $J(R)$ is nilpotent.

Set $A_i = \mathrm{soc}^i(R_R)$ for $i = 0,1,\ldots,n$. These A_i form a chain $A_0 \leq A_1 \leq \ldots \leq A_n$ of right ideals of R such that $A_0 = 0$ and $A_n = R$. Each of the factors A_i/A_{i-1} is a semisimple module and so is a direct sum of simple modules, by Proposition 3.1.

Suppose that one of the factors A_i/A_{i-1} is a direct sum of an infinite family $\mathcal{B}$ of simple modules. Choose distinct $B_1,B_2,\ldots$ in $\mathcal{B}$, and for $k = 1,2,\ldots$ let C_k be the direct sum of the family $\{B_k,B_{k+1},\ldots\}$. Then $C_1 > C_2 > \ldots$ is a strictly descending chain of submodules of A_i/A_{i-1}, whence A_i/A_{i-1} is not artinian. As R_R is artinian, this is impossible.

Thus A_i/A_{i-1} is a finite direct sum of simple modules. As simple modules are noetherian, Corollary 1.4 shows that A_i/A_{i-1} is noetherian. Using Proposition 1.3, we conclude that each A_i is noetherian. Therefore, since $R_R = A_n$, the ring R is right noetherian. □

COROLLARY 3.16. *For a right artinian ring R, the Jacobson radical equals the prime radical.*

Proof. Since every primitive ideal is prime, the intersection of the primitive ideals contains the intersection of the prime ideals, so that the Jacobson radical contains the prime radical in any ring. Conversely, Theorem 3.15 shows that the Jacobson radical of a right artinian ring is nilpotent, and it is thus contained in the prime radical by Corollary 2.9. □

COROLLARY 3.17. [Noether] *For a ring R, the following conditions are equivalent:*

(a) R is right artinian and semiprime.

(b) R is left artinian and semiprime.

(c) R is semisimple.

Proof. Combine Theorem 3.13 and Corollary 3.16. □

EXERCISE 3K. Show that any right noetherian, left artinian ring is also right artinian. □

COROLLARY 3.18. *For a ring R, the following conditions are equivalent:*

(a) R is prime and right artinian.

(b) R is prime and left artinian.

(c) R is simple and right artinian.

(d) R is simple and left artinian.

(e) R is simple and semisimple.

(f) $R \cong M_n(D)$ for some positive integer n and some division ring D.

Proof. (a) ⇒ (f) by Corollary 3.17 and Theorem 3.4, (f) ⇒ (e) by Exercise 3F, (e) ⇒ (c) by Theorem 3.13, and (c) ⇒ (a) is clear. By symmetry, (b), (d), and (f) are also equivalent. □

Because of the symmetry in Corollary 3.18, the rings characterized there are referred to as *simple artinian rings.*

PROPOSITION 3.19. *If R is a nonzero right or left artinian ring, then all prime ideals in R are maximal.*

Proof. If R contains a non–maximal prime ideal P, then R/P is a prime right or left artinian ring which is not simple, contradicting Corollary 3.18. □

PROPOSITION 3.20. *If R is a commutative noetherian ring, then R is artinian if and only if all prime ideals in R are maximal.*

Proof. Assume that all prime ideals in R are maximal. By Theorem 2.4, there are (minimal) prime ideals $P_1,\dots,P_n$ in R such that $P_1P_2\cdots P_n = 0$. If $I_0 = R$ and $I_j = P_1P_2\cdots P_j$ for $j = 1,\dots,n$, then each of the factors I_{j-1}/I_j is a finitely generated module over R/P_j. Moreover, since P_j is maximal R/P_j is a field and hence artinian. It follows from Corollary 3.7 that each I_{j-1}/I_j is artinian, and then we conclude from Proposition 3.5 that R is artinian. □

♦ ESSENTIAL EXTENSIONS

Over a commutative domain, the torsion part of a module consists of those elements whose annihilators are nonzero ideals of the ring, and we take the point of view that nonzero ideals in a commutative domain are "large". (Consider $\mathbb{Z}$: while $\mathbb{Z}$ is infinite, $\mathbb{Z}$ modulo any nonzero ideal is finite.) Over a more general ring, choosing an analog of torsion amounts to choosing a specific type of "largeness"; torsion submodules then consist of elements whose annihilators in the ring are "large" in the chosen sense. For our purposes, an appropriate type of "largeness" is embodied in the notion of an "essential" right or left ideal. This notion also works well for submodules, and will prove useful in other contexts, such as the study of injective modules. Hence, we next devote a short section to a general discussion of "essential" submodules, after which we turn to our analogs of torsion and torsionfree modules ("singular" and "nonsingular" modules).

DEFINITION. An *essential* (or *large*) *submodule* of a module B is any submodule A which has nonzero intersection with every nonzero submodule of B. We write $A \leq_e B$ to denote this situation, and we also say that B is an *essential extension* of A.

If A is a submodule of a right module B over a ring R, then $A \leq_e B$ if and only if for each nonzero element $b \in B$ there exists $r \in R$ such that $br \neq 0$ and $br \in A$.

Since any two nonzero $\mathbb{Z}$–submodules of $\mathbb{Q}$ have nonzero intersection, all nonzero $\mathbb{Z}$–submodules of $\mathbb{Q}$ are essential. For instance, $\mathbb{Z} \leq_e \mathbb{Q}$. Given a prime integer p and a positive integer n, all nonzero submodules of $\mathbb{Z}/p^n\mathbb{Z}$ are essential. At the other extreme, the only essential submodule (subspace) of a vector space V is V itself.

EXERCISE 3L. If I is a nonzero ideal in a prime ring R, show that I is both an essential right ideal and an essential left ideal of R. □

DEFINITION. Let A and B be modules. An *essential monomorphism* from A to

B is any monomorphism $f : A \to B$ such that $f(A) \leq_e B$.

Note that a submodule A of a module B is essential if and only if the inclusion map $A \to B$ is an essential monomorphism.

PROPOSITION 3.21. *(a) Let $A, B,$ and C be modules with $A \leq B \leq C$. Then $A \leq_e C$ if and only if both $A \leq_e B$ and $B \leq_e C$.*

(b) Let A_1, A_2, B_1, B_2 be submodules of a module C. If $A_1 \leq_e B_1$ and $A_2 \leq_e B_2$, then $A_1 \cap A_2 \leq_e B_1 \cap B_2$.

(c) Let A be a submodule of a module C, and let $f : B \to C$ be a homomorphism. If $A \leq_e C$, then $f^{-1}(A) \leq_e B$.

(d) Let $\{A_i \mid i \in I\}$ and $\{B_i \mid i \in I\}$ be collections of submodules of a module C. If the A_i are independent and each $A_i \leq_e B_i$, then the B_i are independent and $\oplus A_i \leq_e \oplus B_i$.

Proof. (a) If $A \leq_e C$, then obviously $A \leq_e B$. Since any nonzero submodule of C has nonzero intersection with A, it also has nonzero intersection with B. Thus $B \leq_e C$.

Conversely, assume that $A \leq_e B \leq_e C$. Given any nonzero submodule $M \leq C$, we have $B \cap M \neq 0$ because $B \leq_e C$. Then $B \cap M$ is a nonzero submodule of B, and so $A \cap B \cap M \neq 0$ because $A \leq_e B$. Thus $A \cap M \neq 0$, proving that $A \leq_e C$.

(b) Given any nonzero submodule $M \leq B_1 \cap B_2$, we have $A_2 \cap M \neq 0$ because $A_2 \leq_e B_2$. Then $A_2 \cap M$ is a nonzero submodule of B_1, whence $A_1 \cap A_2 \cap M \neq 0$ because $A_1 \leq_e B_1$. Thus $A_1 \cap A_2 \leq_e B_1 \cap B_2$.

(c) Let M be any nonzero submodule of B. If $f(M) = 0$, then $M \leq f^{-1}(A)$ and hence $f^{-1}(A) \cap M \neq 0$. If $f(M) \neq 0$, then $A \cap f(M) \neq 0$ because $A \leq_e C$, whence $f^{-1}(A) \cap M \neq 0$. Thus $f^{-1}(A) \leq_e B$.

(d) If I consists of a single index, there is nothing to prove. Next, assume that $I = \{1,2\}$. Since $A_1 \cap A_2 = 0$, it follows from (b) that $0 \leq_e B_1 \cap B_2$, whence $B_1 \cap B_2 = 0$. Thus B_1 and B_2 are independent. Applying (c) to the projection maps $B_1 \oplus B_2 \to B_i$, we find that $A_1 \oplus B_2 \leq_e B_1 \oplus B_2$ and $B_1 \oplus A_2 \leq_e B_1 \oplus B_2$. Then

$$A_1 \oplus A_2 = (A_1 \oplus B_2) \cap (B_1 \oplus A_2) \leq_e B_1 \oplus B_2$$

by (b). Thus (d) holds for a 2–element index set. By induction, it follows that (d) holds for all finite index sets.

In the general case, all finite collections of the B_i are independent, whence the B_i are independent. Given any nonzero submodule $M \leq \oplus B_i$, there exists a finite subset $J \subseteq I$ such that $M \cap (\bigoplus_{i \in J} B_i) \neq 0$. Since $\bigoplus_{i \in J} A_i \leq_e \bigoplus_{i \in J} B_i$, it follows that

$M \cap (\bigoplus_{i \in J} A_i) \neq 0$. Therefore $\oplus A_i \leq_e \oplus B_i$. □

Part (c) of Proposition 3.21 is often used with respect to essential right ideals in a ring R. Namely, if I is an essential right ideal of R and $J = \{r \in R \mid tr \in I\}$ for some $t \in R$, then J is the inverse image of I under the homomorphism $R_R \to R_R$ given by left multiplication by t, whence J is an essential right ideal of R.

EXERCISE 3M. Let $C = \mathbb{Z} \oplus (\mathbb{Z}/2\mathbb{Z})$, and let A, B_1, B_2 denote the cyclic subgroups of C generated by (2,0), (1,0), $(1,1+2\mathbb{Z})$. Show that $A \leq_e B_1$ and $A \leq_e B_2$ but $A \not\leq_e B_1 + B_2$. □

PROPOSITION 3.22. *Let A be a submodule of a module C, and B a submodule of C which is maximal with respect to the property $A \cap B = 0$. Then $A \oplus B \leq_e C$ and $(A \oplus B)/B \leq_e C/B$.*

Proof. If M is a submodule of C such that $(A \oplus B) \cap M = 0$, then A,B,M are independent, whence $A \cap (B \oplus M) = 0$. Then $B \oplus M = B$ by the maximality of B, and so $M = 0$. Thus $A \oplus B \leq_e C$.

Any nonzero submodule of C/B has the form D/B for some submodule D of C which properly contains B. Then $A \cap D \neq 0$ by the maximality of B, whence $(A \oplus B) \cap D \neq B$ and so $[(A \oplus B)/B] \cap (D/B) \neq 0$. Thus $(A \oplus B)/B \leq_e C/B$. □

COROLLARY 3.23. *Any submodule of a module C is a direct summand of an essential submodule of C.*

Proof. Given a submodule A of C, Zorn's Lemma guarantees the existence of a submodule B of C maximal with respect to the property $A \cap B = 0$. □

Corollary 3.23 guarantees a large supply of essential submodules in any module C, which can sometimes be used to reduce problems about arbitrary submodules to the essential case. For instance, to prove that C is noetherian it suffices to show that all essential submodules of C are finitely generated.

A module C always has at least one essential submodule, namely C itself, but in the extreme case C may not have any other essential submodules. This occurs precisely when C is semisimple, as follows.

COROLLARY 3.24. *A module C is semisimple if and only if C has no proper essential submodules.*

Proof. Assume first that C is semisimple. If A is any proper submodule of C, then Proposition 3.2 implies that $C = A \oplus B$ for some submodule B. Since $B \cap A = 0$, we infer that $A \not\leq_e C$.

Conversely, if C has no proper essential submodules, then Corollary 3.23 shows that every submodule of C is a direct summand. Therefore, using Proposition 3.2 again, C is semisimple. □

EXERCISE 3N. Show that in any module C, the intersection of the essential submodules of C equals soc(C). More generally, show that any submodule of C containing soc(C) equals the intersection of some collection of essential submodules of C. □

♦ NONSINGULAR MODULES

LEMMA 3.25. *Given a right module A over a ring R, the set*

$$Z(A) = \{x \in A \mid xI = 0 \text{ for some } I \leq_e R_R\} = \{x \in A \mid ann(x) \leq_e R_R\}$$

is a submodule of A.

Proof. As R is an essential right ideal of itself, we at least have $0 \in Z(A)$. Given any $x,y \in Z(A)$, there are essential right ideals I,J in R such that $xI = yJ = 0$. Since $I \cap J$ is an essential right ideal of R and $(x \pm y)(I \cap J) = 0$, we obtain $x \pm y \in Z(A)$. For any $t \in R$, the right ideal $K = \{r \in R \mid tr \in I\}$ is essential by Proposition 3.21, and $xtK \leq xI = 0$, whence $xt \in Z(A)$. Thus Z(A) is a submodule of A. □

DEFINITION. The submodule Z(A) described in Lemma 3.25 is called the *(maximal) singular submodule* of A. If $Z(A) = A$ then A is called a *singular module,* while if $Z(A) = 0$ then A is called a *nonsingular module*. For left R–modules, the singular submodule is defined analogously, using essential left ideals.

For example, suppose that R is a commutative domain. Then the essential ideals of R are exactly the nonzero ideals, and so the singular submodule of any R–module is just its torsion submodule. In this case the nonsingular R–modules are exactly the torsionfree R–modules.

PROPOSITION 3.26. *A module A is singular if and only if $A \cong B/C$ for some module B and some essential submodule C of B.*

Proof. We may assume that A is a right module over a ring R. First suppose that $A \cong B/C$ for some right R–modules $C \leq_e B$. Given any $b \in B$, the right ideal $I = \{r \in R \mid br \in C\}$ is essential in R by Proposition 3.21, and $(b + C)I = 0$. Thus B/C, and hence A, is singular.

Conversely, assume that A is singular, and write $A \cong F/K$ for some free right R–module F and some submodule $K \leq F$. Choose a basis $\{x_j \mid j \in J\}$ for F. For each $j \in J$, there is an essential right ideal I_j in R such that $x_jI_j \leq K$, because F/K is singular. By Proposition 3.21, $\oplus\, x_jI_j \leq_e \oplus\, x_jR = F$, and thus $K \leq_e F$. □

PROPOSITION 3.27. *Let A be a submodule of a nonsingular module B. Then B/A is singular if and only if $A \leq_e B$.*

Proof. We may assume that B is a right module over a ring R. If $A \leq_e B$, then B/A is singular by Proposition 3.26. Conversely, assume that B/A is singular. Given a nonzero submodule $C \leq B$, choose a nonzero element $x \in C$. Since B/A is singular, there is some $I \leq_e R_R$ such that $xI \leq A$. As B is nonsingular, $xI \neq 0$, whence $A \cap C \neq 0$. Thus $A \leq_e B$. □

Proposition 3.27 does not hold for arbitrary modules B. For instance, if B is a nonzero singular module, then $B/0$ is singular yet $0 \not\leq_e B$.

PROPOSITION 3.28. *(a) All submodules, factor modules, and sums (direct or not) of singular modules are singular.*

(b) All submodules, direct products, and essential extensions of nonsingular modules are nonsingular.

(c) Let B be a submodule of a module A. If B and A/B are both nonsingular, then A is nonsingular.

Proof. (a) It is clear that all submodules and factor modules of singular modules are singular. If $\{A_i \mid i \in I\}$ is a family of submodules of a module C, and each A_i is singular, then each A_i is contained in $Z(C)$, whence $\sum A_i \leq Z(C)$ and so $\sum A_i$ is singular.

(b) It is clear that all submodules of nonsingular modules are nonsingular. Given a family $\{A_i \mid i \in I\}$ of modules, observe that each of the projections $\prod A_i \to A_j$ maps $Z(\prod A_i)$ into $Z(A_j)$. Thus if each A_j is nonsingular, then $Z(\prod A_i)$ is contained in the kernels of all the projections $\prod A_i \to A_j$, whence $Z(\prod A_i) = 0$.

If A is a submodule of a module B, then $A \cap Z(B) = Z(A)$. Thus if A is nonsingular, then $A \cap Z(B) = 0$, whence if $A \leq_e B$ we infer that $Z(B) = 0$.

(c) The quotient map $A \to A/B$ carries $Z(A)$ into $Z(A/B)$. Since $Z(A/B) = 0$, we infer that $Z(A) \leq B$, and since B is nonsingular, it follows that $Z(A) = 0$. □

In general, essential extensions of singular modules need not be singular. For example, if $R = \mathbb{Z}/4\mathbb{Z}$ and $M = 2R$, then M is a singular R–module and R is an essential extension of M, but R is not a singular R–module (even though it is singular as a $\mathbb{Z}$–module). Also, R/M is a singular R–module.

We have already observed in some of the proofs above that any module homomorphism $A \to B$ carries $Z(A)$ into $Z(B)$. In particular, in any ring R left multiplication by elements of R carries $Z(R_R)$ into itself. Thus $Z(R_R)$ is an ideal of R, and similarly $Z({}_RR)$ is an ideal.

DEFINITION. The *right singular ideal* of a ring R is the ideal $Z_r(R) = Z(R_R)$, and the *left singular ideal* of R is the ideal $Z_l(R) = Z({}_RR)$.

EXERCISE 3O. Let k be a field, let x be an indeterminate, and set

$$R = \begin{pmatrix} k[x]/x^2k[x] & k[x]/x^2k[x] \\ 0 & k \end{pmatrix}$$

Show that $Z_r(R) = 0$ while $Z_l(R)$ is an essential left ideal of R. □

DEFINITION. A *right (left) nonsingular ring* is any ring whose right (left) singular ideal is zero. Of course, a *nonsingular ring* is a ring which is both right and left nonsingular.

For instance, every domain is a nonsingular ring. Also, every semisimple ring R is nonsingular. (Since R has no proper essential one–sided ideals, all R–modules are nonsingular.)

EXERCISE 3P. Show that $\begin{pmatrix} \mathbb{Q} & \mathbb{Q} \\ 0 & \mathbb{Z} \end{pmatrix}$ is a nonsingular ring. □

EXERCISE 3Q. Given an integer $n > 1$, show that $\mathbb{Z}/n\mathbb{Z}$ is a nonsingular ring if and only if n is square–free. □

EXERCISE 3R. Show that a commutative ring is nonsingular if and only if it is semiprime. □

PROPOSITION 3.29. *Let R be a right nonsingular ring.*

(a) For every right R–module A, the factor module $A/Z(A)$ is nonsingular.

(b) If A is a submodule of a right R–module B such that A and B/A are both singular, then B is singular.

(c) All essential extensions of singular right R–modules are singular.

Proof. (a) Let $B/Z(A) = Z(A/Z(A))$. We first claim that $Z(A) \leq_e B$. If M is a submodule of B such that $Z(A) \cap M = 0$, then M is nonsingular. On the other hand, M embeds in the singular module $B/Z(A)$, whence M is singular. Consequently, $M = 0$, proving that $Z(A) \leq_e B$.

Now consider any $x \in B$, and set $I = \text{ann}(x)$. If J is any right ideal of R for which $I \cap J = 0$, then $J \cong xJ$. Since $Z(xJ) = xJ \cap Z(A) \leq_e xJ \cap B = xJ$, we see that $Z(J) \leq_e J$. However, $Z(J) = 0$ because R_R is nonsingular, whence $J = 0$. Thus $I \leq_e R_R$, and so $x \in Z(A)$.

Therefore $B/Z(A) = 0$, and hence $A/Z(A)$ is nonsingular.

(b) Since A is singular, $A \leq Z(B)$, whence B/A maps onto $B/Z(B)$, and so $B/Z(B)$ is singular. On the other hand, $B/Z(B)$ is nonsingular by (a), and hence $B/Z(B) = 0$. Thus B is singular.

(c) Let A be an essential submodule of a right R–module B, and suppose that A is singular. By Proposition 3.26, B/A is singular, and so (b) shows that B is singular. □

That the nonsingularity hypothesis is crucial in Proposition 3.29 may be seen from the example $R = \mathbb{Z}/4\mathbb{Z}$. First, $Z(R) = 2R$ and $R/Z(R)$ is singular rather than nonsingular. Second, $2R$ and $R/2R$ are both singular, yet R is not. Finally, R is also an essential extension of the singular module $2R$.

EXERCISE 3S. If $I_1,\ldots,I_n$ are essential right ideals in a right nonsingular ring R, show that $I_1I_2\cdots I_n$ is an essential right ideal. □

♦ ADDITIONAL EXERCISES

3T. If N is a nilpotent ideal in a ring R, show that $\text{l.ann}(N)$ is an essential right ideal of R. □

3U. If I is any ideal in a ring R, show that the ideal $I + \text{l.ann}(I)$ is essential as a right ideal of R. □

3V. Show that a $\mathbb{Z}$–module A has essential socle if and only if A is a torsion module. □

3W. Show that any nonsingular right module over a ring R has an essential submodule which is a direct sum of submodules isomorphic to principal right ideals of R. □

3X. If R is a semiprime ring and $\text{soc}(R_R)$ is an essential right ideal of R, show that R is right and left nonsingular. □

3Y. Let $J = \mathrm{soc}(R_R)$ for some ring R.

(a) Assuming that $J \leq_e R_R$, show that $Z_r(R) = 0$ if and only if $J^2 = J$.

(b) Show that $J \leq Z_r(R)$ if and only if $J^2 = 0$, if and only if J is nil, if and only if J contains no nonzero idempotents. □

3Z. Show that a semisimple right module over a right nonsingular ring is nonsingular if and only if it is projective. □

♦ NOTES

Socles. This concept was introduced by Krull [1928c, Definition 5, p. 64]. The name was first used by Remak to label the product of all minimal normal subgroups of a group [1930, p. 4], and was later adopted by Dieudonné to define the right and left socles of a ring [1942, pp. 47, 51].

Structure of Semisimple Rings. Noether proved that a ring R is (right) semisimple if and only if all finitely generated (right) R–modules are semisimple, if and only if R is isomorphic to a finite direct product of matrix rings over division rings [1929, §§ 13, 14, 18].

Zassenhaus's Lemma. Lemma 3.9 (sometimes called the Butterfly Lemma) was proved by Zassenhaus [1934, p. 107] in simplifying Schreier's original proof of the refinement theorem.

Schreier Refinement Theorem. This was first proved by Schreier for normal series of subgroups [1928, Satz 1], and he remarks in a footnote that his proof also works for groups with operators.

Jordan–Hölder Theorem. This was first developed for composition series of a finite group G. Jordan proved that for any two composition series of G, the list of the orders of the composition factors in one series is a permutation of the corresponding list for the other series [1869a, p. 140; 1869b, §§ 19–21]. That any two composition series for G are isomorphic was proved by Hölder [1889, § 10].

Wedderburn–Artin Theorem. Matrix representations were first considered for finite–dimensional algebras. Molien proved (essentially) that any finite–dimensional simple algebra over $\mathbb{C}$ must be isomorphic to a matrix algebra over $\mathbb{C}$ [1893, Satz 30]; later, Cartan explicitly proved this, along with the result that any finite–dimensional simple algebra over $\mathbb{R}$ must be isomorphic to a matrix algebra over $\mathbb{R}$, $\mathbb{C}$, or $\mathbb{H}$ [1898, §§ 71, 85]. Wedderburn proved that any finite–dimensional semiprime algebra over a field F is isomorphic to a finite direct product of matrix algebras over finite–dimensional division algebras over F [1908, Theorems 10, 17, 22, 23]. Artin developed the generalization to rings satisfying both ACC and DCC for right ideals, but only gave the proofs up to the point where Wedderburn's arguments could be used [1927, cf. Satz 11]. The ACC hypothesis was not removed until Hopkins and Levitzki obtained Theorem 3.15.

ACC and Radical Nilpotence in Artinian Rings. The nilpotence of the radical in a right artinian ring was proved independently by Hopkins [1938, (3.1); 1939, (1.4)] and Levitzki [1939, Theorem 6]. Hopkins obtained the consequence that every right artinian ring is right noetherian [1939, (6.4)].

Semisimplicity of Semiprime Artinian Rings. That a ring is semisimple if and only if it is (right) artinian and semiprime was proved by Noether [1929, § 13].

Essential Submodules. The concept was introduced by Johnson [1951, p. 891], the name by Eckmann and Schopf [1953, (4.1)].

Singular Submodules. The right singular ideal of a ring was introduced by Johnson [1951, p. 894], and the singular submodule of a module later [1957, p. 537].

4 INJECTIVE HULLS

Injective modules may be regarded as modules that are "complete" in the following algebraic sense: any "partial" homomorphism (from a submodule of a module B) into an injective module A can be "completed" to a "full" homomorphism (from all of B) into A. (Other types of completeness often entail similar extension properties. For instance: (a) If X,Y are metric spaces with X complete, then any uniformly continuous map from a dense subspace of Y to X extends to a uniformly continuous map from Y to X; (b) If Y is a normed linear space, then any bounded linear map from a linear subspace of Y to $\mathbb{R}$ extends to a bounded linear map from Y to $\mathbb{R}$; (c) If X,Y are boolean algebras with X complete, then any boolean homomorphism from a subalgebra of Y to X extends to a boolean homomorphism from Y to X.)

In topological and order–theoretic contexts, incomplete objects can be investigated by enlarging them to their completions. Following this pattern, one way to study a module A is to "complete" it to an injective module, i.e., to embed A in an injective module in some canonical, minimal fashion. The resulting module, called the "injective hull" of A, is introduced in this chapter. After studying general properties of injective hulls, we show how injective hulls may be used to develop a useful notion of "finite rank" for modules, and we show how the injective hull of a nonsingular ring may be made into a ring itself which in the finite rank case is a semisimple ring. This new ring will play a key role in constructing semisimple rings of fractions in the following chapter. In later chapters we shall see that injective hulls are intimately bound up with prime ideals in noetherian rings.

Injective modules first appeared in the context of abelian groups ($\mathbb{Z}$–modules). Zippin observed in 1935 that an abelian group is divisible if and only if it is a direct summand of any larger group containing it as a subgroup, and that divisible abelian groups can be completely described. The general notion for modules was first investigated by Baer in 1940. It is interesting in retrospect that the theory of these modules was investigated long before the dual notion of projective modules was considered. The "injective" and "projective" terminology originated with Cartan and Eilenberg in 1956.

♦ INJECTIVE MODULES

This introductory section covers the basic properties of injective modules. Since this is completely standard material, we merely sketch it; readers who have not met injective modules before should fill in the details for themselves or consult a text on homological algebra.

DEFINITION. A right (left) module A over a ring R is *injective* provided that for any right (left) R–module B and any submodule C of B, all homomorphisms $C \to A$ extend to homomorphisms $B \to A$.

For example, over a semisimple ring all modules are injective (Exercise 3G). On the other hand, $\mathbb{Z}$ is not an injective $\mathbb{Z}$–module, since the homomorphism $f : 2\mathbb{Z} \to \mathbb{Z}$ given by the rule $f(2n) = n$ cannot be extended to a homomorphism from $\mathbb{Z}$ to $\mathbb{Z}$.

Observe that all direct summands and direct products of injective modules are injective.

PROPOSITION 4.1. [Baer's Criterion] *Let A be a right module over a ring R. Then A is injective if and only if for every right ideal I of R and every $f \in Hom_R(I,A)$, there exists $a \in A$ such that $f(r) = ar$ for all $r \in I$.*

Proof. If A is injective, then given $I \leq R_R$, any $f \in \mathrm{Hom}_R(I,A)$ extends to some $f_1 \in \mathrm{Hom}_R(R,A)$, and $f(r) = f_1(r) = f_1(1)r$ for all $r \in I$. Conversely, assume that A satisfies the given condition, and consider right R–modules $C \leq B$ together with a homomorphism $f : C \to A$.

Let X be the set of all pairs (C_1,f_1) where C_1 is a submodule of B containing C and f_1 is a homomorphism from C_1 to A extending f. Define a relation $\leq$ on X by declaring that $(C_1,f_1) \leq (C_2,f_2)$ if and only if $C_1 \leq C_2$ and f_2 extends f_1. One then checks that this relation is a partial order on X, and that every nonempty chain in X has an upper bound. By Zorn's Lemma, there is a maximal element (C^*,f^*) in X, and if $C^* = B$ we are done.

If not, choose $b \in B - C^*$ and set $I = \{r \in R \mid br \in C^*\}$. The rule $r \mapsto f^*(br)$ defines a homomorphism $I \to A$, and hence, by assumption, there exists $a \in A$ such that $f^*(br) = ar$ for all $r \in I$. One checks that there is a well–defined homomorphism $f_1 : C^* + bR \to A$ such that $f_1(c + br) = f^*(c) + ar$ for all $c \in C^*$, $r \in R$. However, this contradicts the maximality of (C^*,f^*). □

Recall that a $\mathbb{Z}$–module A is *divisible* provided $nA = A$ for all nonzero $n \in \mathbb{Z}$.

PROPOSITION 4.2. *(a) A $\mathbb{Z}$–module A is injective if and only if it is divisible.*
(b) Every $\mathbb{Z}$–module is a submodule of a divisible module.

Proof. (a) An element $a \in A$ is divisible by a nonzero integer n if and only if the homomorphism $n\mathbb{Z} \to A$ sending n to a extends to $\mathbb{Z}$.

(b) Any $\mathbb{Z}$–module is isomorphic to one of the form F/K where F is free and $K \leq F$. Now F is a direct sum of copies of $\mathbb{Z}$. If D is the corresponding direct sum of copies of $\mathbb{Q}$, then D and D/K are divisible, and $F/K \leq D/K$. □

Recall that if R is a ring and D is an abelian group, $\mathrm{Hom}_{\mathbb{Z}}(R,D)$ may be made into either a right or a left R–module. If $f \in \mathrm{Hom}_{\mathbb{Z}}(R,D)$ and $r \in R$ then fr is defined by the rule $(fr)(x) = f(rx)$ for all $x \in R$.

LEMMA 4.3. *If R is a ring and D is a divisible $\mathbb{Z}$–module, then $H = \mathrm{Hom}_{\mathbb{Z}}(R,D)$ is an injective right (or left) R–module.*

Proof. If $I \leq R_R$ and $f \in \mathrm{Hom}_R(I,H)$, the rule $r \mapsto f(r)(1)$ defines a $\mathbb{Z}$–module homomorphism from I to D. Since D is an injective $\mathbb{Z}$–module, this extends to a $\mathbb{Z}$–module homomorphism $g : R \to D$. Now $g \in H$, and one checks that $f(r) = gr$ for all $r \in I$. □

THEOREM 4.4. [Baer] *Every module is a submodule of an injective module.*

Proof. Consider a right module A over a ring R. Viewed as a $\mathbb{Z}$–module, A is a submodule of a divisible $\mathbb{Z}$–module D. Then $\mathrm{Hom}_{\mathbb{Z}}(R,D)$ is an injective right R–module, and there are right R–module embeddings

$$A \cong \mathrm{Hom}_R(R,A) \leq \mathrm{Hom}_{\mathbb{Z}}(R,A) \leq \mathrm{Hom}_{\mathbb{Z}}(R,D). \ \square$$

COROLLARY 4.5. [Baer] *A module A is injective if and only if A is a direct summand of every module that contains it.*

Proof. If A is injective and $A \leq B$, the identity map on A extends to a homomorphism $f : B \to A$, and then $B = A \oplus \ker(f)$. Conversely, if the direct summand condition holds then A is a direct summand of an injective module. □

♦ INJECTIVE HULLS

By an injective hull for a module A is meant, roughly, an injective module containing A which is as small as possible. Alternatively, an injective hull for A turns out to be an essential extension of A which is as large as possible. In order to construct injective hulls, and to exhibit this alternate description of them, we first look at the relationships between injectivity and essential extensions.

DEFINITION. A *proper essential extension* of a module A is any module B such that $A \leq_e B$ while $B > A$. Note that A has a proper essential extension if and only if there exists an essential monomorphism $f : A \to C$ such that $f(A) < C$.

PROPOSITION 4.6. [Eckmann–Schopf] *A module A is injective if and only if A has no proper essential extensions.*

Proof. First assume that A is injective, and consider an essential extension $A \leq_e B$. By Corollary 4.5, $B = A \oplus C$ for some submodule C. Since $A \cap C = 0$, we obtain $C = 0$ because $A \leq_e B$, whence $B = A$.

Conversely, if A is not injective then by Corollary 4.5 there exists a module $C \geq A$ such that A is not a direct summand of C. Choose a submodule $B \leq C$ maximal with respect to the property $A \cap B = 0$, and note that $A \oplus B < C$. By Proposition 3.22, $(A \oplus B)/B \leq_e C/B$, whence the map $A \to C/B$ (given by composing the inclusion map $A \to C$ with the quotient map $C \to C/B$) is an essential monomorphism. Thus A has a proper essential extension. □

DEFINITION. Let C be a module and A a submodule. We say that A is *essentially closed* in C provided A has no proper essential extensions within C, that is, the only submodule B of C for which $A \leq_e B$ is A. In short, A is essentially closed in C if and only if $A \leq_e B \leq C$ implies $B = A$.

EXERCISE 4A. If a module A is a direct summand of a module C, show that A is essentially closed in C. □

EXERCISE 4B. Let $A \leq C$ be modules. Show that A is essentially closed in C if and only if there exists a submodule $B \leq C$ such that A is maximal with respect to the property $A \cap B = 0$, if and only if there exists a submodule $B \leq C$ such that $A \cap B = 0$ and $(A + B)/A \leq_e C/A$. □

PROPOSITION 4.7. *Let A be a submodule of an injective module E. Then A is injective if and only if A is essentially closed in E.*

Proof. If A is injective, then by Proposition 4.6 A is essentially closed in any module containing A.

Conversely, assume that A is essentially closed in E, and consider any essential extension $A \leq_e B$. The inclusion map $A \to B$ extends to a homomorphism $f : B \to E$. Since $A \cap \ker(f) = 0$ and $A \leq_e B$, we obtain $\ker(f) = 0$, and so f provides an isomorphism of B onto $f(B)$. Then $A = f(A) \leq_e f(B) \leq E$, and so $f(B) = A$ (because A

is essentially closed in E), whence $B = A$. Thus A has no proper essential extensions. By Proposition 4.6, A is injective. □

DEFINITION. An *injective hull* (or *injective envelope*) for a module A is any injective module which is an essential extension of A.

For example, $\mathbb{Q}$ is an injective hull for $\mathbb{Z}$.

EXERCISE 4C. If R is any commutative domain, show that the quotient field of R is an injective hull for R_R. □

THEOREM 4.8. [Baer, Eckmann–Schopf] *Let A be a module.*

(a) Any injective module containing A contains an injective hull for A. In particular, there exist injective hulls for A.

Let E be any injective hull for A, and let $j : A \to E$ be the inclusion map.

(b) Given any essential monomorphism $f : A \to B$, there exists a monomorphism $g : B \to E$ such that $gf = j$.

(c) Given any monomorphism $f : A \to E'$ with E' injective, there exists a monomorphism $g : E \to E'$ such that $gj = f$.

Proof. (a) Consider any injective module $F \geq A$. By Zorn's Lemma, there exists a submodule $E \leq F$ such that $A \leq E$ and E is maximal with respect to the property $A \leq_e E$. If E' is any submodule of F for which $E \leq_e E'$, then $A \leq_e E'$, and hence $E' = E$ by maximality of E. Thus E is essentially closed in F, and so E is injective, by Proposition 4.7. Therefore E is an injective hull for A.

(b) Since E is injective, there exists a homomorphism $g : B \to E$ such that $gf = j$. Then

$$f(A) \cap \ker(g) = f(\ker(j)) = 0,$$

whence $\ker(g) = 0$, because $f(A) \leq_e B$. Thus g is a monomorphism.

(c) Since E' is injective, there exists a homomorphism $g : E \to E'$ such that $gj = f$. Then $A \cap \ker(g) = \ker(f) = 0$, whence $\ker(g) = 0$, because $A \leq_e E$. Thus g is a monomorphism. □

Parts (b) and (c) of Theorem 4.8 may be summarized by saying that an injective hull for a module A is a "maximal essential extension" of A as well as a "minimal injective extension" of A.

Injective hulls need not be unique as sets, even within a given injective module, as the following example shows. Let $R = \mathbb{Z}/4\mathbb{Z}$, and observe that R_R is injective. Set $F = R \oplus R$ and $A = (2,0)R \leq F$. Then $(1,0)R$ and $(1,2)R$ are submodules of F isomorphic to R, and so they are injective R–modules. Moreover, $A \leq_e (1,0)R$ and $A \leq_e (1,2)R$, so that $(1,0)R$ and $(1,2)R$ are both injective hulls for A.

However, injective hulls are unique up to isomorphism, as follows.

PROPOSITION 4.9. *If E and E' are injective hulls for a module A, the identity map on A extends to an isomorphism of E onto E'.*

Proof. Let $j : A \to E$ and $j' : A \to E'$ be the inclusion maps. By Theorem 4.8, there exists a monomorphism $g : E \to E'$ such that $gj = j'$. Then $A = g(A) \leq g(E)$ and so $g(E) \leq_e E'$, that is, g is an essential monomorphism. However E, being injective, has no proper essential extensions (Proposition 4.6), and hence $g(E) = E'$. Thus g is an isomorphism. □

DEFINITION. Given a module A, we use the notation E(A) for an injective hull of A. As this is only unique up to isomorphism, general assertions about E(A) must be valid for all injective hulls of A. The equation "B = E(A)" should only be used as an abbreviation for "B is an injective hull for A".

For instance, given modules A_1 and A_2 we may state that

$$E(A_1) \oplus E(A_2) = E(A_1 \oplus A_2),$$

since the direct sum of any injective hull for A_1 with any injective hull for A_2 is an injective hull for $A_1 \oplus A_2$.

EXERCISE 4D. Given any torsionfree $\mathbb{Z}$–module A, show that $E(A) \cong A \otimes_{\mathbb{Z}} \mathbb{Q}$. □

EXERCISE 4E. Let I be an ideal in a ring R, let A be a right (R/I)–module, and let E be an injective hull for A_R. Show that the (R/I)–module $\mathrm{ann}_E(I)$ is an injective hull for $A_{R/I}$. □

♦ MODULES OF FINITE RANK

In abelian group theory, the "rank" of a torsionfree abelian group A is defined as the dimension of the vector space $A \otimes_{\mathbb{Z}} \mathbb{Q}$. While this notion extends immediately to torsionfree modules over commutative domains, it is not readily apparent how to usefully adapt it to torsion modules, or to modules over other rings. We proceed by relating the vector space viewpoint to injective hulls. Recall from Exercise 4D that the injective hull of a torsionfree abelian group A is isomorphic to $A \otimes_{\mathbb{Z}} \mathbb{Q}$. Under this isomorphism, as is easily checked, one–dimensional subspaces of $A \otimes_{\mathbb{Z}} \mathbb{Q}$ correspond to nonzero indecomposable direct summands of E(A). Hence, the dimension of $A \otimes_{\mathbb{Z}} \mathbb{Q}$, which may be calculated as the number of summands in a decomposition of $A \otimes_{\mathbb{Z}} \mathbb{Q}$ as a direct sum of one–dimensional subspaces, equals the number of nonzero summands in a decomposition of E(A) as a direct sum of indecomposable submodules.

To adapt this idea to arbitrary modules, two obstacles must be overcome. First, not every injective module is a direct sum of indecomposable modules (Exercise 4F). Second, if an injective module is a finite direct sum of indecomposable submodules in two different ways, we need to know whether the number of nonzero summands is the same in both decompositions. The first obstacle we finesse by restricting attention to modules whose injective hulls are finite direct sums of indecomposable submodules, and within this context we prove that the second obstacle vanishes.

EXERCISE 4F. Let $R = (\prod F_n)/(\oplus F_n)$ for some fields $F_1, F_2, \ldots$. Show that $E(R_R)$ has no nonzero indecomposable direct summands. [Hint: Show that no nonzero principal ideal of R is indecomposable.] □

DEFINITION. A module A has *finite rank* provided E(A) is a finite direct sum of indecomposable submodules. (We shall consider a value for the rank of A later.) In the literature, a module of finite rank is sometimes called a *finite–dimensional* module. Moreover, one or the other of the equivalent conditions given in Proposition 4.11 and Theorem 4.13 is often taken as the definition of finite rank.

As indicated in the discussion above, this notion of finite rank coincides with the traditional one when applied to torsionfree abelian groups. Thus, for instance, every subgroup of a finite–dimensional vector space over $\mathbb{Q}$ is a $\mathbb{Z}$–module of finite rank. Moreover, we shall prove shortly that every noetherian module has finite rank (Corollary 4.14).

The obvious building blocks for finite rank modules should be those modules whose injective hulls are indecomposable. We approach this property from within the modules, as follows.

DEFINITION. A *uniform module* is a nonzero module A such that the intersection of any two nonzero submodules of A is nonzero, or, equivalently, such that every nonzero submodule of A is essential in A.

Note that all nonzero submodules and all essential extensions of uniform modules are uniform. For an example, the quotient field of a commutative domain R is a uniform R–module.

EXERCISE 4G. Show that the finitely generated uniform $\mathbb{Z}$–modules are (up to isomorphism) exactly $\mathbb{Z}$ and $\mathbb{Z}/p^n\mathbb{Z}$ for prime integers p and positive integers n. □

LEMMA 4.10. *A nonzero module A is uniform if and only if $E(A)$ is indecomposable.*

Proof. First suppose that A is uniform, and that $E(A) = B \oplus C$ for some submodules B, C. As $(B \cap A) \cap (C \cap A) = 0$, either $B \cap A = 0$ or $C \cap A = 0$ (because A is uniform), whence either $B = 0$ or $C = 0$ (because $A \leq_e E(A)$). Thus $E(A)$ is indecomposable.

Conversely, if A is not uniform it has nonzero submodules B and C such that $B \cap C = 0$. Then $E(A)$ has a nonzero submodule E which is an injective hull for B, and $E \cap C = 0$ because $B \leq_e E$, whence $E \neq E(A)$. Thus E is a nontrivial direct summand of $E(A)$, and so $E(A)$ is not indecomposable. □

In particular, Lemma 4.10 shows that an injective module is uniform if and only if it is nonzero and indecomposable. Thus the terms "uniform injective module" and "nonzero indecomposable injective module" are synonymous, and since the former is shorter we shall generally use it.

We now show that the condition of finite rank can be characterized internally, using uniform submodules.

PROPOSITION 4.11. *A module A has finite rank if and only if A has an essential submodule which is a finite direct sum of uniform submodules.*

Proof. First assume that A contains independent uniform submodules $A_1,\ldots,A_n$ such that $A_1 \oplus \ldots \oplus A_n \leq_e A$. Then $A_1 \oplus \ldots \oplus A_n \leq_e E(A)$, whence

$$E(A) = E(A_1 \oplus \ldots \oplus A_n) \cong E(A_1) \oplus \ldots \oplus E(A_n).$$

Since each $E(A_i)$ is indecomposable by Lemma 4.10, A has finite rank.

Conversely, if A has finite rank then $E(A) = E_1 \oplus \ldots \oplus E_n$ for some indecomposable submodules E_i, and we may assume that each $E_i \neq 0$. Each E_i is uniform by Lemma 4.10. Each of the submodules $A_i = A \cap E_i$ is nonzero because $A \leq_e E(A)$, whence A_i is uniform. The A_i are clearly independent submodules of A, and each $A_i \leq_e E_i$ because E_i is uniform. Finally,

$$A_1 \oplus \ldots \oplus A_n \leq_e E_1 \oplus \ldots \oplus E_n = E(A),$$

and consequently $A_1 \oplus \ldots \oplus A_n \leq_e A$. □

Modules of finite rank need not be direct sums of uniform submodules, as the following examples show.

EXERCISE 4H. If $R = \{(x,y) \in \mathbb{Z} \times \mathbb{Z} \mid x \equiv y \pmod 2\}$, show that R_R has finite rank but is not a direct sum of uniform submodules. □

EXERCISE 4I. Let $S = k[x,y]$ be a polynomial ring over a field k, and set $R = S/(xS + yS)^2$. Show that R_R has finite rank but is not a direct sum of uniform submodules. □

LEMMA 4.12. *If a module E is a finite direct sum of n uniform submodules, E does not contain any direct sums of $n + 1$ nonzero submodules.*

Proof. If $n = 0$ then $E = 0$, while if $n = 1$ then E is uniform; in either case the conclusion is clear. Now let $n > 1$ and assume the lemma holds for direct sums of $n - 1$ uniform modules.

We are given that $E = E_1 \oplus \dots \oplus E_n$ with each E_i uniform. Suppose that E contains a direct sum $A_1 \oplus \dots \oplus A_{n+1}$ of $n + 1$ nonzero submodules. Set $A = A_1 \oplus \dots \oplus A_n$. If $A \cap E_1 = 0$, then A embeds in $E_2 \oplus \dots \oplus E_n$ (via the natural projection $E \to E_2 \oplus \dots \oplus E_n$), whence $E_2 \oplus \dots \oplus E_n$ contains a direct sum of n nonzero submodules, contradicting the induction hypothesis. Thus $A \cap E_1 \neq 0$, and similarly $A \cap E_i \neq 0$ for all i.

Since the E_i are uniform, $A \cap E_i \leq_e E_i$ for all i. Then

$$(A \cap E_1) \oplus \dots \oplus (A \cap E_n) \leq_e E_1 \oplus \dots \oplus E_n = E,$$

whence $A \leq_e E$. However, as $A \cap A_{n+1} = 0$, this is impossible. Therefore E does not contain a direct sum of $n + 1$ nonzero submodules. □

THEOREM 4.13. [Goldie] *A module A is of finite rank if and only if A contains no infinite direct sums of nonzero submodules.*

Proof. If A is of finite rank, $E(A)$ is a direct sum of n uniform submodules for some $n \in \mathbb{Z}^+$. By Lemma 4.12, $E(A)$ cannot contain a direct sum of more than n nonzero submodules, and hence neither can A.

If A is not of finite rank, then $A \neq 0$ and $E(A)$ is not a finite direct sum of indecomposable submodules. Set $C_0 = E(A)$. Since C_0 is not indecomposable, $C_0 = M \oplus N$ for some nonzero submodules M and N; moreover, M and N cannot both be finite direct sums of indecomposable submodules. Hence, $C_0 = B_1 \oplus C_1$ for some nonzero submodules B_1 and C_1 such that C_1 is not a finite direct sum of indecomposable submodules.

Repeat this argument with respect to C_1, and continue inductively. We obtain submodules $B_1, C_1, B_2, C_2, \dots$ such that each $C_{n-1} = B_n \oplus C_n$ with B_n nonzero and C_n not a finite direct sum of indecomposable submodules. Since $B_k \leq C_n$ whenever $k > n$, we find that

$$B_n \cap \left(\sum_{k=n+1}^{\infty} B_k \right) \leq B_n \cap C_n = 0$$

for all $n \in \mathbb{N}$, from which it follows that $B_1, B_2, \ldots$ are independent submodules of $E(A)$.

Now $B_1 \cap A$, $B_2 \cap A, \ldots$ is an infinite sequence of independent submodules of A, and as $A \leq_e E(A)$, it follows that $B_n \cap A \neq 0$ for each $n \geq 1$. Therefore A contains an infinite direct sum of nonzero submodules. □

COROLLARY 4.14. *Any noetherian module A has finite rank.*

Proof. If not, then by Theorem 4.13, A contains an infinite direct sum of nonzero submodules, and hence A contains an infinite sequence $A_1, A_2, \ldots$ of independent nonzero submodules. But then

$$A_1 < A_1 \oplus A_2 < A_1 \oplus A_2 \oplus A_3 < \ldots$$

is a strictly ascending infinite chain of submodules of A, contradicting our noetherian hypothesis. □

COROLLARY 4.15. *Every nonzero noetherian module has a uniform submodule.* □

EXERCISE 4J. Show that a module A has finite rank if and only if every submodule of A is an essential extension of a finitely generated submodule. □

EXERCISE 4K. Show that a module A has finite rank if and only if A satisfies the ACC on essentially closed submodules. □

♦ UNIFORM RANK

DEFINITION. If A is a module of finite rank, there exists a nonnegative integer n such that $E(A)$ is a direct sum of n uniform submodules. Moreover, because of Lemma 4.12 any other decomposition of $E(A)$ into a direct sum of uniform submodules has exactly n summands. Thus n is uniquely determined by A. We shall call this integer the *uniform rank*, or just the *rank*, of A, and denote it by $\operatorname{rank}(A)$. (In the literature, the rank of A is also called the *Goldie rank*, the *Goldie dimension*, the *uniform dimension*, or the *dimension* of A.)

Observe that if $A_1,\ldots,A_n$ are modules of finite rank, then $A_1 \oplus \ldots \oplus A_n$ has finite rank, and

$$\operatorname{rank}(A_1 \oplus \ldots \oplus A_n) = \operatorname{rank}(A_1) + \ldots + \operatorname{rank}(A_n).$$

EXERCISE 4L. Let R be a commutative domain with quotient field K, and let A be a torsionfree R–module of finite rank. Show that $\text{rank}(A) = \dim_K(A \otimes_R K)$. □

PROPOSITION 4.16. *Let A be a module and n a nonnegative integer. Then the following conditions are equivalent:*

(a) A has finite rank n.

(b) A has an essential submodule which is a direct sum of n uniform submodules.

(c) A contains a direct sum of n nonzero submodules, but no direct sums of $n + 1$ nonzero submodules.

Proof. (a) ⇒ (c): By assumption, $E(A) = E_1 \oplus \ldots \oplus E_n$ for some uniform submodules E_i. Lemma 4.12 then shows that $E(A)$ contains no direct sums of $n + 1$ nonzero submodules, and hence A does not either. On the other hand, A contains the direct sum of the n nonzero submodules $E_1 \cap A, \ldots, E_n \cap A$.

(c) ⇒ (b): Let $A_1,\ldots,A_n$ be n independent nonzero submodules of A. There cannot be two nonzero submodules $B,C \leq A_1$ with $B \cap C = 0$, for then A would contain the direct sum $B \oplus C \oplus A_2 \oplus \ldots \oplus A_n$ of $n + 1$ nonzero submodules. Thus A_1 is uniform, and similarly all the A_i are uniform. If $A_1 \oplus \ldots \oplus A_n$ is not essential in A, there is a nonzero submodule $A_{n+1} \leq A$ such that $(A_1 \oplus \ldots \oplus A_n) \cap A_{n+1} = 0$, but then A would contain the direct sum $A_1 \oplus \ldots \oplus A_{n+1}$ of $n + 1$ nonzero submodules. Therefore $A_1 \oplus \ldots \oplus A_n \leq_e A$.

(b) ⇒ (a): See the proof of Proposition 4.11. □

COROLLARY 4.17. *Let A be a module of finite rank and B a submodule of A. Then B has finite rank and $\text{rank}(B) \leq \text{rank}(A)$. Moreover, $\text{rank}(B) = \text{rank}(A)$ if and only if $B \leq_e A$.*

Proof. By Proposition 4.16, B contains no direct sums of $\text{rank}(A) + 1$ nonzero submodules. Hence, applying the proposition to B, we find that B has finite rank and that $\text{rank}(B) \leq \text{rank}(A)$.

Proposition 4.16 also says that B has an essential submodule C which is a direct sum of $\text{rank}(B)$ uniform submodules. If $B \leq_e A$, then $C \leq_e A$, whence the proposition shows that $\text{rank}(A) = \text{rank}(B)$. Conversely, if $\text{rank}(A) = \text{rank}(B)$ the proposition says that A does not contain a direct sum of $\text{rank}(B) + 1$ nonzero submodules, from which we conclude that $C \leq_e A$, and therefore $B \leq_e A$. □

COROLLARY 4.18. *Let A be a module with finite rank. If $f : A \to A$ is a monomorphism, then $f(A) \leq_e A$.*

Proof. As f(A) is isomorphic to A, it has the same rank as A. □

EXERCISE 4M. Prove Corollary 4.18 using only the property that A contains no infinite direct sums of nonzero submodules. □

EXERCISE 4N. If A is a module of finite rank and B is an essentially closed submodule of A, show that rank(A) = rank(B) + rank(A/B). (This additivity usually fails in case B is not essentially closed, as seen in the next exercise.) □

EXERCISE 4O. Given a positive integer n, find a nonzero ideal B of $\mathbb{Z}$ such that rank($\mathbb{Z}$/B) = n. Note in particular that rank($\mathbb{Z}$) < rank(B) + rank($\mathbb{Z}$/B). □

EXERCISE 4P. If $R = M_n(D)$ for some $n \in \mathbb{N}$ and some division ring D, show that $\text{rank}(R_R) = n$. □

♦ DIRECT SUMS OF INJECTIVE MODULES

THEOREM 4.19. [Papp, Bass] *A ring R is right noetherian if and only if every direct sum of injective right R–modules is injective.*

Proof. First assume that R is right noetherian, and let E be the direct sum of a family $\mathscr{E}$ of injective right R–modules. Let J be a right ideal of R, and let $f : J \to E$ be a homomorphism.

Choose generators $x_1,...,x_n$ for J. Each $f(x_k)$ lies in $\oplus \mathscr{E}_k$ for some finite subfamily $\mathscr{E}_k \subseteq \mathscr{E}$. If $E' = \oplus (\mathscr{E}_1 \cup ... \cup \mathscr{E}_n)$, then each $f(x_k) \in E'$, whence $f(J) \leq E'$. Since $\mathscr{E}_1 \cup ... \cup \mathscr{E}_n$ is finite, E' is injective, and so f extends to a homomorphism $R_R \to E' \leq E$. Therefore E is injective.

Conversely, assume that all direct sums of injective right R–modules are injective, and let $I_1 \leq I_2 \leq ...$ be an ascending chain of right ideals of R. Set

$$I = \bigcup_{n=1}^{\infty} I_n \qquad \text{and} \qquad E = \bigoplus_{n=1}^{\infty} E(R/I_n).$$

Then I is a right ideal of R, and E is an injective right R–module. Define a homomorphism

$$f : I \to \prod_{n=1}^{\infty} E(R/I_n)$$

so that $f(x)_n = x + I_n$ for all $x \in I$ and all $n \in \mathbb{N}$. For any $x \in I$, we have $x \in I_k$ for some k, whence $x + I_n = 0$ for all $n \geq k$. Hence, $f(I) \leq E$.

As E is injective, there exists $z \in E$ such that $f(x) = zx$ for all $x \in I$. Then $z_k = 0$ for some k. For all $x \in I$, we have

$$x + I_k = f(x)_k = (zx)_k = z_k x = 0,$$

and so $x \in I_k$. Thus $I = I_k$, whence $I_n = I_k$ for all $n \geq k$.

Therefore R is right noetherian. □

EXERCISE 4Q. (This is for those readers who have met direct limits somewhere.) If R is a right noetherian ring, show that all direct limits of injective right R–modules are injective. □

COROLLARY 4.20. [Matlis, Papp] *If R is a right noetherian ring, every injective right R–module is a direct sum of uniform injective modules.*

Proof. Let E be a nonzero injective right R–module. Let $\mathscr{A}$ be a maximal independent family of nonzero finitely generated submodules of E. If $\oplus \mathscr{A} \nleq_e E$, then E has a nonzero submodule B such that $(\oplus \mathscr{A}) \cap B = 0$. Since B contains a nonzero finitely generated submodule, we may assume that B itself is finitely generated. But then $\mathscr{A} \cup \{B\}$ is independent, contradicting the maximality of $\mathscr{A}$. Thus $\oplus \mathscr{A} \leq_e E$.

By Theorem 4.8, each $A \in \mathscr{A}$ has an injective hull $E_A \leq E$. As $A \leq_e E_A$, Proposition 3.21 says that the E_A are independent. Hence, $\oplus E_A$ is an essential submodule of E. On the other hand, $\oplus E_A$ is injective by Theorem 4.19, and so $E = \oplus E_A$.

Each $A \in \mathscr{A}$ is a noetherian module and hence has finite rank, by Corollary 4.14. Thus each E_A is a (finite) direct sum of indecomposable submodules, all of which must be injective. Therefore E is a direct sum of uniform injective submodules. □

Thus over a right noetherian ring finding all the injective right modules reduces to finding all the uniform ones. This is easy to do over a commutuative noetherian ring, as follows.

PROPOSITION 4.21. [Matlis] *Let R be a commutative noetherian ring, and let E be an injective R–module. Then E is uniform if and only if $E \cong E((R/P)_R)$ for some prime ideal P of R.*

Proof. If $E \cong E((R/P)_R)$ for some prime ideal P, then since R/P is a uniform R–module, so is E.

Conversely, assume that E is uniform. By Proposition 2.12, E has an associated prime P, and the submodule $A = \text{ann}_E(P)$ is a fully faithful (R/P)–module. Choose a nonzero element $x \in A$, and note that $\text{ann}(x) = \text{ann}(xR) = P$, whence $xR \cong R/P$. Now E contains a submodule E' which is an injective hull for xR, and E' is a nonzero direct summand of E. As E is uniform, E' = E, and therefore $E = E(xR) \cong E((R/P)_R)$. □

For example, Proposition 4.21 shows that every uniform injective $\mathbb{Z}$–module is isomorphic either to $E(\mathbb{Z})$ or to $E(\mathbb{Z}/p\mathbb{Z})$ for some prime integer p. On one hand, $E(\mathbb{Z}) = \mathbb{Q}$, while on the other hand $E(\mathbb{Z}/p\mathbb{Z}) \cong \mathbb{Z}(p^\infty)$ (which may be described as the p–torsion subgroup of $\mathbb{Q}/\mathbb{Z}$). Thus every uniform injective $\mathbb{Z}$–module is isomorphic to one of $\mathbb{Q}, \mathbb{Z}(2^\infty), \mathbb{Z}(3^\infty), \mathbb{Z}(5^\infty), \ldots$.

By Corollary 4.20, every injective $\mathbb{Z}$–module is isomorphic to a direct sum of some copies of these modules.

♦ ASSASSINATOR PRIMES

We may view Proposition 4.21 as parametrizing the uniform injective modules over a commutative noetherian ring R via the prime ideals of R. An obvious question is whether this parametrization is a bijection, i.e., whether each prime ideal P is uniquely determined by the injective module $E((R/P)_R)$. The answer is positive, and to prove that we make use of the concept of assassinator.

LEMMA 4.22. *Let U be a uniform right module over a right noetherian ring R. Then there is a unique prime ideal P in R such that P equals the annihilator of some nonzero submodule of U and P contains the annihilators of all nonzero submodules of U. Moreover, P is the unique associated prime of U, and $ann_U(P)$ is a fully faithful (R/P)–module.*

Proof. Choose a nonzero submodule A of U such that the ideal $P = ann(A)$ is maximal among annihilators of nonzero submodules of U. Then P is an associated prime of U, and $ann_U(P)$ is a fully faithful (R/P)–module (Proposition 2.12).

If B is a nonzero submodule of U, then $A \cap B \neq 0$ because U is uniform, and $P \leq ann(A \cap B)$ because $AP = 0$. Then $ann(A \cap B) = P$ by the maximality of P, whence $ann(B) \leq P$. Thus P contains the annihilators of all nonzero submodules of U. The uniqueness of P in the first conclusion of the lemma is now clear.

Given an associated prime Q of U, there is a nonzero submodule C in U such that $ann(C) = Q$ and C is a fully faithful (R/Q)–module. Since $A \cap C \neq 0$ (because U is uniform), and since A is a fully faithful (R/P)–module, we conclude that $P = ann(A \cap C) = Q$. Therefore P is the only associated prime of U. □

DEFINITION. If U is a uniform right module over a right noetherian ring, the unique associated prime of U is called the *assassinator* of U.

EXERCISE 4R. Identify the annihilator and the assassinator of each uniform submodule of the $\mathbb{Z}$–module $\mathbb{Q}/\mathbb{Z}$. □

EXERCISE 4S. If $U \geq V$ are uniform right modules over a right noetherian ring, show that they have the same assassinator. □

EXERCISE 4T. If A is a right module over a right noetherian ring, show that Ass(A) equals the set of assassinators of uniform submodules of A. □

LEMMA 4.23. *Let P be a prime ideal in a right noetherian ring R, and let U be a uniform right ideal of R/P. Then $E(U_R)$ is a uniform injective right R–module, and its assassinator is P.*

Proof. Since U is uniform, so is $E(U_R)$. Observe that U is a fully faithful (R/P)–module, whence P is an associated prime of $E(U_R)$. Therefore P is the assassinator of $E(U_R)$. □

PROPOSITION 4.24. [Matlis] *If R is a commutative noetherian ring, then the rule $P \mapsto E((R/P)_R)$ provides a bijection between the set of prime ideals of R and the set of isomorphism classes of uniform injective R–modules. The inverse bijection is given by the rule $E \mapsto$ (assassinator of E).*

Proof. Proposition 4.21 and Lemma 4.23. □

In the noncommutative case, a noetherian ring may have more uniform injective modules than prime ideals. For instance, a simple noetherian ring R has only one prime ideal, namely 0, yet R may have more than one uniform injective module. This holds in particular for Weyl algebras over division rings of characteristic zero, as follows.

EXERCISE 4U. Let R be any simple noetherian domain which is not a division ring, and let A be any simple right R–module. Show that $E(R_R)$ and E(A) are uniform and not isomorphic. □

♦ INJECTIVE HULLS OF NONSINGULAR RINGS

The injective hull of a commutative domain R is just the quotient field of R (Exercise 4C), which is itself a ring. More generally, we shall see in this section that the injective hull of any nonsingular ring R can be given a unique ring structure consistent with its R–module structure. This larger ring E is called the *maximal quotient ring* of the nonsingular ring R (see Exercise 4ZI), and if R has finite rank then E is a semisimple ring.

While it is possible to define this ring structure directly and to verify the axioms of a ring, it is more convenient to identify the new ring with a ring we already have available, namely an endomorphism ring. To see why this should be expected, observe that if a ring R is a subring of a ring S, then left multiplication by any element of S defines a right R–module endomorphism of the module S_R. By this means, we may identify S with a subring of $End_R(S_R)$. On the other hand, if $Hom_R((S/R)_R, S_R) = 0$, it is not hard to see that then $S = End_R(S_R)$. This condition is certainly fulfilled in case R is right nonsingular and $R_R \leq_e S_R$, since then S/R is a singular right R–module.

We conclude from this discussion that if R is a right nonsingular ring and if the injective hull $E(R_R)$ can be given a ring structure consistent with its R–module structure, then the new ring must be isomorphic to the endomorphism ring of $E(R_R)$. In particular, the endomorphism ring of $E(R_R)$ must then be at least additively isomorphic to $E(R_R)$, and our first step is to show that this is indeed the case.

PROPOSITION 4.25. *Let R be a right nonsingular ring, and set $E = E(R_R)$ and $Q = End_R(E)$. Then the rule $f \mapsto f(1)$ defines an abelian group isomorphism of Q onto E.*

Proof. Let $\phi : Q \to E$ be the group homomorphism defined by the rule $\phi(f) = f(1)$.

Since R_R is nonsingular, and $R_R \leq_e E$, Proposition 3.28 shows that E is a nonsingular right R–module. On the other hand, Proposition 3.26 shows that E/R is a singular R–module.

If $f \in \ker(\phi)$, then $f(1) = 0$ and so $f(R) = 0$. Then f(E) is an epimorphic image of the singular module E/R and so is a singular module. As E is nonsingular, $f(E) = 0$. Thus ϕ is injective.

Given $x \in E$, define a homomorphism $g : R_R \to E$ so that $g(1) = x$. Then g extends to a homomorphism $f : E \to E$, giving $f \in Q$ with $\phi(f) = x$. Thus ϕ is surjective. □

The isomorphism given in Proposition 4.25 will be used to carry the ring structure of Q over to a ring structure on E. To avoid ambiguity, we must show that those sums and products already available in E (from its R–module structure) don't differ from the corresponding sums and products of the new ring structure.

DEFINITION. Let R be a ring, and let E be a right R–module containing R_R as a submodule. A ring structure on E is *compatible with the right R–module structure* provided that

(a) The ring addition in E is the same as the R–module addition;

(b) The ring product of any $x \in E$ with any $r \in R$ is the same as the R–module product of x with r.

PROPOSITION 4.26. *If R is a right nonsingular ring and $E = E(R_R)$, there is a unique ring structure on E compatible with its right R–module structure. Using this ring* *structure on* E, *the map* $f \mapsto f(1)$ *gives a ring isomorphism of* $End_R(E)$ *onto* E.

Proof. Set $Q = End_R(E)$. By Proposition 4.25, there is an abelian group isomorphism $\phi : Q \to E$ given by the rule $\phi(f) = f(1)$. Hence, there exists a ring structure on E consisting of the original addition together with a multiplication $*$ following the rule

$$x * y = \phi(\phi^{-1}(x)\phi^{-1}(y)) = [\phi^{-1}(x)](y).$$

For all $x \in E$ and $r \in R$, we compute that

$$x * r = [\phi^{-1}(x)](r) = [[\phi^{-1}(x)](1)]r = xr.$$

Thus this ring structure on E is compatible with the right R–module structure.

Now consider any other compatible ring structure on E, using the original addition and a multiplication $**$. If $x \in E$ then left $**$–multiplication by x is a right R–module endomorphism of E taking 1 to x (since $x ** ar = x ** a ** r = (x ** a)r$ for all $a \in E$, $r \in R$). By Proposition 4.25, the only such endomorphism is $\phi^{-1}(x)$. Thus for any y in E, we have $x ** y = [\phi^{-1}(x)](y) = x * y$. Therefore $**$ coincides with $*$. □

Given a right nonsingular ring R, we shall assume whenever convenient that $E(R_R)$ has been equipped with the ring structure obtained from Proposition 4.26. Because of the compatibility with the R–module structure, there is no ambiguity in writing xy for any product in $E(R_R)$.

For example, if R is any commutative domain, the ring structure on $E(R_R)$ compatible with the R–module structure is the ring structure of the quotient field of R.

DEFINITION. Let $R \subseteq Q$ be rings, and let A be a right R–module. A right Q–module structure on A is *compatible with the right* R*–module structure* provided that

(a) The Q–module addition in A is the same as the R–module addition;

(b) The Q–module product of any $x \in A$ with any $r \in R$ is the same as the R–module product of x with r.

PROPOSITION 4.27. *Let* R *be a right nonsingular ring, set* $Q = E(R_R)$, *and let* A *be any nonsingular injective right* R*–module. Then there is a unique right* Q*–module structure on* A *compatible with its right* R*–module structure.*

Proof. Set $Q' = End_R(Q_R)$ and $A' = Hom_R(Q_R, A)$, and let $\phi : Q' \to Q$ and $\psi : A' \to A$ be the group homomorphisms given by the rules $\phi(f) = f(1)$ and $\psi(g) = g(1)$. By Proposition 4.26, ϕ is a ring isomorphism of Q' onto Q. By the same method as in Proposition 4.25, ψ is an abelian group isomorphism of A' onto A.

Now A' is naturally a right Q'–module, with composition as the module product. Hence, there is a right Q–module structure on A using the original addition together with a module product $*$ following the rule $x * q = \psi(\psi^{-1}(x)\phi^{-1}(q))$. As in Proposition 4.26, this

Q–module structure on A is compatible with the R–module structure and is unique. □

THEOREM 4.28. [Gabriel] *If R is a right nonsingular ring such that R_R has finite rank, then $E(R_R)$ is a semisimple ring.*

Proof. Set $Q = E(R_R)$. Since R_R has finite rank, $Q_R = E_1 \oplus \dots \oplus E_n$ where each E_i is a uniform injective right R–module. According to Proposition 4.27, each E_i has a right Q–module structure, and Q_R is thus given a right Q–module structure with respect to which it is the direct sum of the Q–submodules E_i. The uniqueness statement of Proposition 4.27 shows that this Q–module structure agrees with the usual structure on Q_Q and hence that the E_i are right ideals of Q.

Now consider any nonzero element $z \in E_i$, and set $J = r.ann_Q(z)$. Then $Q/J \cong zQ$ as right Q–modules and as right R–modules, and so Q/J is a nonsingular right R–module. Consequently, J is an essentially closed R–submodule of Q. (Namely, if J' is any right R–submodule of Q such that $J \leq_e J'$, then J'/J is singular, whence $J'/J = 0$.) By Proposition 4.7, J_R is injective. Then J_R is a direct summand of Q_R, whence $(Q/J)_R$ is isomorphic to a direct summand of Q_R and so is injective. Thus $(zQ)_R$ is injective and hence is a direct summand of the uniform R–module E_i. Consequently, $zQ = E_i$, proving that E_i is a simple right Q–module.

Therefore Q_Q is a semisimple module. □

EXERCISE 4V. If R is a right nonsingular ring, R_R has finite rank, and $Q = E(R_R)$, show that $\text{rank}(Q_Q) = \text{rank}(R_R)$. □

EXERCISE 4W. If R is a right nonsingular ring such that $E(R_R)$ is a semisimple ring, show that R_R has finite rank. □

EXERCISE 4X. Let $R \subseteq Q$ be rings such that $R_R \leq_e Q_R$ and Q is a semisimple ring. Show that R is a right nonsingular ring and that $Q = E(R_R)$. [Hint: First show that $IQ = Q$ for each $I \leq_e R_R$. Then show that Q is a subring of $E(R_R)$.] □

EXERCISE 4Y. If $R = \begin{pmatrix} \mathbb{Z} & \mathbb{Q} \\ 0 & \mathbb{Q} \end{pmatrix}$ and $Q = M_2(\mathbb{Q})$, show that $Q = E(R_R)$ and $Q = E({}_RR)$. □

♦ ADDITIONAL EXERCISES

4Z. Let R be a commutative noetherian ring, J an ideal of R, and A a finitely generated right R–module. If A has an essential submodule B such that $BJ = 0$, show that there exists $n \in \mathbb{N}$ such that $AJ^n = 0$. [Hint: Find n such that $\text{ann}_A(J^n) = \text{ann}_A(J^{n+1})$. If $a \in A$ and $aJ^n \neq 0$, find $r \in R$ such that $0 \neq aJ^n r \subseteq B$.] □

4ZA. If R is a commutative noetherian ring, J is an ideal of R, and $E = E((R/J)_R)$, show that $E = \bigcup_{n=0}^{\infty} \mathrm{ann}_E(J^n)$. □

4ZB. Let R be a commutative noetherian ring and A a finitely generated right R–module.

(a) Show that there exist $P_1,\dots,P_n \in \mathrm{Ass}(A)$ (repetitions allowed) such that $AP_1P_2\cdots P_n = 0$. [Hint: Exercises 2J and 4Z.]

(b) If P is any annihilator prime for A, show that $P \in \mathrm{Ass}(A)$. [Hint: Apply (a) to the module $\mathrm{ann}_A(P)$.] Conclude using Exercise 2L that $\mathrm{Ass}(A)$ equals the set of affiliated primes for A as well as the set of annihilator primes for A. □

4ZC. Given modules A and B, show that $E(A) \cong E(B)$ if and only if there exist essential submodules $A' \leq_e A$ and $B' \leq_e B$ such that $A' \cong B'$. □

4ZD. Given a right module A over a ring R, show that A is injective if and only if every homomorphism from an essential right ideal of R to A extends to a homomorphism from R to A. □

4ZE. If $A \leq F$ are modules with F nonsingular and injective, show that F contains a unique injective hull for A. □

4ZF. Let $A \leq B \leq C$ be modules. If A is essentially closed in B and B is essentially closed in C, show that A is essentially closed in C. [Hint: Choose submodules $A' \leq B$ and $B' \leq C$ maximal with respect to $A \cap A' = 0$ and $B \cap B' = 0$, and show that $(A \oplus A' \oplus B')/A \leq_e C/A$.] □

4ZG. Show that a module A has finite rank if and only if A has DCC on essentially closed submodules. □

4ZH. Show that any module A with finite rank n has submodules $A_1,\dots,A_n$ such that each A/A_i is uniform and $A_1 \cap \dots \cap A_n = 0$. □

4ZI. Let R be a right nonsingular ring. A *right quotient ring* of R is any ring extension $S \supseteq R$ such that $R_R \leq_e S_R$. Show that the identity map on R extends to a ring embedding of S into $E(R_R)$. (For this reason, $E(R_R)$ is called the *maximal right quotient ring* of R.) □

4ZJ. If R is a right nonsingular ring and $Q = E(R_R)$, show that every nonsingular injective right R–module is also injective as a right Q–module. (In particular, Q is a right self–injective ring.) □

4ZK. Let R be a right nonsingular ring such that R_R has finite rank.

(a) Show that any direct sum of nonsingular injective right R–modules is injective. [Hint: Use the proof of Theorem 4.19 and Exercise 4J. Observe that if M is a module and $N \leq_e M$, and $f: M \to A \oplus B$ is a homomorphism in which A and B are nonsingular and $f(N) \leq A$, then $f(M) \leq A$.]

(b) Conclude that the nonsingular injective right R–modules are precisely the right Q–modules, where $Q = E(R_R)$. Since Q is semisimple, this provides a classification of the nonsingular injective right R–modules similar to what we have for commutative integral domains. □

♦ NOTES

Existence of Injectives. Baer worked with what he called "complete" modules over a ring R, namely modules A such that every homomorphism from a one–sided ideal of R to A extends to a homomorphism from R to A. He proved that every module is a submodule of a complete module [1940, Theorem 3], and that a module is complete if and only if it is a direct summand of every module that contains it [1940, Theorem 1]. The method we have given for embedding modules into injectives is due to Eckmann and Schopf [1953, § 2].

Injectivity Versus Inessentiality. That a module is injective if and only if it has no proper essential extensions is due to Eckmann and Schopf [1953, § 4].

Injective Hulls. The concept was developed by Eckmann and Schopf [1953, § 4]. The name "injective envelope" appeared in a paper of Matlis [1958, p. 512], the name "injective hull" in a paper of Rosenberg and Zelinsky [1959, p. 373].

Existence of Injective Hulls. Baer proved that any module A can be embedded in a complete module E with the property that every monomorphism from A to a complete module E' extends to a monomorphism from E to E' [1940, Theorem 4]. Eckmann and Schopf proved that any module has a maximal essential extension (in the sense of Theorem 4.8(b)), which is also a minimal injective extension (in the sense of Theorem 4.8(c)) [1953, § 4].

Finite Rank. This was first developed from the standpoint of modules containing no infinite direct sums of nonzero submodules. The latter concept was introduced for rings by Goldie [1958b, p. 590], and later for modules [1960, p. 202], under the name "finite–dimensional modules".

Uniform Modules. Uniform right ideals were introduced by Goldie [1958b, p. 590], and later uniform modules [1960, p. 201].

Finite Rank Versus Infinite Direct Sums. Goldie proved that a module has finite rank if and only if it contains no infinite direct sums of nonzero submodules [1960, Theorem (1.1)].

Uniform Rank. This was introduced, under the name "dimension", by Goldie [1960, p. 202].

Injectivity of Direct Sums. That a ring R is right noetherian if and only if every direct sum of injective right R–modules is injective was proved independently by Papp [1959, Theorem 1] and Bass [1962, Theorem 1.1] (his result was actually donated to an earlier paper of Chase [1960, Proposition 4.1]).

Uniform Decompositions of Injectives. That every injective right module over a right noetherian ring is a direct sum of uniform injectives was proved independently by Matlis [1958, Theorem 2.5] and Papp [1959, Theorem 2].

Uniform Injectives over Commutative Noetherian Rings. Matlis developed the bijection between prime ideals of a commutative noetherian ring R and isomorphism classes of uniform injective R–modules [1958, Proposition 3.1].

Ring Structures on Nonsingular Injective Hulls. A ring structure on the injective hull of a right nonsingular ring R was described by Gabriel [1962, Théorème 1, p. 418]. That R_R has finite rank if and only if the ring $E(R_R)$ is semisimple is equivalent to a result of Gabriel [1962, Lemme 6, p. 418].

5 SEMISIMPLE RINGS OF FRACTIONS

A very useful technique in commutative ring theory is to pass from a commutative ring R to a prime factor ring R/P and then to the quotient field of R/P. In the noncommutative case we could ask whether it is possible to pass to a factor ring from which a division ring may be built from fractions. Since noncommutative noetherian rings need not have any factor rings which are domains, this is rather restrictive. Instead, we look for factor rings from which simple artinian rings can be built using fractions. The main result is Goldie's theorem which says in particular that if P is a prime ideal in a noetherian ring then the factor R/P has such a ring of fractions. It turns out to be no extra work to investigate rings from which semisimple rings of fractions can be built. We must be careful to distinguish between fractions with denominators on the right, such as ab^{-1}, and those with denominators on the left, such as $c^{-1}d$.

The semisimple rings of fractions constructed in this chapter are special cases of the general theory of rings of fractions, which we have placed in Chapter 9. Some readers may prefer to have the core of this general theory at hand in order to simplify the work of the present chapter (in particular, to avoid our use of Proposition 4.26 and Theorem 4.28). Readers with this inclination may wish to study Chapter 9 before the present chapter, and to use the results of Chapter 9 wherever appropriate.

♦ ORDERS IN SEMISIMPLE RINGS

DEFINITION. A *regular element* in a ring R is any non–zero–divisor, i.e., any element $x \in R$ such that $\text{r.ann}_R(x) = 0$ and $\text{l.ann}_R(x) = 0$.

Note that if $R \subseteq Q$ are rings and x is any element of R which is invertible in Q, then x must be a regular element of R.

When we come to consider regular elements in factor rings, it will be convenient to have the following notation available.

DEFINITION. Let I be an ideal in a ring R. An element $x \in R$ is said to be *regular modulo* I provided the coset $x + I$ is a regular element of the ring R/I. The set of all such x is denoted by $\mathscr{C}(I)$, or by $\mathscr{C}_R(I)$ if the ring R needs emphasis. Thus, the set of all regular elements in R may be denoted $\mathscr{C}_R(0)$.

DEFINITION. Let Q be a ring. A *right order in* Q is any subring $R \subseteq Q$ such that

(a) Every regular element of R is invertible in Q;

(b) Every element of Q has the form ab^{-1} for some $a \in R$ and some regular $b \in R$.

Left orders are defined analogously, using fractions of the form $b^{-1}a$. In case both rings are commutative, the adjectives "right" and "left" are not needed.

For example, any commutative domain is an order in its quotient field. For another example, the ring $\{(a,b) \in \mathbb{Z} \times \mathbb{Z} \mid a \equiv b \pmod 2\}$ is an order in the ring $\mathbb{Q} \times \mathbb{Q}$.

EXERCISE 5A. Let R be a commutative domain with quotient field F, and let $n \in \mathbb{N}$. Show that the ring of all upper triangular $n \times n$ matrices over R is a right and left order in the ring of all upper triangular $n \times n$ matrices over F (but not a right or left order in $M_n(F)$, unless $n = 1$). □

EXERCISE 5B. If R is a right artinian ring, show that R is a right and left order in itself. □

LEMMA 5.1. *Let R be a right order in a ring Q.*

(a) Given $a,b \in R$ with b regular, there exist $c,d \in R$ with d regular such that $ad = bc$.

(b) Given any regular elements $d_1,\ldots,d_n \in R$, there exists a regular element in $d_1R \cap \ldots \cap d_nR$.

(c) Given any $x_1,\ldots,x_n \in Q$, there exist $a_1,\ldots,a_n,b \in R$ with b regular such that each $x_i = a_ib^{-1}$.

Proof. (a) Since R is a right order in Q, the regular element b is invertible in Q, and $b^{-1}a = cd^{-1}$ for some $c,d \in R$ with d regular. Then $ad = bc$.

(b) By induction on n, we need only consider the case that $n = 2$. By (a), we have $d_1d = d_2c$ for some $c,d \in R$ with d regular, whence $d_1d \in d_1R \cap d_2R$. As d_1 and d are regular, so is d_1d.

(c) Each $x_i = c_id_i^{-1}$ for some $c_i,d_i \in R$ with d_i regular. By (b), there exists a regular element b in $d_1R \cap \ldots \cap d_nR$. For each $i = 1,\ldots,n$, we have $b = d_ie_i$ for some $e_i \in R$, and hence $x_i = c_ie_ib^{-1}$. □

DEFINITION. A *right (left) annihilator* in a ring R is any right (left) ideal of R which equals the right (left) annihilator of some subset of R.

Note that a right ideal I of R is a right annihilator if and only if $I = \text{r.ann}(\text{l.ann}(I))$. Namely, if $I = \text{r.ann}(X)$ for some $X \subseteq R$, then $X \subseteq \text{l.ann}(I)$, whence

$$I = \text{r.ann}(X) \geq \text{r.ann}(\text{l.ann}(I)) \geq I.$$

PROPOSITION 5.2. *If a ring R is a right order in a semisimple ring Q, then R is a semiprime ring, R_R has finite rank, and R has ACC on right annihilators.*

Proof. Let $I_1 \leq I_2 \leq \dots$ be an ascending chain of right annihilators in R. Set $J_n = \text{l.ann}(I_n)$ for all n. Then $J_1 \geq J_2 \geq \dots$, and each $I_n = \text{r.ann}(J_n)$. In Q, we have

$$\text{r.ann}_Q(J_1) \leq \text{r.ann}_Q(J_2) \leq \dots .$$

Since Q is semisimple, it is right noetherian, and so there is a positive integer m such that $\text{r.ann}_Q(J_n) = \text{r.ann}_Q(J_m)$ for all $n \geq m$. Consequently,

$$I_n = R \cap \text{r.ann}_Q(J_n) = R \cap \text{r.ann}_Q(J_m) = I_m$$

for all $n \geq m$. Therefore R has ACC on right annihilators.

If R_R does not have finite rank, it contains an infinite sequence $A_1, A_2, \dots$ of independent nonzero right ideals (Theorem 4.13). Choose a nonzero element $x_i \in A_i$ for each i. Since Q is right noetherian, the right ideals $x_1Q, x_2Q, \dots$ cannot be independent, and so there exist $q_1, \dots, q_n \in Q$ such that $x_1q_1 + \dots + x_nq_n = 0$ but $x_nq_n \neq 0$. By Lemma 5.1, there exist $a_1, \dots, a_n, b \in R$ with b regular such that each $q_i = a_ib^{-1}$. Then

$$x_1a_1 + \dots + x_na_n = (x_1q_1 + \dots + x_nq_n)b = 0.$$

Since $A_1, \dots, A_n$ are independent, $x_na_n = 0$. But then $x_nq_n = x_na_nb^{-1} = 0$, a contradiction. Thus R_R does have finite rank.

Finally, consider any ideal $N \leq R$ such that $N^2 = 0$. Set $L = \text{l.ann}(N)$, and note that L is an ideal of R. By Exercise 3T, L is essential as a right ideal in R. Next, we show that LQ is an essential right ideal of Q. Given any nonzero right ideal $B \leq Q$, choose a nonzero element $x \in B$, and write $x = ab^{-1}$ for some $a, b \in R$ with $a \neq 0$ and b regular. Then $a = xb$ lies in $R \cap B$, whence $R \cap B$ is a nonzero right ideal of R. Consequently, $R \cap B \cap L \neq 0$, and hence $B \cap LQ \neq 0$. Thus $LQ \leq_e Q_Q$.

As Q_Q is semisimple, $LQ = Q$, by Corollary 3.24. Hence, there exist $y_1, \dots, y_n \in L$ and $q_1, \dots, q_n \in Q$ such that $y_1q_1 + \dots + y_nq_n = 1$. By Lemma 5.1, there exist $a_1, \dots, a_n, b \in R$ with b regular such that each $q_i = a_ib^{-1}$. Then

$$b = (y_1q_1 + \dots + y_nq_n)b = y_1a_1 + \dots + y_na_n \in L,$$

whence $bN = 0$. Since b is regular, $N = 0$.

Therefore R is semiprime, by Corollary 2.8. □

DEFINITION. A *right Goldie ring* is any ring R such that R_R has finite rank and R has ACC on right annihilators.

For example, every right noetherian ring is right Goldie.

Proposition 5.2 says that any ring which is a right order in a semisimple ring must be a semiprime right Goldie ring. Proving the converse statement occupies a major portion of this chapter.

EXERCISE 5C. Let $k(x)$ be a rational function field over a field k, let α be the k–algebra endomorphism of $k(x)$ given by the rule $\alpha(f) = f(x^2)$, and set $R = k(x)[\theta;\alpha]$. Show that R is left Goldie but not right Goldie. [Hint: Show that $\theta R \cap x\theta R = 0$.] □

♦ TECHNICALITIES

PROPOSITION 5.3. [Mewborn–Winton] *If R is a ring with ACC on right annihilators, then $Z_r(R)$ is nilpotent.*

Proof. Set $J = Z_r(R)$, and note that $r.ann(J) \leq r.ann(J^2) \leq \dots$. As R has ACC on right annihilators, $r.ann(J^k) = r.ann(J^{k+1})$ for some $k \in \mathbb{N}$. We claim that $J^k = 0$.

If not, then we may choose an x in $R - r.ann(J^k)$ such that $r.ann(x)$ is as large as possible. If $a \in J$ then $r.ann(a) \cap xR \neq 0$ since $r.ann(a) \leq_e R_R$. Hence, there is some $s \in R$ such that $axs = 0$ while $xs \neq 0$. Then $r.ann(x) < r.ann(ax)$, and so (by the maximality of $r.ann(x)$), we obtain $ax \in r.ann(J^k)$. Hence $J^k ax = 0$, and since this holds for all $a \in J$, we infer that $x \in r.ann(J^{k+1})$. Since $r.ann(J^{k+1}) = r.ann(J^k)$, this contradicts our choice of x, and therefore $J^k = 0$. □

COROLLARY 5.4. *If R is a semiprime ring with ACC on right annihilators, then R is a right nonsingular ring.* □

Now if R is a semiprime right Goldie ring, then R is right nonsingular and so $E(R_R)$ is a ring; moreover, $E(R_R)$ is a semisimple ring because R_R has finite rank (Theorem 4.28). Thus to prove the converse of Proposition 5.2, it suffices to show that R is a right order in $E(R_R)$. We first work toward showing that all regular elements of R are invertible in $E(R_R)$.

LEMMA 5.5. *Let R be a ring such that R_R has finite rank. If $x \in R$ and $r.ann(x) = 0$, then $xR \leq_e R_R$.*

Proof. Since left multiplication by x defines a monomorphism from R_R to itself, this follows from Corollary 4.18. □

COROLLARY 5.6. *If R is a ring such that R_R has finite rank, then any right or left invertible element of R is invertible.*

Proof. Given $x,y \in R$ such that $yx = 1$, we must show that $xy = 1$. Observe that $r.ann(x) = 0$, whence Lemma 5.5 shows that $xR \leq_e R_R$. Since $xyx = x$, we see that $xR = xyR$. Moreover, $xyxy = xy$, so that xy is idempotent, and hence xR is a direct summand of R_R. As $xR \leq_e R_R$, we obtain $xR = R$. It follows that $l.ann(x) = 0$. Since $(xy - 1)x = 0$, we conclude that $xy - 1 = 0$. □

LEMMA 5.7. *Let R be a right nonsingular ring such that R_R has finite rank, and set $Q = E(R_R)$. For any $x \in R$, the following conditions are equivalent:*

(a) x is a regular element of R.

(b) $r.ann_R(x) = 0$.

(c) $xR \leq_e R_R$.

(d) x is invertible in Q.

Proof. Obviously (d) ⇒ (a) ⇒ (b), and (b) ⇒ (c) by Lemma 5.5.

(c) ⇒ (d): Since $R_R \leq_e Q_R$, we have $xR \leq_e Q_R$, and hence $xQ \leq_e Q_R$. It follows that $xQ \leq_e Q_Q$. As Q is a semisimple ring (Theorem 4.28), it has no proper essential right ideals (Corollary 3.24). Thus $xQ = Q$, and so x is right invertible in Q. Since Q is right noetherian, Q_Q has finite rank (Corollary 4.14). Therefore Corollary 5.6 shows that x is invertible in Q. □

It remains to show that when R is a semiprime right Goldie ring, every element $x \in E(R_R)$ has the form ab^{-1} for some $a,b \in R$ with b regular. Thus we need a regular element $b \in R$ satisfying $xb \in R$. As $\{r \in R \mid xr \in R\}$ is an essential right ideal of R, we may accomplish this by showing that every essential right ideal of R contains a regular element.

LEMMA 5.8. *If R is a semiprime right Goldie ring, then R has DCC on right annihilators.*

Proof. Because of Corollary 5.4 and Theorem 4.28, R is a subring of a semisimple ring Q. From this, DCC on right annihilators follows as in the first part of the proof of Proposition 5.2. □

PROPOSITION 5.9. [Goldie] *Let R be a semiprime right Goldie ring, and let I be a right ideal of R. Then I is an essential right ideal if and only if I contains a regular element.*

Proof. If there is a regular element $x \in I$, then as $xR \leq_e R_R$ by Lemma 5.7, we must have $I \leq_e R_R$.

Conversely, assume that $I \leq_e R_R$. By Lemma 5.8, R has DCC on right annihilators. Hence, there is some $x \in I$ such that the right ideal $A = r.ann(x)$ is minimal among right annihilators of elements of I. We claim that $xR \leq_e I$.

Thus consider any right ideal $B \leq I$ for which $B \cap xR = 0$. Given any $b \in B$, we have $bR \cap xR = 0$, from which it follows that

$$r.ann(b + x) = r.ann(b) \cap r.ann(x) \leq A.$$

As $b + x \in I$, the minimality of A forces

$$A = r.ann(b + x) = r.ann(b) \cap r.ann(x) \leq r.ann(b),$$

whence $bA = 0$. Thus $BA = 0$, and then $(AB)^2 = 0$. Since R is semiprime, $AB = 0$. Now $(RB \cap A)^2 \leq AB = 0$, and hence $RB \cap A = 0$.

Since $RB \cap r.ann(x) = 0$, left multiplication by x provides a monomorphism from $(RB)_R$ to itself. As $(RB)_R$ has finite rank, Corollary 4.18 shows that $xRB \leq_e RB$. However, we also have $B \cap xRB = 0$, and consequently $B = 0$.

Therefore $xR \leq_e I$, as claimed. Since $I \leq_e R_R$, it follows that $xR \leq_e R_R$. By Corollary 5.4 and Lemma 5.7, x is regular. □

EXERCISE 5D. Let R be a semiprime right Goldie ring. Show that R is left Goldie if and only if $_RR$ has finite rank. □

♦ GOLDIE'S THEOREMS

THEOREM 5.10. [Goldie] *A ring R is a right order in a semisimple ring if and only if R is a semiprime right Goldie ring.*

Proof. Necessity is given by Proposition 5.2. Conversely, assume that R is a semiprime right Goldie ring. By Corollary 5.4, R is a right nonsingular ring. Let Q denote the ring $E(R_R)$, which is semisimple by Theorem 4.28.

Lemma 5.7 shows that all regular elements of R are invertible in Q. Now consider any $x \in Q$, and set $I = \{r \in R \mid xr \in R\}$. Since $R_R \leq_e Q_R$, Proposition 3.21 shows that $I \leq_e R_R$. There exists a regular element $b \in I$ by Proposition 5.9, whence $xb \in R$ and $x = xbb^{-1}$. Therefore R is a right order in Q. □

LEMMA 5.11. *Let R be a right order in a semisimple ring Q. Then Q is a simple ring if and only if R is a prime ring.*

Proof. We first note that $R_R \leq_e Q_R$, since any nonzero submodule $A \leq Q_R$ contains ab^{-1} for some nonzero element $a \in R$ and some regular element $b \in R$, and $a \in A \cap R$.

Suppose that R is prime. If I is any nonzero ideal of Q, then as $R_R \leq_e Q_R$ we see that $I \cap R$ is a nonzero ideal of R. Then $I \cap R \leq_e R_R$ (Exercise 3L), and so $I \cap R \leq_e Q_R$. As a result, $I_R \leq_e Q_R$, and hence $I_Q \leq_e Q_Q$. Since Q is semisimple, $I = Q$. Therefore Q is simple.

Conversely, assume that Q is simple, and consider any ideals $A,B \leq R$ such that $AB = 0$ but $A \neq 0$. Then QAQ is a nonzero ideal of Q, whence $QAQ = Q$ and so

$$x_1a_1y_1 + \ldots + x_na_ny_n = 1$$

for some $x_i,y_i \in Q$ and some $a_i \in A$. By Lemma 5.1, there exist $b_1,\ldots,b_n,c$ in R with c regular such that each $y_i = b_ic^{-1}$. Hence

$$c = (x_1a_1y_1 + \ldots + x_na_ny_n)c = x_1a_1b_1 + \ldots + x_na_nb_n \in QA,$$

and so $cB \leq QAB = 0$. As c is regular, $B = 0$. Therefore R is prime. □

THEOREM 5.12. [Goldie, Lesieur–Croisot] *A ring R is a right order in a simple artinian ring if and only if R is a prime right Goldie ring.*

Proof. Theorem 5.10 and Lemma 5.11. □

DEFINITION. Let R be a semiprime right Goldie ring. Any semisimple ring in which R is a right order is called a *right Goldie quotient ring of* R.

EXERCISE 5E. Let $R \subseteq Q$ be rings such that R is semiprime and Q is semisimple. If $R_R \leq_e Q_R$, show that R is a right Goldie ring and that Q is a right Goldie quotient ring of R. [Hint: Exercise 4X.] □

PROPOSITION 5.13. *Let R be a semiprime right Goldie ring.*

(a) If Q is a right Goldie quotient ring of R, then Q_R is an injective hull for R_R. In particular, $R_R \leq_e Q_R$.

(b) If Q_1 and Q_2 are right Goldie quotient rings of R, the identity map on R extends to a ring isomorphism of Q_1 onto Q_2.

Proof. (a) As in the proof of Lemma 5.11, $R_R \leq_e Q_R$.

Consider a right ideal $I \leq R$ and a homomorphism $f : I \to Q_R$. We claim that whenever $x_1,\ldots,x_n \in I$ and $q_1,\ldots,q_n \in Q$ with $\Sigma\, x_iq_i = 0$, then $\Sigma\, f(x_i)q_i = 0$. By Lemma 5.1, there exist $a_1,\ldots,a_n,b \in R$ with b regular such that each $q_i = a_ib^{-1}$. Then $\Sigma\, x_ia_i = (\Sigma\, x_iq_i)b = 0$, whence $\Sigma\, f(x_i)a_i = 0$ and so $\Sigma\, f(x_i)q_i = 0$, as claimed.

Consequently, f induces a well–defined Q–module homomorphism $g : IQ \to Q_Q$. As Q is semisimple, IQ is a direct summand of Q_Q, and so g extends to a homomorphism $h : Q_Q \to Q_Q$. The restriction of h to R provides a homomorphism $R_R \to Q_R$ extending f.

Therefore Q_R is injective.

(b) By (a), $(Q_1)_R$ and $(Q_2)_R$ are injective hulls for R_R, and hence the identity map on R extends to a right R–module isomorphism $\phi : Q_1 \to Q_2$. Consequently, Q_1 has a ring structure using the given addition together with a multiplication $*$ following the rule $x * y = \phi^{-1}(\phi(x)\phi(y))$. Observe that this ring structure on Q_1 is compatible with its right R–module structure, as is the original ring structure on Q_1.

By Corollary 5.4, R is a right nonsingular ring. Hence, Proposition 4.26 shows that Q_1 has only one ring structure compatible with its right R–module structure. Thus $*$ coincides with the original multiplication on Q_1, whence ϕ is a ring isomorphism. □

Because of the uniqueness expressed in Proposition 5.13, we may speak of *the* right Goldie quotient ring of a semiprime right Goldie ring.

PROPOSITION 5.14. *If R is a semiprime right and left Goldie ring, then every right (left) Goldie quotient ring of R is also a left (right) Goldie quotient ring of R.*

Proof. Let Q be a left Goldie quotient ring of R. By definition, all regular elements of R are invertible in Q. Any element of Q has the form $b^{-1}a$ for some $a,b \in R$ with b regular. Since R is a right order in a (semisimple) ring, Lemma 5.1 says that $ad = bc$ for some $c,d \in R$ with d regular. Hence, $b^{-1}a = cd^{-1}$ in Q. Therefore R is a right order in Q, and so Q is a right Goldie quotient ring of R.

The reverse implication follows by symmetry. □

Because of Proposition 5.14, when R is a semiprime right and left Goldie ring we refer just to *the Goldie quotient ring* of R.

EXERCISE 5F. Let R be a semiprime right Goldie ring. If R is a right order in a ring Q, prove that Q must be a semisimple ring (whence Q is a right Goldie quotient ring of R). □

EXERCISE 5G. If a ring R is a right order in a ring Q, and R_R has finite rank, show that Q_Q has finite rank and that $\text{rank}(Q_Q) = \text{rank}(R_R)$. [Compare Exercise 4V.] Conclude in particular that if R is a semiprime right Goldie ring with right Goldie quotient ring Q, then $\text{length}(Q_Q) = \text{rank}(R_R)$. □

♦ DIVISION RINGS OF FRACTIONS

Returning to the analogy with the quotient field of a commutative domain, we consider the question of when a noncommutative domain R can be a right order in a division ring D. By Theorem 5.12, R must at least be a prime right Goldie ring. Primeness is no restriction, because R is a domain. Neither is ACC on right annihilators a restriction, because 0 and R are the only right annihilators in R. Thus it remains to consider the finite rank condition, which, as the following example shows, is indeed a restriction.

EXERCISE 5H. If R is the domain described in Exercise 5C, show that R_R does not have finite rank. □

DEFINITION. A *right Ore domain* is any domain R in which every two nonzero elements have a nonzero common right multiple, i.e., for each nonzero $x,y \in R$ there exist $r,s \in R$ such that $xr = ys \neq 0$.

For example, every commutative domain is right Ore.

LEMMA 5.15. *For a domain R, the following conditions are equivalent:*

(a) R is a right Ore domain.

(b) R_R is uniform.

(c) R_R has finite rank.

Proof. (a) ⇒ (b): The right Ore condition is equivalent to saying that the intersection of any two nonzero principal right ideals of R is nonzero, from which (b) is immediate.

(b) ⇒ (c): This is clear.

(c) ⇒ (a): Since every nonzero element of R is regular, Lemma 5.5 shows that every nonzero principal right ideal of R is essential in R. Thus the intersection of any two nonzero principal right ideals of R is nonzero. □

COROLLARY 5.16. *Every right noetherian domain is right Ore.*

Proof. Corollary 4.14 and Lemma 5.15. □

In particular, Lemma 5.15 shows that the domain R of Exercise 5C/H is not right Ore. However, R is a principal left ideal domain by Theorem 1.11, and so Corollary 5.16 shows that R is left Ore.

EXERCISE 5I. If R is a *right Bezout domain* (i.e., a domain in which every finitely generated right ideal is principal), show that R is right Ore. □

THEOREM 5.17. [Ore] *A ring R is a right order in a division ring if and only if R is a right Ore domain.*

Proof. If R is a right order in a division ring D, then obviously R is a domain. As D is semisimple, R_R has finite rank by Proposition 5.2. Then Lemma 5.15 shows that R is right Ore.

Conversely, assume that R is a right Ore domain. Then R is a prime ring with ACC on right annihilators, and Lemma 5.15 shows that R_R has finite rank. By Theorem 5.12, R is a right order in a simple artinian ring Q. Any nonzero element of Q has the form ab^{-1} for some nonzero elements $a,b \in R$. Since R is a domain, a is regular in R and hence invertible in Q, whence ab^{-1} is invertible in Q. Therefore Q is a division ring. □

Theorem 5.17 shows that any right Ore domain R is a semiprime right Goldie ring with a right Goldie quotient ring which is a division ring. By Proposition 5.13, all right Goldie quotient rings of R are division rings. The right Goldie quotient ring of R is usually called the *right Ore quotient ring* because Ore constructed his quotient rings 30 years before Goldie constructed his.

Of course for a right and left Ore domain R, every right (left) Ore quotient ring is also a left (right) Ore quotient ring, because of Proposition 5.14, and so we refer just to *the Ore quotient ring* of R in this case.

EXERCISE 5J. Let D be a division ring with center C, let δ be a derivation on D, and set $R = D[\theta; \delta]$. By Theorem 1.11, R is a right and left principal ideal domain, and so R is right and left Ore. Let Q be the Ore quotient ring of R.

(a) Show that $\delta(C) \subseteq C$. Then show that the set $k = \{\alpha \in C \mid \delta(\alpha) = 0\}$ is a subfield of C. (It is called the *subfield of central constants of* D.) Show that k is contained in the center of Q.

(b) Given any $x \in Q$, write $x = ab^{-1}$ for some $a,b \in R$ with $b \neq 0$ and set $\deg(x) = \deg(a) - \deg(b)$. Show that $\deg(x)$ is well–defined. For all $x,y \in Q$, show that $\deg(xy) = \deg(x) + \deg(y)$ and that $\deg(x \pm y) \leq \max\{\deg(x), \deg(y)\}$.

(c) If $\operatorname{char}(D) = 0$ and δ is an outer derivation, show that k equals the center of Q. [Hint: Given $x \in Q - D$, write $x = \beta\theta^n + \gamma\theta^{n-1} + y$ for some $\beta, \gamma \in D$ with $\beta \neq 0$, some $n \in \mathbb{Z}$, and some $y \in Q$ with $\deg(y) \leq n - 2$. Then show that $x\alpha \neq \alpha x$ for some $\alpha \in D$. (The case $n = 0$ must be treated separately.)] □

EXERCISE 5K. Let k be a field of characteristic zero, and let D be the Ore quotient ring of $A_1(k)$. Show that the center of D is k, and that D is transcendental over k. [Hint: Write $A_1(k)$ as $k[x][\theta; d/dx]$ and show that D is also the Ore quotient ring of $k(x)[\theta; d/dx]$.] □

♦ NIL SUBSETS

Goldie's Theorem gives us the option of trying to study a noetherian ring R by working first within a semisimple ring (the Goldie quotient ring of R/N, where N is the prime radical of R), trying to pull information back from there to R/N, and finally trying to lift information from R/N to R. Later we will see this principle in action a number of times. Here, as a digression, we show how this method may be used to give an easy proof of a classical result – that nil subrings (non–unital subrings, of course!) of noetherian rings must be nilpotent. (Actually, the details go more smoothly if we work with nil multiplicatively closed subsets.)

Recall that a subset S of a ring R is nil provided all elements of S are nilpotent. To say that S itself is nilpotent, on the other hand, means that $S^n = 0$ for some $n \in \mathbb{N}$. Now in standard ring–theoretic notation, S^n denotes the set of all sums of products of n elements from S. Note, however, that $S^n = 0$ if and only if all products of n elements from S equal zero.

THEOREM 5.18. [Levitzki, Hopkins, Goldie] *If R is a right noetherian ring and S is a nil multiplicatively closed subset of R, then S is nilpotent.*

Proof. If N is the prime radical of R, then since N is nilpotent, it suffices to show that the image of S in R/N is nilpotent. Moreover, by Goldie's Theorem R/N is a subring of an artinian ring, and so it is enough to show that nil multiplicatively closed subsets of artinian rings are nilpotent. Thus we may assume, without loss of generality, that R is artinian.

Let $\ell = \text{length}(R_R)$. We first show that every nilpotent multiplicatively closed subset N of R satisfies $N^\ell = 0$. Consider the descending chain

$$R \geq NR \geq N^2R \geq \dots$$

of right ideals of R. Since R_R has length ℓ, there must be a nonnegative integer $m \leq \ell$ such that $N^mR = N^{m+1}R$. It follows (by induction) that $N^mR = N^kR$ for all $k > m$. Then since N is nilpotent, $N^mR = 0$, and thus $N^\ell = 0$, as claimed.

Using Zorn's Lemma, there is a multiplicatively closed subset $N \subseteq S$ maximal with respect to the property $N^\ell = 0$. In view of the previous paragraph, N is maximal among nilpotent multiplicatively closed subsets of S. Note that $0 \in N$.

If $S \neq N$, we claim that there is an element $s \in S - N$ such that $sN \subseteq N$. Otherwise, there are elements $s_0 \in S - N$ and $n_1, n_2, \dots \in N$ such that $s_0n_1n_2 \cdots n_k \notin N$ for $k = 1,2,\dots$. But then $s_0n_1n_2 \cdots n_\ell \neq 0$, contradicting the fact that $N^\ell = 0$. Thus there does exist $s \in S - N$ such that $sN \subseteq N$, as claimed.

There is a positive integer m such that $s^m = 0$. Set

$$T = \{s^i \mid i = 1,2,\dots,m-1\} \cup \{ns^i \mid n \in N \text{ and } i = 0,1,\dots,m-1\},$$

and observe that T is a multiplicatively closed subset of S properly containing N. Any product of m elements of T either equals 0 or contains a factor of the form ns^i. Hence, all products of m elements from T belong to the set

$$U = \{ns^i \mid n \in N \text{ and } i = 0,1,\dots,m-1\}.$$

Now any product of ℓ elements from U has the form

$$n_1 s^{i(1)} n_2 s^{i(2)} \cdots n_\ell s^{i(\ell)}$$

for some $n_j \in N$ and $i(j) \in \{0,1,\ldots,m-1\}$, and such a product is zero because

$$n_1,\ s^{i(1)}n_2,\ s^{i(2)}n_3, \ldots,\ s^{i(\ell-1)}n_\ell \in N.$$

Thus $U^\ell = 0$, whence $T^{m\ell} = 0$. However, this contradicts the maximality of N.

Therefore $S = N$, and S is nilpotent. □

One is perhaps more likely to encounter nil one–sided ideals than nil multiplicatively closed subsets in a noetherian ring R. The nilpotence of these, a result known as *Levitzki's Theorem*, may be obtained without using Goldie's Theorem, in the following manner.

EXERCISE 5L. Show (without using Goldie's Theorem) that any nil right or left ideal in a right noetherian ring R must be nilpotent. [Hints: First reduce to the case that R is semiprime. Next observe that if xR is a nil right ideal of R, then Rx is a nil left ideal. Finally, if I is a nonzero nil left ideal of R, look at a nonzero element of I whose right annihilator is as large as possible.] □

♦ ADDITIONAL EXERCISES

5M. If a ring R is a right order in a ring Q, show that $M_n(R)$ is a right order in $M_n(Q)$ for each $n \in \mathbb{N}$. □

5N. Let R be a semiprime right Goldie ring, and let $x,y \in R$. Show that xy is regular if and only if x and y are both regular. □

5O. If R is a prime right Goldie ring, $x \in R$, and $I \leq_e R_R$, show that $x + I$ contains a regular element of R. [Hint: Choose $a \in I$ with $\text{r.ann}(x + a)$ minimal. If $B \leq R_R$ with $B \cap (x + a)R = 0$, show that $(B \cap I)\text{r.ann}(x + a) = 0$.] □

5P. Let k be a field of characteristic zero, let $T = A_1(k) = k[x][\theta; d/dx]$, and let $R = \begin{pmatrix} k & T/\theta T \\ 0 & T \end{pmatrix}$. Observe that R is right noetherian, by Proposition 1.9. Show that R does not have DCC on right annihilators. [Hint: Consider the elements $\begin{pmatrix} 0 & x^n+\theta T \\ 0 & 0 \end{pmatrix}$ for $n \in \mathbb{N}$.] Show that R is not isomorphic to a subring of a left noetherian ring. □

♦ NOTES

Nilpotence of Singular Ideals. That the right singular ideal is nilpotent in a ring with ACC on right annihilators comes from a proof of Mewborn and Winton [1969, Theorem 2.5].

Regular Elements in Essential Ideals. The existence of regular elements in essential right ideals was proved by Goldie first for prime right and left Goldie rings [1958a, Theorem 7; 1958b, Theorem 10], and later for semiprime right Goldie rings [1960, Theorem 3.9].

Goldie's Theorems. Initially, Goldie proved that a ring R is a right and left order in a simple artinian ring if and only if R is a prime right and left Goldie ring [1958a, Theorem A; 1958b, Theorem 13]. His methods were then modified by Lesieur and Croisot to prove a one–sided version of this result [1959a, Théorème 7; 1959b, Théorèmes 9, 10, 11]. After that, Goldie proved the characterization of right orders in semisimple rings [1960, Theorems 4.1, 4.4].

Orders in Division Rings. Ore proved that a ring is a right order in a division ring if and only if it is a right Ore domain [1931, Theorems 1, II].

Nilpotence of Nil Subrings. This was first obtained by Levitzki for a ring with both ACC and DCC on right ideals [1931, Hauptsatz, p. 625]. For right artinian rings, this was later obtained independently by Hopkins [1938, (3.3); 1939, (1.7)] and Levitzki [1939, Theorem 12]. Later still, Levitzki proved that nil subrings of a right and left noetherian ring are nilpotent [1945, Theorem 6], and that nil one–sided ideals in a right noetherian ring are nilpotent [1945, Theorem 5]. Finally, the nilpotence of nil subrings of right noetherian rings was obtained by Goldie [1960, Theorem 6.1].

6 MODULES OVER SEMIPRIME GOLDIE RINGS

Many investigations of the structure of a right module A over a right noetherian ring R involve related modules over prime or semiprime factor rings of R. For instance, if A is finitely generated then by using an affiliated series we may view A as built from a chain of subfactors each of which is a fully faithful module over a prime factor ring of R (see Proposition 2.13). Alternatively, we may relate the structure of A to the structure of the (R/N)–module A/AN, where N is the prime radical of R. Thus we need a good grasp of the structure of modules over prime or semiprime noetherian rings. The fundamentals of such structure can be obtained with little extra effort for modules over prime or semiprime Goldie rings.

♦ MINIMAL PRIME IDEALS

PROPOSITION 6.1. *Let R be a semiprime right Goldie ring, and let Q be the right Goldie quotient ring of R.*

(a) There are only finitely many distinct maximal ideals in Q, say $M_1,\dots,M_n$, and $M_1 \cap \dots \cap M_n = 0$.

(b) Each of the ideals $P_i = R \cap M_i$ is a minimal prime ideal of R, and $P_1,\dots,P_n$ are distinct. Each of the factor rings R/P_i is right Goldie, and Q/M_i is (via the natural embedding $R/P_i \to Q/M_i$) the right Goldie quotient ring of R/P_i.

(c) The only minimal prime ideals of R are $P_1,\dots,P_n$.

(d) Each $P_i = r.ann_R(\bigcap_{j \neq i} P_j)$.

Proof. (a) Since Q is semisimple, we may identify it with a finite direct product of simple artinian rings, say $Q = Q_1 \times \dots \times Q_n$, so that elements of Q may be written as n–tuples. The maximal ideals of Q are just the ideals $M_i = \{x \in Q \mid x_i = 0\}$, for $i = 1,\dots,n$, and obviously $M_1 \cap \dots \cap M_n = 0$.

(b) If $A_i = \bigcap_{j \neq i} M_j = \{x \in Q \mid x_j = 0 \text{ for all } j \neq i\}$, then $R \cap A_i = \bigcap_{j \neq i} P_j$. Since $R_R \leq_e Q_R$ (Proposition 5.13), $R \cap A_i \neq 0$. On the other hand,

$$R \cap A_i \cap P_i = P_1 \cap \ldots \cap P_n = 0,$$

and so $R \cap A_i \nleq P_i$, whence $P_j \nleq P_i$ for all $j \neq i$. In particular, $P_j \neq P_i$ for all $j \neq i$.

Note that P_i is the kernel of the natural map $v_i : R \to Q \to Q_i$ sending each element $r \in R$ to its i–th. component r_i. We claim that $v_i(R)$ is a right order in Q_i. If $v_i(c)$ is a regular element of $v_i(R)$, then $c \in \mathscr{C}_R(P_i)$ and so $(cR + P_i)/P_i$ is an essential right ideal of R/P_i (Lemma 5.7), whence $cR + P_i \leq_e R_R$. Hence, $cR + P_i$ contains a regular element d (Proposition 5.9), and since d is invertible in Q it follows that $(cR + P_i)Q = Q$, whence $cQ + M_i = Q$. Consequently, $v_i(c)$ is right invertible in Q_i, and hence invertible (Corollary 5.6). Thus all regular elements of $v_i(R)$ are invertible in Q_i. Any element of Q_i has the form x_i for some $x \in Q$, and $x = ab^{-1}$ for some $a,b \in R$ with b regular. Then b_i is invertible in Q_i and is regular in $v_i(R)$. Since $xb = a$, we have $x_i b_i = a_i$ and hence $x_i = a_i b_i^{-1}$. Therefore $v_i(R)$ is a right order in Q_i, as claimed.

By Theorem 5.12, $v_i(R)$ is a prime right Goldie ring, that is, P_i is a prime of R and R/P_i is right Goldie. That Q_i is the right Goldie quotient ring of $v_i(R)$ is the same as to say that Q/M_i is the right Goldie quotient ring of R/P_i.

By Proposition 2.3, there is a minimal prime ideal $P \leq P_i$. Observe that

$$P_1 P_2 \cdots P_n \leq M_1 \cap M_2 \cap \ldots \cap M_n = 0,$$

whence $P_1 P_2 \cdots P_n \leq P$. Then some $P_k \leq P \leq P_i$. As $P_j \nleq P_i$ for all $j \neq i$, we must have $k = i$, and so $P_i = P$. Therefore P_i is a minimal prime.

(c) Given a minimal prime P in R, we observe (as above) that some $P_k \leq P$. Since P_k is prime, $P = P_k$ by minimality.

(d) Set $B_i = \bigcap_{j \neq i} P_j$. Since $B_i P_i \leq P_1 \cap P_2 \cap \ldots \cap P_n = 0$, we see that $P_i \leq r.ann_R(B_i)$. On the other hand, as $P_1,\ldots,P_n$ are distinct minimal primes, $B_i \nleq P_i$. Then, since $B_i \cdot (r.ann_R(B_i)) = 0 \leq P_i$, we conclude that $r.ann_R(B_i) \leq P_i$. □

LEMMA 6.2. *If A is an ideal in a semprime ring R, then l.ann(A) = r.ann(A).*

Proof. Note that l.ann(A) and r.ann(A) are ideals of R. Since

$$(A \cdot l.ann(A))^2 = A(l.ann(A) \cdot A) l.ann(A) = 0,$$

we obtain $A \cdot l.ann(A) = 0$ and so $l.ann(A) \leq r.ann(A)$. The reverse inclusion is obtained symmetrically. □

PROPOSITION 6.3. *Let P be a prime ideal in a semiprime right Goldie ring R. Then the following conditions are equivalent:*

(a) P is a minimal prime.

(b) P is a right annihilator.

(c) P is a left annihilator.

(d) P contains no regular elements.

Proof. (a) $\Rightarrow$ (b) by Proposition 6.1.

(b) $\Rightarrow$ (c): If $L = \text{l.ann}(P)$, then L is an ideal of R and $P = \text{r.ann}(L)$. By Lemma 6.2, $P = \text{l.ann}(L)$ as well.

(c) $\Rightarrow$ (d): By assumption, $P = \text{l.ann}(X)$ for some $X \subseteq R$, and $X \nsubseteq \{0\}$ because P is a proper ideal. Thus $\text{l.ann}(X)$ contains no regular elements.

(d) $\Rightarrow$ (a): If P properly contains a prime ideal P_0, then P/P_0 is an essential right ideal of R/P_0 (Exercise 3L), and hence also an essential right R–submodule. But then P is an essential right ideal of R (Proposition 3.21), and so P contains a regular element (Proposition 5.9), contradicting (d). □

EXERCISE 6A. Show that in a semiprime right Goldie ring R, the minimal prime ideals are exactly those ideals P maximal with respect to the property that P contains no regular elements. □

LEMMA 6.4. *If R is a semiprime right Goldie ring and $P_1,\dots,P_n$ are its minimal primes, then the set of regular elements of R equals $\mathscr{C}(P_1) \cap \dots \cap \mathscr{C}(P_n)$.*

Proof. If $x \in \mathscr{C}(P_1) \cap \dots \cap \mathscr{C}(P_n)$, then $\text{r.ann}_R(x) \leq P_1 \cap \dots \cap P_n = 0$ (because x is regular modulo each of $P_1,\dots,P_n$), and similarly $\text{l.ann}_R(x) = 0$. Hence, x is regular.

Conversely, assume that x is regular. By Proposition 6.3 and Lemma 6.2, each

$$P_i = \text{l.ann}_R(A_i) = \text{r.ann}_R(A_i)$$

for some ideal A_i. If $r \in R$ and $xr \in P_i$, then $xrA_i = 0$, whence $rA_i = 0$ (because x is regular), and so $r \in P_i$. Similarly, $rx \in P_i$ implies $r \in P_i$, and therefore $x \in \mathscr{C}(P_i)$. □

PROPOSITION 6.5. *Let N be a proper semiprime ideal in a right noetherian ring R, and let $P_1,\dots,P_n$ be the prime ideals of R minimal over N. Then $P_1,\dots,P_n$ are precisely the prime ideals of R that contain N but are disjoint from $\mathscr{C}(N)$, and*

$$\mathscr{C}(N) = \mathscr{C}(P_1) \cap \dots \cap \mathscr{C}(P_n).$$

Proof. After reducing to R/N, we may assume that $N = 0$. Then the first conclusion follows from Proposition 6.3 and the second from Lemma 6.4. □

PROPOSITION 6.6. *If $P_1,\dots,P_n$ are the distinct minimal prime ideals of a semiprime right Goldie ring R, then the natural map $f: R \to (R/P_1) \oplus \dots \oplus (R/P_n)$ is an essential monomorphism of right (or left) R–modules.*

Proof. That f is a monomorphism is clear, since $P_1 \cap \ldots \cap P_n = 0$. For $j = 1,\ldots,n$, set $A_j = \bigcap_{i \neq j} P_i$. Since $P_i \nsubseteq P_j$ for all $i \neq j$, we must have $A_j \nsubseteq P_j$, whence $(A_j + P_j)/P_j$ is a nonzero ideal of R/P_j. Then $(A_j + P_j)/P_j$ is essential as a right or left ideal of R/P_j (Exercise 3L), and thus also essential as a right or left R–submodule. Since

$$f(A_1 + \ldots + A_n) = ((A_1 + P_1)/P_1) \oplus \ldots \oplus ((A_n + P_n)/P_n),$$

we conclude that $f(\Sigma A_j)$ is essential as a right or left R–submodule of $\oplus (R/P_j)$, and therefore the same is true of $f(R)$. □

COROLLARY 6.7. *If N is a proper semiprime ideal in a right noetherian ring R and $P_1,\ldots,P_n$ are the distinct prime ideals of R minimal over N, then*

$$E((R/N)_R) \cong E((R/P_1)_R) \oplus \ldots \oplus E((R/P_n)_R).$$

Proof. By Proposition 6.6, R/N is isomorphic to an essential right R–submodule of $(R/P_1) \oplus \ldots \oplus (R/P_n)$. □

EXERCISE 6B. Give alternative proofs (and interpretations) of Lemma 6.4 and Proposition 6.6 using the product decomposition of the Goldie quotient ring Q, as in the proof of Proposition 6.1. In Lemma 6.4, note that

$$\mathscr{C}_R(P_i) = \{x \in R \mid x_i \text{ is invertible in } Q_i\}. \ \square$$

PROPOSITION 6.8. *Let P be a minimal prime ideal in a semiprime right Goldie ring R. Then R/P is nonsingular as either a right R–module or a right (R/P)–module, and $Z(A_{R/P}) = Z(A_R)$ for all right (R/P)–modules A.*

Proof. By Proposition 6.6, R is isomorphic to an essential right R–submodule of $(R/P_1) \oplus \ldots \oplus (R/P_n)$, where $P_1,\ldots,P_n$ are the distinct minimal primes of R. As R is a right nonsingular ring (Corollary 5.4), it follows that $(R/P_1) \oplus \ldots \oplus (R/P_n)$ is a nonsingular right R–module, and thus $(R/P)_R$ is nonsingular.

Given $x \in Z(A_{R/P})$, there is an essential right ideal I of R/P such that $xI = 0$. Then $I = J/P$ for some essential right ideal J of R, and $xJ = 0$, whence $x \in Z(A_R)$. Thus $Z(A_{R/P}) \leq Z(A_R)$. In particular, it follows that $Z((R/P)_{R/P}) = 0$.

Now consider any $y \in Z(A_R)$, and set $K = \text{ann}_R(y)$. Then K is a right ideal of R that contains P, and $(R/P)/(K/P) \cong R/K \cong yR$, whence $(R/P)/(K/P)$ is a singular right R–module. As R/P is a nonsingular right R–module, $K/P \leq_e (R/P)_R$ by Proposition 3.27. Thus K/P is an essential right ideal of R/P, and so from $y(K/P) = 0$ we conclude that $y \in Z(A_{R/P})$. Therefore $Z(A_R) \leq Z(A_{R/P})$. □

♦ TORSION

PROPOSITION 6.9. *If A is a right module over a semiprime right Goldie ring R, then*

$$Z(A) = \{a \in A \mid ax = 0 \text{ for some regular element } x \in R\}.$$

Proof. This is immediate from Proposition 5.9. □

Proposition 6.9 shows that over a semiprime right Goldie ring R, the singular submodule of any right R–module A is a direct analog of the torsion submodule of a module over a commutative domain. Because of this analogy, it is common to use torsion terminology in this situation. Thus Z(A) is called the *torsion submodule* of A, while A is a *torsion module* if and only if Z(A) = A, and A is a *torsionfree module* if and only if Z(A) = 0. Warning: this torsion terminology, and the description of singular submodules as consisting of elements annihilated by regular elements, is only to be used for modules over semiprime Goldie rings (see Exercises 6C, 6D).

Note that given any finite subset $X \subseteq Z(A)$, there exists a regular element $y \in R$ such that $Xy = \{0\}$ (either using Proposition 5.1 or observing that $\mathrm{ann}_R(X)$ is a finite intersection of essential right ideals and so is essential).

The general properties of singular and nonsingular modules derived in Chapter 3 may be restated for torsion and torsionfree R–modules as in the following proposition. Since R is a right nonsingular ring (Corollary 5.4), Propositions 3.28 and 3.29 are both applicable to R–modules.

PROPOSITION 6.10. *For right modules over a semiprime right Goldie ring, the following properties hold:*

(a) All submodules, factor modules, sums, and essential extensions of torsion modules are torsion.

(b) If B is a submodule of a module A such that B and A/B are both torsion, then A is torsion.

(c) All submodules, direct products, and essential extensions of torsionfree modules are torsionfree.

(d) If B is a submodule of a module A such that B and A/B are both torsionfree, then A is torsionfree. □

One should beware of using torsion terminology loosely when there is ambiguity about coefficient rings. In particular, if R is a semiprime right noetherian ring and N is a semiprime ideal of R, then R and R/N are both semiprime right Goldie rings, and the torsion submodule of a right (R/N)–module A may well differ from the torsion submodule of A viewed as an R–module. For example, any vector space over $\mathbb{Z}/2\mathbb{Z}$ is torsionfree as a $(\mathbb{Z}/2\mathbb{Z})$–module but torsion as a $\mathbb{Z}$–module.

EXERCISE 6C. Let k be a field and k[x] a polynomial ring, and let $R = \begin{pmatrix} k[x] & k[x] \\ 0 & k \end{pmatrix}$. If A is the right R–module $(k[x]/xk[x] \;\; k[x]/xk[x])$, show that the set

$$\{a \in A \mid ay = 0 \text{ for some regular element } y \in R\}$$

is not a submodule of A. □

EXERCISE 6D. Let R be a ring. If

$$Z(A) = \{a \in A \mid ax = 0 \text{ for some regular element } x \in R\}$$

for all right R–modules A, show that R is a semiprime right Goldie ring. [Hint: Show that R is a right order in a semisimple ring.] □

PROPOSITION 6.11. *If U is a uniform right module over a semiprime right Goldie ring, then U is either torsion or torsionfree.*

Proof. If U is not torsionfree then $Z(U) \neq 0$, whence $Z(U) \leq_e U$ (because U is uniform), and therefore U must be torsion. □

EXERCISE 6E. Let R be a semiprime right Goldie ring with right Goldie quotient ring Q, let A be a right R–module, and let $f : A \to A \otimes_R Q$ be the natural map given by the rule $f(a) = a \otimes 1$. Note that f is a right R–module homomorphism.

(a) Show that $\ker(f) = Z(A)$. [Hint: Proposition 4.27.] Conclude that A is torsion if and only if $A \otimes_R Q = 0$.

(b) Show that $A/Z(A)$ has finite rank if and only if $A \otimes_R Q$ has finite length (as a right Q–module), and that in this case $\text{rank}(A/Z(A)) = \text{length}(A \otimes_R Q)$. □

♦ TORSIONFREE INJECTIVE MODULES

DEFINITION. A right module A over a ring R is *divisible* provided $Ax = A$ for all regular elements $x \in R$.

For example, every injective module is divisible. Over $\mathbb{Z}$, or more generally over any principal right ideal domain, the divisible right modules are exactly the injective right modules. Over other domains, however, divisible modules need not be injective, as the following example shows.

EXERCISE 6F. Let $R = k[x,y]$ be a polynomial ring over a field k, and let $F = k(x,y)$ be the quotient field of R. Show that $F/(xR + yR)$ is a divisible R–module which is not injective. [Hint: Look at homomorphisms from $(xR + yR)/(xR + yR)^2$ to $R/(xR + yR)$.] □

However, over a commutative domain all *torsionfree* divisible modules are injective, and the same result holds over semiprime Goldie rings, as follows.

PROPOSITION 6.12. [Gentile, Levy] *Let A be a torsionfree right module over a semiprime right Goldie ring R. Then A is divisible if and only if it is injective.*

Proof. First assume that A is injective. Given $a \in A$ and a regular element $x \in R$, there is a well–defined homomorphism $f : xR \to A$ such that $f(xr) = ar$ for all $r \in R$. Then f extends to a homomorphism $g : R \to A$, and $g(1)x = f(x) = a$. Thus A is divisible.

Conversely, assume that A is divisible, let I be a right ideal of R, and let $f : I \to A$ be a homomorphism. By Corollary 3.23, I is a direct summand of an essential right ideal J, and f extends to a homomorphism $g : J \to A$. By Proposition 5.9, J contains a regular element x, and $xR \leq_e R_R$. In particular, J/xR is a torsion module.

Since A is divisible, $g(x) = ax$ for some $a \in A$. Define a homomorphism $h : R \to A$ so that $h(1) = a$, and observe that $h(x) = ax = g(x)$. Now $(h - g)(xR) = 0$, whence $(h - g)(J)$ is an epimorphic image of the torsion module J/xR. As A is torsionfree, $(h - g)(J) = 0$. Thus h is an extension of g, and hence an extension of f. Therefore A is injective. □

EXERCISE 6G. If R is a semiprime right Goldie ring, show that all direct sums of torsionfree injective right R–modules are injective. □

PROPOSITION 6.13. *Let R be a semiprime right Goldie ring, and let Q be the right Goldie quotient ring of R.*

(a) All right Q–modules are torsionfree and injective as right R–modules, and their Q–submodules are exactly their divisible R–submodules.

(b) Every torsionfree divisible right R–module has a unique right Q–module structure compatible with its right R–module structure.

(c) A right Q–module is uniform as an R–module if and only if it is isomorphic to a minimal right ideal of Q.

Proof. (a) Since all regular elements of R are invertible in Q, any right Q–module A is obviously torsionfree and divisible as an R–module. By Proposition 6.12, A_R is injective.

Obviously any Q–submodule of A is divisible as an R–module. Conversely, let B be any divisible R–submodule of A. Given $b \in B$ and $q \in Q$, write $q = xy^{-1}$ for some $x,y \in R$ with y regular. There exists $b' \in B$ such that $b'y = bx$, whence $(bq - b')y = 0$. As A is torsionfree, $bq - b' = 0$, and so $bq \in B$. Thus B is a Q–submodule of A.

(b) By Corollary 5.4 and Proposition 5.13, R is a right nonsingular ring and $Q_R = E(R_R)$. If A is any torsionfree divisible right R–module, then A is nonsingular and Proposition 6.12 shows that A is injective. Hence, Proposition 4.27 applies.

(c) Let A be a right Q–module. If A is uniform as an R–module, it is indecomposable, and hence also indecomposable as a Q–module. Then since Q is semisimple A must be simple, whence A is isomorphic to a minimal right ideal of Q.

Conversely, assume that A is isomorphic to a minimal right ideal of Q, so that A is a simple right Q–module. If $A = B \oplus C$ for some R–submodules B and C, then B and C are divisible and so are Q–submodules of A, by (a). As A_Q is simple, either $B = 0$ or $C = 0$. Thus A is indecomposable as an R–module, and therefore uniform (because it is injective). □

COROLLARY 6.14. *Let R be a semiprime right Goldie ring, and let Q be the right Goldie quotient ring of R. Up to isomorphism, the torsionfree uniform injective right R–modules are exactly the minimal right ideals of Q, and there are only finitely many isomorphism classes of them.* □

EXERCISE 6H. Let R be a semiprime right Goldie ring with right Goldie quotient ring Q, let A be a right R–module, let $f : A \to A \otimes_R Q$ be the natural map given by the rule $f(a) = a \otimes 1$, and let $g : A \to A/Z(A) \to E(A/Z(A))$ be the natural map. Show that there is an isomorphism (of right R–modules) $h : A \otimes_R Q \to E(A/Z(A))$ such that $hf = g$. □

EXERCISE 6I. If R is a semiprime right Goldie ring, show that its right Goldie quotient ring Q is a flat left R–module [i.e., given any monomorphism $t : A \to B$ of right R–modules, the induced map $t \otimes 1 : A \otimes_R Q \to B \otimes_R Q$ is also a monomorphism]. □

THEOREM 6.15. *If R is a semiprime right Goldie ring, there is a central idempotent $e \in R$ such that $eR = soc(R_R)$. This yields a ring decomposition $R \cong S \times T$ (where $S = eR$ and $T = (1-e)R$) such that S is a semisimple ring and $soc(T_T) = 0$.*

Proof. Set $J = soc(R_R)$, and choose a right ideal K of R such that $J \oplus K \leq_e R_R$. Note that $KJ \leq K \cap J = 0$.

By Proposition 5.9, there exists a regular element $x \in J \oplus K$. Since $xJ \leq J$ and left multiplication by x is a monomorphism, Corollary 4.18 shows that $xJ \leq_e J$. Thus $xJ = J$, because J_R is semisimple.

Write $x = y + z$ for some $y \in J$ and $z \in K$. As $xJ = J$, we must have $y = xe$ for some $e \in J$. For all $r \in J$, we have $zr \in KJ = 0$, whence $xr = yr + zr = yr = xer$ and so $r = er$, because x is regular. Thus $e = e^2$ and $J = eR$.

Since J is an ideal, $(1-e)Re \subseteq (1-e)J = 0$. Then

$$(eR(1-e)R)^2 = eR[(1-e)Re]R(1-e)R = 0$$

and hence $eR(1-e)R = 0$, by semiprimeness. For all $s \in R$, we obtain

$$(1-e)se = es(1-e) = 0$$

and so $se = ese = es$. Therefore e is central.

Now $S = eR$ and $T = (1-e)R$ are rings with unit, and $R \cong S \times T$. As $soc(R_R) = S$, we conclude that $soc(S_S) = S$ while $soc(T_T) = 0$. □

COROLLARY 6.16. *Let R be a prime right Goldie ring. If $soc(R_R) \neq 0$, then R is a simple artinian ring.*

Proof. Since R is prime, the only central idempotents in R are 0 and 1. Hence, Theorem 6.15 implies that $soc(R_R) = R$, that is, R is a semisimple ring. As R is prime, it must be simple artinian. □

EXERCISE 6J. If R is a semiprime right Goldie ring, show that all torsionfree semisimple right R–modules are injective. [Hint: In case $soc(R_R) = 0$, show that every maximal right ideal of R is essential, and consequently all semisimple right R–modules are torsion.] □

EXERCISE 6K. Let R be the ring of all upper triangular 2×2 matrices over a field k. Show that R is a right and left nonsingular artinian ring, and that $soc(R_R)$ is not a direct summand of R_R. Show that up to isomorphism, R has only one nonsingular simple right module A, and that A is not injective. □

♦ TORSIONFREE UNIFORM MODULES

LEMMA 6.17. *Let R be a semiprime right Goldie ring, and let A be a right R–module. If A is not torsion, then A has a uniform submodule isomorphic to a right ideal of R.*

Proof. Choose an element $x \in A$ which is not torsion. Then $ann(x)$ is a non–essential right ideal of R, and hence there is a nonzero right ideal I in R such that $I \cap ann(x) = 0$. Since R_R has finite rank, I must contain a uniform right ideal J, by Proposition 4.11. As $J \cap ann(x) = 0$, we see that $J \cong xJ$, and therefore xJ is the desired uniform submodule of A. □

Lemma 6.17 guarantees that nonzero torsionfree right modules over a semiprime right Goldie ring R have uniform submodules isomorphic to right ideals of R. However, it does not say that all torsionfree uniform right R–modules are necessarily isomorphic to right ideals of R. For instance, if R is right noetherian, non–finitely–generated uniform R–modules cannot be isomorphic to right ideals of R. (For example, $\mathbb{Q}$ is a uniform torsionfree $\mathbb{Z}$–module which is not isomorphic to an ideal of $\mathbb{Z}$.) More strictly, even finitely generated torsionfree uniform right R–modules need not be isomorphic to right ideals of R, as the following (left–handed) example shows. We shall see shortly that if R is both left and right Goldie, then finitely generated torsionfree uniform R–modules do always embed in R (Corollary 6.20).

EXERCISE 6L. Let $k(x)$ be a rational function field over a field k, let α be the k–algebra endomorphism of $k(x)$ given by the rule $\alpha(f) = f(x^2)$, and let $R = k(x)[\theta; \alpha]$. Then R is a principal left ideal domain (Theorem 1.11), whence R is a left Ore domain. Let Q be the left Ore quotient ring of R, and let A be the left R–submodule $R + R\theta^{-1}x\theta \leq Q$. Show that A is a torsionfree uniform left R–module which is not isomorphic to a left ideal of R. [Hint: As mentioned in Exercise 5C, $\theta R \cap x\theta R = 0$.] □

PROPOSITION 6.18. *If A is a torsionfree right module over a semiprime right Goldie ring R, then A has an essential submodule which is a direct sum of uniform submodules isomorphic to right ideals of R.*

Proof. Let $\mathscr{A}$ be the collection of those uniform submodules of A which are isomorphic to right ideals of R. Choose a maximal independent subfamily $\mathscr{A}_0 \subseteq \mathscr{A}$, and set $A_0 = \oplus \mathscr{A}_0$. If B is a submodule of A satisfying $B \cap A_0 = 0$, then by the maximality of $\mathscr{A}_0$ we see that B has no uniform submodules isomorphic to right ideals of R. This forces B to be zero, by Lemma 6.17. Therefore $A_0 \leq_e A$. □

Of course the direct sum of uniform submodules obtained in Proposition 6.18 is finite if A has finite rank.

PROPOSITION 6.19. [Gentile, Levy] *If R is a semiprime right and left Goldie ring and A is a finitely generated torsionfree right R–module, then A can be embedded in a finitely generated free right R–module.*

Proof. Let Q be the (right and left) Goldie quotient ring of R.

Since $E(A)$ is an essential extension of a torsionfree module, it is torsionfree. Hence, Proposition 6.13 shows that $E(A)$ can be made into a right Q–module. Now AQ is a finitely generated right Q–module, and so it is a finite direct sum of simple right Q–modules, each of which is isomorphic to a right ideal of Q. Consequently, A is isomorphic to a right R–submodule of Q^n for some $n \in \mathbb{N}$, and there is no loss of generality in assuming that $A \leq Q^n$.

Choose generators $x_1,\ldots,x_t$ for A, and write each $x_i = (x_{i1},\ldots,x_{in})$ for some $x_{ij} \in Q$. By the left–hand version of Lemma 5.1, there exist $b, c_{ij} \in R$ with b regular such that each $x_{ij} = b^{-1}c_{ij}$. Then $bx_i \in R^n$ for $i = 1,\ldots,t$, whence $bA \leq R^n$. As b is invertible in Q, we conclude that $A \cong bA$. □

COROLLARY 6.20. *If R is a semiprime right and left Goldie ring and U is a finitely generated torsionfree uniform right R–module, then U is isomorphic to a right ideal of R.*

Proof. In view of Proposition 6.19, there exist $f_1,\ldots,f_n \in \mathrm{Hom}_R(U,R)$ such that

$$\ker(f_1) \cap \ldots \cap \ker(f_n) = 0.$$

Since U is uniform, $\ker(f_i)$ must be zero for some i. □

PROPOSITION 6.21. *Let R be a semiprime right Goldie ring, and let U be a torsionfree uniform right R–module. Then ann(U) is a minimal prime ideal of R, and E(U) is a fully faithful right (R/ann(U))–module.*

Proof. Let Q be the right Goldie quotient ring of R. Since U is torsionfree, so is E(U), and hence Corollary 6.14 shows that E(U) is isomorphic to a minimal right ideal I of Q. Since Q is a semisimple ring, $r.\mathrm{ann}_Q(I)$ is a maximal ideal of Q. Hence if

$$P = \mathrm{ann}_R(E(U)) = R \cap r.\mathrm{ann}_Q(I),$$

then P is a minimal prime ideal of R, by Proposition 6.1. Note that E(U) is a faithful right (R/P)–module. By Proposition 6.8, E(U) must be nonsingular as an (R/P)–module.

Given any nonzero R–submodule B in E(U), note that $P \leq \mathrm{ann}_R(B)$. Since B is a nonsingular (R/P)–module, it follows as in the proof of Lemma 6.17 that B has a nonzero submodule isomorphic to a right ideal J of R/P. Then $J \cdot \mathrm{ann}_R(B) = 0$, whence $\mathrm{ann}_R(B) = P$. Therefore E(U) is a fully faithful (R/P)–module. In particular, $\mathrm{ann}_R(U) = P$. □

♦ TORSIONFREE MODULES OVER PRIME GOLDIE RINGS

Restricting attention to a prime right Goldie ring R, we show that all torsionfree uniform right R–modules are "essentially the same", and consequently that the rank of any torsionfree right R–module can be measured using copies of a single uniform right ideal of R.

LEMMA 6.22. *Let A be a nonzero torsionfree right module over a prime right Goldie ring R. Then A is fully faithful, and every uniform right ideal U in R is isomorphic to a submodule of A.*

Proof. If B is a nonzero submodule of A, then by Lemma 6.17, B has a uniform submodule isomorphic to a right ideal V in R. Since R is prime, V is a faithful R–module, whence B is faithful. Thus A is fully faithful.

Since A is faithful, there exists an element $a \in A$ such that $aU \neq 0$. Set $I = \mathrm{ann}_U(a)$, and note that $U/I \cong aU$. Since A is torsionfree, U/I must be torsionfree, and so $I \not\leq_e U$. Then $I = 0$ because U is uniform, and therefore $U \cong aU$. □

PROPOSITION 6.23. *If P is a prime ideal in a right noetherian ring R and U,V are torsionfree uniform right (R/P)–modules, then $E(U_R) \cong E(V_R)$.*

Proof. Lemma 6.22 shows that U is isomorphic to a submodule of V, whence $E(U)$ is isomorphic to a nonzero direct summand of $E(V)$. Since V is uniform, $E(V)$ is indecomposable, and therefore $E(U) \cong E(V)$. □

PROPOSITION 6.24. *Let R be a prime right Goldie ring, U a uniform right ideal of R, and A a torsionfree right R–module of finite rank. If $n = \mathrm{rank}(A)$, then A has an essential submodule isomorphic to $\oplus^n U$.*

Proof. Since A has rank n, it has an essential submodule of the form $V_1 \oplus \dots \oplus V_n$ where each V_i is uniform. By Lemma 6.22, each V_i has a submodule U_i isomorphic to U, and $U_i \leq_e V_i$ because V_i is uniform. Therefore $U_1 \oplus \dots \oplus U_n$ is an essential submodule of A, isomorphic to $\oplus^n U$. □

COROLLARY 6.25. *Let R be a prime right Goldie ring, U a uniform right ideal of R, and $n = \mathrm{rank}(R_R)$. Then each essential right ideal of R contains an essential right ideal isomorphic to $\oplus^n U$, and R_R is isomorphic to an essential submodule of $\oplus^n U$.*

Proof. Since all essential right ideals of R have rank n, the first conclusion follows immediately from Proposition 6.24. In particular, R has at least one essential right ideal I which is isomorphic to $\oplus^n U$. By Proposition 5.9, there exists a regular element $x \in I$, and $xR \leq_e R$. Therefore $R_R \cong xR \leq_e I \cong \oplus^n U$. □

COROLLARY 6.26. *Let R be a prime right Goldie ring and $n = \mathrm{rank}(R_R)$.*

(a) If A is any torsionfree right R–module, then A^n has an essential free submodule, which will be finitely generated in case A has finite rank.

(b) If A is an arbitrary right R–module, then A^n has a free submodule F such that A^n/F is torsion, and F will be finitely generated in case $A/Z(A)$ has finite rank.

Proof. (a) By Proposition 6.18, A has an essential submodule of the form $\oplus A_i$ where the A_i are uniform. Pick a uniform right ideal U in R; then Lemma 6.22 shows that U is isomorphic to an essential submodule of each A_i. Hence, some direct sum of copies of $\oplus^n U$ is isomorphic to an essential submodule of $\oplus A_i^n$, which is an essential submodule of A^n. By Corollary 6.25, R_R is isomorphic to an essential submodule of $\oplus^n U$, and thus any direct sum of copies of $\oplus^n U$ has an essential free submodule. This yields an essential free submodule F in A^n, and if A has finite rank F must also have finite rank and hence be finitely generated.

(b) Choose a submodule $B \leq A$ such that $B \oplus Z(A) \leq_e A$. Then $A/(B \oplus Z(A))$ is torsion, whence A/B is torsion. Since B is torsionfree, (a) provides us with an essential free submodule F in B^n. Then B^n/F is torsion, and so A^n/F is torsion. Since F embeds in $(A/Z(A))^n$, it must be finitely generated in case $A/Z(A)$ has finite rank. □

EXERCISE 6M. Let R be a prime right and left Goldie ring, and let A,B be finitely generated torsionfree right R–modules.

(a) Show that A and B have finite rank. [Hint: Exercise 4N or Exercise 6E.]

(b) Show that $\mathrm{rank}(A) \leq \mathrm{rank}(B)$ if and only if A is isomorphic to a submodule of B. [Hint: Inside $E(A)$, show that A is contained in a direct sum of $\mathrm{rank}(A)$ finitely generated uniform modules.]

(c) Show that $A^{\mathrm{rank}(B)}$ and $B^{\mathrm{rank}(A)}$ are isomorphic to essential submodules of each other. In particular, if $n = \mathrm{rank}(R_R)$ then A^n is isomorphic to an essential submodule of a free module. □

EXERCISE 6N. If U is a uniform right ideal in a prime right Goldie ring R, show that $\mathrm{End}_R(U)$ is a right Ore domain. [Hint: Let Q be the right Goldie quotient ring of R, and show that $\mathrm{End}_R(U)$ is isomorphic to a right order in $\mathrm{End}_Q(UQ)$.] □

EXERCISE 6O. Prove that a prime ring in which every right ideal is principal is isomorphic to a matrix ring over a right Ore domain. (This was proved by Goldie in 1962). [Hint: Corollary 6.25 and Exercise 6N.] Generalize this to show that a semiprime ring in which every right ideal is principal is isomorphic to a direct product of such matrix rings. Generalize both results to semiprime right Goldie rings in which every finitely generated right ideal is principal. (This was done by Robson in 1967.) □

♦ NOTES

Much of this chapter consists of folklore, worked out by many hands in extending and applying Goldie's Theorems. Thus we do not give many specific attributions for these results.

Divisibility Versus Injectivity. Gentile proved that over a right Ore domain, a torsionfree right module is injective if and only if it is divisible [1960, Proposition 1.1]. That the same holds over a semiprime right Goldie ring follows from a result of Levy [1963, Theorem 3.3].

Embedding Torsionfree Modules in Free Modules. For a right Ore domain R, Gentile proved that all finitely generated torsionfree right R–modules can be embedded in free right R–modules if and only if R is also left Ore [1960, Proposition 4.1]. The analogous equivalence over a semiprime right Goldie ring follows from results of Levy [1963, Theorems 5.2, 5.3].

7 BIMODULES AND AFFILIATED PRIME IDEALS

Bimodules have become of increasing importance in the ideal theory of a noetherian ring R, particularly ideal factors I/J where $I \geq J$ are ideals of R, and overrings $S \supseteq R$, viewed as (S,R)–bimodules. To make the notation more convenient in both cases, we study bimodules in general. In this chapter, we investigate the structure of bimodules over noetherian rings, particularly bimodules which are noetherian or artinian on at least one side, and we illustrate the results by indicating a number of applications, particularly to the relationships between prime ideals of a ring and prime ideals of a subring. While the latter sections of the chapter are designed to show the reader a number of contexts where bimodules have been successfully used, only the first three sections are strictly needed for later chapters of the book.

♦ NOETHERIAN BIMODULES

By a "noetherian bimodule" is usually meant a bimodule ${}_RA_S$ which not only has the ACC on sub–bimodules, but also is noetherian as a left R–module and as a right S–module. (Some authors mean in addition that R and S are noetherian rings.) We state our results under somewhat more general hypotheses, while thinking mainly of applications to noetherian bimodules.

Our first result, although elementary, is fundamental to the entire development of noetherian bimodules. In particular, as the exercise following it indicates, it gives severe restrictions on which modules can appear as one–sided submodules of noetherian bimodules.

LEMMA 7.1. *Let ${}_RA_S$ be a bimodule, and let B be a right S–submodule of A such that ${}_R(RB)$ is finitely generated. (The latter hypothesis is automatically satisfied if ${}_RA$ is noetherian.) If $I = r.ann_S(B)$, there exists $n \in \mathbb{N}$ such that S/I is isomorphic to a (right) submodule of B^n.*

Proof. There is a finite set of generators for ${}_R(RB)$, each of which is a finite sum of products rb with $r \in R$ and $b \in B$. Hence, there exist $b_1,\dots,b_n \in R$ such that $RB = Rb_1 + \dots + Rb_n$. Consequently,

$$I = r.ann_S(B) = r.ann_S(RB) = r.ann_S(b_1) \cap \ldots \cap r.ann_S(b_n).$$

Thus the rule $s \mapsto (b_1 s, \ldots, b_n s)$ defines a right S–module homomorphism $S \to B^n$ with kernel I. □

EXERCISE 7A. Let R be a left noetherian ring and B a minimal (nonzero) right ideal of R (that is, B_R is simple). Show that $R/r.ann_R(B)$ is a simple artinian ring. (In other words, the right primitive ideal of R that annihilates B must be "co–artinian".) □

PROPOSITION 7.2. *Let ${}_RA_S$ be a bimodule, and let B be a right S–submodule of A such that ${}_R(RB)$ is finitely generated. (The latter hypothesis is automatically satisfied if ${}_RA$ is noetherian.) Set $I = r.ann_S(B)$. If B_S is artinian (noetherian), then S/I is right artinian (right noetherian).*

Proof. Since any finite direct sum of copies of B is an artinian (noetherian) right (S/I)–module, this is immediate from Lemma 7.1. □

LEMMA 7.3. *Let ${}_RA_S$ be a bimodule such that ${}_RA$ is noetherian and S is a prime right noetherian ring. Let C be a sub–bimodule of A, and B a right S–submodule containing C. If B/C is torsion as a right S–module, then there is a nonzero ideal I of S such that $BI \leq C$.*

Proof. We apply Lemma 7.1 to B/C. If $I = r.ann_S(B/C)$, the lemma implies that S/I embeds in a finite direct sum of copies of $(B/C)_S$, and hence S/I is torsion as a right S–module. It follows that $I \neq 0$. □

PROPOSITION 7.4. *Let ${}_RA_S$ be a bimodule, where R is a prime left noetherian ring and S is a prime right noetherian ring, and suppose that the modules ${}_RA$ and A_S are both finitely generated and torsionfree. Let B and C be sub–bimodules with $B \geq C$. Then the following conditions are equivalent:*

(a) B/C is torsion as a right S–module.

(b) B/C is torsion as a left R–module.

(c) There is a nonzero ideal I of S such that $BI \leq C$ (that is, B/C is an unfaithful right S–module).

(d) There is a nonzero ideal J of R such that $JB \leq C$ (that is, B/C is an unfaithful left R–module).

(e) C is essential as a right S–submodule of B.

(f) C is essential as a left R–submodule of B.

Proof. (a) $\Rightarrow$ (c): Lemma 7.3.

(c) $\Rightarrow$ (f): Since I is a nonzero ideal in the prime right Goldie ring S, it contains a regular element c. Then $Bc \leq C$, and we note that since B is torsionfree as a right S–module, right multiplication by c is an injective map of B into C. Right multiplication by c is also a homomorphism of left R–modules. By Corollary 4.18, $Bc \leq_e {}_RB$, whence ${}_RC \leq_e {}_RB$.

(f) $\Rightarrow$ (b) is trivial, and finally (b) $\Rightarrow$ (d) $\Rightarrow$ (e) $\Rightarrow$ (a) by symmetry. □

We next consider affiliated submodules and affiliated primes (as defined in Chapter 2) in the context of bimodules. If ${}_RA_S$ is a bimodule, an affiliated submodule for the module A_S is a nonzero submodule B such that the ideal $P = \text{r.ann}_S(B)$ is maximal among right annihilators of nonzero submodules of A, and $B = \text{l.ann}_A(P)$. Observe that B is a sub–bimodule of A, and hence that P is maximal among right annihilators of nonzero sub–bimodules of A. Conversely, if Q is an ideal of S maximal among right annihilators of nonzero sub–bimodules of A, then Q is also maximal among right annihilators of nonzero submodules of A (because $\text{r.ann}_S(C) = \text{r.ann}_S(RC)$ for any right S–submodule $C \leq A$), and so $\text{l.ann}_A(Q)$ is an affiliated submodule for A_S.

Thus we may speak of "affiliated sub–bimodules" of a bimodule, as follows.

DEFINITION. Let ${}_RA_S$ be a bimodule. A *right affiliated sub–bimodule* of A is any affiliated submodule for A_S, a *right affiliated series* for A is any affiliated series

$$A_0 = 0 < A_1 < A_2 < \dots < A_m = A$$

for A_S, and the corresponding affiliated prime ideals $P_i = \text{r.ann}_S(A_i/A_{i-1})$ are called *right affiliated primes* of A. *Left affiliated sub–bimodules*, *series*, and *primes* are defined analogously.

PROPOSITION 7.5. *Let ${}_RA_S$ be a bimodule such that ${}_RA$ is noetherian and S is right noetherian. Then there exists a right affiliated series*

$$A_0 = 0 < A_1 < A_2 < \dots < A_m = A$$

for A. If $P_1,\dots,P_m$ are the corresponding right affiliated primes, then each A_i/A_{i-1} is a torsionfree right (S/P_i)–module.

Proof. Since A has ACC on sub–bimodules, it suffices to show that when A is nonzero, it contains a right affiliated sub–bimodule B, and that B is a torsionfree right module over $S/\text{r.ann}_S(B)$.

As S is right noetherian, there is a nonzero right S–submodule $C \leq A$ such that the ideal $P = \text{r.ann}_S(C)$ is maximal among right annihilators of nonzero right S–submodules of A. If $B = \text{l.ann}_A(P)$, then B is a right affiliated sub–bimodule of A with $P = \text{r.ann}_S(B)$, and P is a prime ideal of S.

Now B is an (R,S/P)–bimodule. Let D be the torsion submodule of B as a right (S/P)–module, and observe that D is an (R,S)–sub–bimodule of A. If $D \neq 0$, then Lemma 7.3 shows that $DI = 0$ for some nonzero ideal I of S/P, contradicting the maximality of P. Therefore $D = 0$, and so B is torsionfree as a right (S/P)–module. □

COROLLARY 7.6. *Let ${}_RA_S$ be a bimodule such that ${}_RA$ and A_S are finitely generated. If R is left noetherian and S is right noetherian, there exist sub–bimodules*

$$A_0 = 0 < A_1 < A_2 < \ldots < A_m = A$$

such that for each $i = 1,\ldots,m$, the ideals $Q_i = l.ann_R(A_i/A_{i-1})$ and $P_i = r.ann_S(A_i/A_{i-1})$ are prime, and A_i/A_{i-1} is torsionfree both as a left (R/Q_i)–module and as a right (S/P_i)–module.

Proof. It suffices to show that when A is nonzero, it contains a nonzero sub–bimodule B such that $l.ann_R(B)$ and $r.ann_S(B)$ are prime ideals and B is torsionfree over $R/l.ann_R(B)$ and over $S/r.ann_S(B)$.

Let C be a right affiliated sub–bimodule of A and B a left affiliated sub–bimodule of C, which exist by Proposition 7.5. Moreover, if $P = r.ann_S(C)$ and $Q = l.ann_R(B)$, then C is a torsionfree right (S/P)–module and B is a torsionfree left (R/Q)–module. In addition, $r.ann_S(B) = P$ by the maximality of P, and since B is contained in C it is a torsionfree right (S/P)–module. □

The noetherian conditions are crucial to Proposition 7.5 and Corollary 7.6, as the following examples show.

EXERCISE 7B. Let $R = A_n(D)$ for some $n \in \mathbb{N}$ and some division ring D of characteristic zero. If A is any simple right R–module and $S = End_R(A)$, show that the (S,R)–bimodule A contains no nonzero right submodule which is a torsionfree module over a prime factor ring of R. Similarly, if Q is the Ore quotient ring of R, show that the (R,R)–bimodule Q/R contains no nonzero right or left submodule which is a torsionfree module over a prime factor ring of R. □

Comparing Proposition 7.5 with Proposition 2.13, we see that the advantage of working with a noetherian bimodule as opposed to a noetherian module is that in the former case we obtain submodules which are torsionfree, rather than just fully faithful, modules over prime factor rings.

In the situation of Corollary 7.6, it is often interesting to study what happens when we take first a right affiliated series for a bimodule, and then refine it further by taking left affiliated series of each of the factors. The next result shows that the right annihilator primes of the resulting sequence of factors are just the original right affiliated primes from the right affiliated series.

PROPOSITION 7.7. *Let* ${}_RA_S$ *be a bimodule, where* R *is a left noetherian ring and* S *is a prime right noetherian ring, and suppose that the modules* ${}_RA$ *and* A_S *are both finitely generated, and that* A_S *is torsionfree. Let*

$$A_0 = 0 < A_1 < \dots < A_m = A$$

be a left affiliated series for A. *Then each factor* A/A_{i-1} *(for* $i = 1,\dots,m$*) is a torsionfree right* S*–module. In particular,* $r.ann_S(A_i/A_{i-1}) = 0$.

Proof. It suffices to show that $(A/A_1)_S$ is torsionfree. Let B/A_1 be the torsion submodule of A/A_1 as a right S–module. By Lemma 7.3, there is a nonzero ideal I of S such that $BI \le A_1$, and consequently there is a regular element $c \in S$ such that $Bc \le A_1$. Hence, as a left R–module B is isomorphic to a submodule of A_1. If $P = l.ann_R(A_1)$, then $PB = 0$, and since A_1 is a left affiliated submodule we conclude that $B \le A_1$. Therefore $B/A_1 = 0$ and A/A_1 is torsionfree as a right S–module. □

♦ ARTINIAN BIMODULES

We turn next to a basic symmetry result – that a noetherian bimodule which is artinian on one side is necessarily artinian on the other. (The corresponding result for a noetherian ring we have already met in Exercise 3K.)

LEMMA 7.8. *Let* ${}_RA_S$ *be a bimodule such that* S *is a semiprime right Goldie ring and* A_S *is torsionfree. Let* Q *be the right Goldie quotient ring of* S.

(a) If ${}_RA$ *has finite length, then* A_S *is divisible.*

(b) If A_S *is divisible, its right* S*–module structure extends to a right* Q*–module structure, and* A *becomes an* (R,Q)*–bimodule.*

Proof. (a) Let c be a regular element of S. Since A_S is torsionfree, right multiplication by c defines an injective map of A to itself. This map is also a left R–module endomorphism of A, whence Ac has the same length as A, and therefore $Ac = A$.

(b) By Proposition 6.13, A has a unique right Q–module structure compatible with its right S–module structure. Given $r \in R$, $a \in A$, $q \in Q$, write $q = sc^{-1}$ for some $s,c \in S$ with c regular. Then

$$[r(aq)]c = r(aqc) = r(as) = (ra)s,$$

whence $r(aq) = [(ra)s]c^{-1} = (ra)q$. Therefore A is an (R,Q)–bimodule. □

EXERCISE 7C. Let R and S be semiprime noetherian rings, and let ${}_RA_S$ be a bimodule which is finitely generated and torsionfree on each side. Let $Q(R)$ and $Q(S)$ be the Goldie quotient rings of R and S, and set $B = Q(R) \otimes_R A$.

(a) Show that B has finite length as a left $Q(R)$–module.

(b) Show that B is torsionfree as a right S–module. [Hint: Every element of B has the form $c^{-1} \otimes a$ where $c \in R$, $a \in A$, and c is regular.] Now by Lemma 7.8, B is a $(Q(R),Q(S))$–bimodule.

(c) Given $c \in R$, $a \in A$ with c regular, show that there exist $b \in A$, $d \in S$ with d regular such that $ad = cb$. [Hint: Corollary 4.18.]

(d) Show that B has finite length as a right $Q(S)$–module. □

LEMMA 7.9. *Let ${}_RA_S$ be a bimodule such that ${}_RA$ and A_S are finitely generated. Assume that S is a semiprime right noetherian ring, and that A is a torsionfree, faithful, divisible right S–module. Then S is a semisimple ring and A_S is a semisimple module with finite length.*

Proof. If Q is the right Goldie quotient ring of S, then in view of Lemma 7.8 we may view A as an (R,Q)–bimodule.

We next observe that A_Q is faithful. Namely, any nonzero $q \in Q$ can be written as sc^{-1} for some $s,c \in S$ with $s \neq 0$ and c regular, and $as \neq 0$ for some $a \in A$, whence $aq \neq 0$. Since ${}_RA$ is finitely generated, Lemma 7.1 shows that Q_Q embeds in some finite direct sum of copies of A_Q. As A_S is noetherian, it follows that Q_S is noetherian.

Now given any regular element $z \in S$, the chain $S \leq z^{-1}S \leq z^{-2}S \leq \ldots$ of right S–submodules of Q must terminate, i.e., there exists $m \in \mathbb{N}$ such that $z^{-m-1}S = z^{-m}S$. On multiplying by z^m, we obtain $z^{-1}S = S$, whence $z^{-1} \in S$. Thus all regular elements of S are invertible in S, whence $Q = S$.

Therefore S is a semisimple ring, and consequently A_S is a semisimple module. As A_S is finitely generated, it must have finite length. □

THEOREM 7.10. [Lenagan] *Let ${}_RA_S$ be a bimodule such that ${}_RA$ has finite length and A_S is noetherian. Then A_S has finite length.*

Proof. We may replace S by $S/\text{r.ann}_S(A)$, in which case Proposition 7.2 shows that S is right noetherian. By Proposition 7.5, there exists a right affiliated series

$$A_0 = 0 < A_1 < A_2 < \ldots < A_m = A$$

for A, and for each $i = 1,\ldots,m$, the ideal $P_i = \text{r.ann}_S(A_i/A_{i-1})$ is a prime ideal of S and A_i/A_{i-1} is a torsionfree right (S/P_i)–module. Each A_i/A_{i-1} has finite length as a left R–module, and it suffices to show that each A_i/A_{i-1} has finite length as a right

(S/P_i)–module.

Thus we may assume, without loss of generality, that S is a prime ring and that A is a torsionfree faithful right S–module. By Lemma 7.9, it suffices to show that A_S is divisible, and this follows from Lemma 7.8. □

COROLLARY 7.11. *Let ${}_RA_S$ be a bimodule which is noetherian on each side. Then ${}_RA$ is artinian if and only if A_S is artinian.* □

COROLLARY 7.12. *Let I be an ideal in a noetherian ring R. Then ${}_RI$ is artinian if and only if I_R is artinian.* □

COROLLARY 7.13. *Let ${}_RA_S$ be a bimodule which is noetherian on each side. There exists a unique sub–bimodule $B \leq A$ such that B is artinian on each side and B contains all artinian right or left submodules of A.*

Proof. Let B be the sum of all the artinian right S–submodules of A, and let C be the sum of all the artinian left R–submodules of A. Since B_S is finitely generated, it must be the sum of finitely many artinian right S–submodules, whence B_S is artinian. Similarly, ${}_RC$ is artinian. For any $x \in R$, observe that $(xB)_S$ is an epimorphic image of B_S and so is artinian, whence $xB \leq B$. Thus B is a sub–bimodule of A, and similarly C is a sub–bimodule.

By Corollary 7.11, ${}_RB$ and C_S are artinian, and therefore $B = C$. In particular, ${}_RB$ is artinian and B contains all the artinian left R–submodules of A. The uniqueness of B is clear. □

DEFINITION. The sub–bimodule B in Corollary 7.13 is called the *artinian radical* of A.

For example, if $R = \begin{pmatrix} k & k[x]/xk[x] \\ 0 & k[x] \end{pmatrix}$ for some field k and indeterminate x, the artinian radical of R (i.e., of the bimodule ${}_RR_R$) is the ideal $\begin{pmatrix} k & k[x]/xk[x] \\ 0 & 0 \end{pmatrix}$.

EXERCISE 7D. Let R be a subring of a ring S such that R is right noetherian and S_R is finitely generated, and let P be a prime ideal of S. Show that S/P is simple artinian if and only if $R/(P \cap R)$ is right artinian. □

Here is a first application of Lenagan's Theorem, which will be used in the next chapter.

THEOREM 7.14. [Ginn–Moss] *Let R be a noetherian ring. If $soc(R_R)$ is essential as either a right or a left ideal of R, then R is an artinian ring.*

Proof. If $I = soc(R_R)$, then since I_R is finitely generated and semisimple it is artinian. By Corollary 7.12, ${}_RI$ is artinian as well. We now have symmetric hypotheses – namely, we have an ideal I of R which is artinian on both sides and essential on (at least) one side. Hence, it is enough to consider the case that $I_R \leq_e R_R$.

If $N = l.ann_R(I)$, then as ${}_RI$ is artinian, we see by Proposition 7.2 that R/N is left artinian. On the other hand, since $NI = 0$ and $I_R \leq_e R_R$ we have $N \leq Z_r(R)$, and so N is nilpotent (Proposition 5.3). For $i = 0,1,\ldots$, note that N^i/N^{i+1} is a finitely generated left (R/N)–module, whence ${}_R(N^i/N^{i+1})$ is artinian. As $N^k = 0$ for some $k \in \mathbb{N}$, we conclude that R is left artinian. By Corollary 7.12, R is right artinian as well. □

Theorem 7.14 does not hold for one–sided noetherian rings, as the following example shows.

EXERCISE 7E. If $R = \begin{pmatrix} \mathbb{Z} & \mathbb{Q} \\ 0 & \mathbb{Q} \end{pmatrix}$, show that R is a right noetherian ring with $soc(R_R) \leq_e R_R$, but that R is neither right nor left artinian. □

EXERCISE 7F. Let I be a right ideal in a noetherian ring R. If I has an essential artinian submodule, show that I is artinian. [Hint: Look at $R/r.ann(I)$.] □

♦ PRIME IDEALS IN FINITE RING EXTENSIONS

In this section, we begin the study of finite ring extensions, to which we will return in Chapters 10 and 12. Suppose that R is a noetherian ring and R is a subring of a ring S such that the modules ${}_RS$ and S_R are finitely generated; this immediately implies that S is noetherian. We can, of course, regard S itself as an (R,S)–bimodule or as an (S,R) bimodule, but there are other bimodules arising from these which give even more information about the relation between the ideal theories of R and S. We will concentrate on properties of prime ideals of R and S. If P is a prime ideal of S, then (unlike the commutative case) $P \cap R$ need not be a prime (or even a semiprime) ideal of R – for instance, let $R = \begin{pmatrix} \mathbb{Q} & \mathbb{Q} \\ 0 & \mathbb{Q} \end{pmatrix}$ and $S = M_2(\mathbb{Q})$, while $P = 0$. In this section we will look at properties of P and see how they compare with properties of the prime ideals of R minimal over $P \cap R$.

LEMMA 7.15. *Let R be a noetherian ring which is a subring of a ring S such that ${}_RS$ and S_R are finitely generated, and let P be a prime ideal of S. For any prime ideal Q of R which is minimal over $P \cap R$, there exist nonzero bimodules ${}_{R/Q}A_{S/P}$ and ${}_{S/P}B_{R/Q}$ which are finitely generated and torsionfree on each side. Similarly, if Q_1 and Q_2 are prime ideals of R which are both minimal over $P \cap R$, there exist nonzero bimodules ${}_{R/Q_1}C_{R/Q_2}$ and ${}_{R/Q_2}D_{R/Q_1}$ which are finitely generated and torsionfree on each side.*

Proof. Since we may factor P out of S and $P \cap R$ out of R, without loss of generality we may assume that $P = 0$, so that S is a prime ring.

Given a minimal prime Q of R, the first bimodule we need to find is an $(R/Q,S)$–bimodule which is torsionfree as a left (R/Q)–module and as a right S–module. We first regard S as an (R,S)–bimodule and choose a left affiliated series

$$0 = S_0 < S_1 < \dots < S_n = S$$

for this bimodule. Propositions 7.5 and 7.7 imply that each factor S_i/S_{i-1} is a torsionfree left (R/L_i)–module, where $L_i = \text{l.ann}_R(S_i/S_{i-1})$, and a torsionfree right S–module. According to Proposition 2.14, since Q is minimal over $\text{l.ann}_R(S)$ we must have $Q = L_j$ for some index j. Thus we obtain the desired bimodule A by choosing $A = S_j/S_{j-1}$. The bimodule B is obtained by a symmetric argument.

Now suppose that Q_1 and Q_2 are minimal primes of R. Let ${}_{R/Q_1}A_S$ be the bimodule obtained in the previous paragraph, and now regard A as an $(R/Q_1,R)$–bimodule. Since A is a faithful right S–module, it is also a faithful right R–module. Let

$$0 = A_0 < A_1 < \dots < A_m = A$$

be a right affiliated series for this bimodule, and let $P_j = \text{r.ann}_R(A_j/A_{j-1})$ for $j = 1,\dots,m$. Using Propositions 7.5 and 7.7 again, each factor A_j/A_{j-1} is torsionfree as a left (R/Q_1)–module and as a right (R/P_j)–module. As Q_2 is minimal over $\text{r.ann}_R(A)$, we again apply Proposition 2.14 to see that $Q_2 = P_k$ for some index k. Therefore we may choose $C = A_k/A_{k-1}$. The corresponding argument (beginning with B as an $(R,R/Q_1)$–bimodule) completes the proof of the lemma. □

THEOREM 7.16. [Jategaonkar, Letzter] *Let R and S be prime noetherian rings and ${}_RB_S$ a nonzero bimodule which is finitely generated and torsionfree on each side.*

(a) If R is semiprimitive, so is S.

(b) If R is right primitive, so is S.

Proof. First observe that since B is finitely generated and torsionfree as a left R–module, ${}_RB$ embeds in a free left R–module (Proposition 6.19). Hence, $\text{Hom}_R({}_RB,{}_RR) \neq 0$. Next let

$$T = \Sigma\ \{f(B) \mid f \in \text{Hom}_R({}_RB,{}_RR)\}$$

and observe that T is a nonzero ideal of R (called the *trace ideal of* $_RB$).

If I is a right ideal of R, we claim that $B/IB = 0$ only if $(R/I)T = 0$. Given any $f \in \mathrm{Hom}_R({}_RB, {}_RR)$, it follows from $B = IB$ that $f(B) = If(B) \le I$. Hence, $T \le I$, and so $(R/I)T = 0$, as claimed.

(a) Let $J = J(S)$. If $J \neq 0$, then by Proposition 7.4 (applied to the sub–bimodules $B \ge BJ$), there is a nonzero ideal K of R such that $KB \le BJ$. Since R is prime, $T \cap K \neq 0$, and then since R is semiprimitive we may choose a maximal right ideal I of R such that I does not contain $T \cap K$. It follows immediately that $K + I = R$, so that $KB + IB = B$. On the other hand, $(R/I)T \neq 0$. By the claim proved above, $B/IB \neq 0$, and so we can find a maximal right submodule C in B such that $IB \le C$. Since B/C is a simple right S–module, $(B/C)J = 0$. Hence, $BJ \le C$ and then $KB \le C$. Thus $B = KB + IB \le C$, which is a contradiction. Therefore $J = 0$, and S is semiprimitive.

(b) There exists a maximal right ideal I in R such that R/I is a faithful simple module. Then $(R/I)T \neq 0$, and the claim above shows that $B/IB \neq 0$. Let C be a maximal right submodule of B containing IB. We claim that the simple right S–module B/C is faithful. If not, then $BJ \le C$ for some nonzero ideal J of S. Applying Proposition 7.4 again, there is a nonzero ideal K of R such that $KB \le BJ$. Since I is a maximal right ideal not containing K (because R/I is faithful), we conclude that $K + I = R$, and so $KB + IB = B$. However, $KB \le BJ \le C$ and $IB \le C$, so this contradicts the fact that $C < B$. Therefore B/C is a faithful simple right S–module, and S is right primitive as required. □

COROLLARY 7.17. *Let R be a noetherian ring which is a subring of a ring S such that $_RS$ and S_R are finitely generated, and assume that S is a prime ring. Then S is right (left) primitive if and only if at least one minimal prime ideal of R is right (left) primitive, if and only if every minimal prime ideal of R is right (left) primitive.*

Proof. Lemma 7.15 and Theorem 7.16. □

DEFINITION. A *Jacobson ring* (sometimes called a *Hilbert ring*) is a ring in which every prime ideal is semiprimitive, i.e., every prime factor ring has zero Jacobson radical.

COROLLARY 7.18. [Cortzen–Small] *Let R be a noetherian ring which is a subring of a ring S such that $_RS$ and S_R are finitely generated. If R is a Jacobson ring, then so is S.*

Proof. Lemma 7.15 and Theorem 7.16. □

EXERCISE 7G. Let R be a noetherian ring which is a subring of a ring S such that ${}_RS$ and S_R are finitely generated. If some minimal prime of R is right primitive, show that some minimal prime of S is right primitive. If all minimal primes of S are right primitive, show that all minimal primes of R are right primitive. □

♦ BIMODULE COMPOSITION SERIES

If R is a subring of a ring S and P is a prime ideal of S, then it is natural (especially when R is noetherian) to study the primes in R minimal over $P \cap R$. However, it turns out that a slightly larger set of "affiliated primes" carries more information about the pair of rings. (When the extension is finite, as in the previous section, these primes can frequently be proved to be precisely the primes minimal over $P \cap R$, but this will not generally be true for infinite extensions.) The idea is to view the Goldie quotient ring C of S/P as a (C,R)–bimodule and to refine an affiliated series for this bimodule even further – to a bimodule composition series, which exists because C has finite length on the left. Thus we begin with another look at bimodules which have finite length on one side.

Most terminology relating to submodules of a module can be directly carried over to sub–bimodules of a bimodule. For instance, a bimodule ${}_TC_R$ is *simple* provided C is nonzero and its only sub–bimodules are 0 and C. A *bimodule composition series* for C is a sequence $C_0 = 0 < C_1 < C_2 < \ldots < C_m = C$ of sub–bimodules of C such that C_i/C_{i-1}, for $i = 1,\ldots,m$, is a simple bimodule. Recall from Exercise 1E that C can be made into a right $(T^{op} \otimes_{\mathbb{Z}} R)$–module so that its $(T^{op} \otimes_{\mathbb{Z}} R)$–submodules are precisely its (T,R)–sub–bimodules. Thus a bimodule composition series for C is precisely a $(T^{op} \otimes_{\mathbb{Z}} R)$–module composition series. One exists if and only if C is both artinian and noetherian as a $(T^{op} \otimes_{\mathbb{Z}} R)$–module, i.e., if and only if C has both DCC and ACC on sub–bimodules. In particular, if C has finite length as either a left T–module or a right R–module, it will have a bimodule composition series.

PROPOSITION 7.19. *Let ${}_TC_R$ be a bimodule such that ${}_TC$ has finite length, and let*

$$C_0 = 0 < C_1 < C_2 < \ldots < C_m = C$$

be a bimodule composition series for C. Let $Q_i = r.ann_R(C_i/C_{i-1})$ for $i = 1,\ldots,m$.

(a) Each Q_i is a prime ideal of R, and if R/Q_i is right Goldie then C_i/C_{i-1} is torsionfree as a right (R/Q_i)–module.

(b) If $D_0 = 0 < D_1 < \ldots < D_n = C$ is another bimodule composition series for C, then $n = m$, and there exists a permutation π of $\{1,2,\ldots,m\}$ such that $r.ann_R(D_i/D_{i-1}) = Q_{\pi(i)}$ for $i = 1,\ldots,m$.

Proof. (a) If I,J are ideals of R not contained in Q_i, then $(C_i/C_{i-1})I$ and

$(C_i/C_{i-1})J$ are nonzero. As these are sub–bimodules of C_i/C_{i-1}, we find that $(C_i/C_{i-1})I = (C_i/C_{i-1})J = C_i/C_{i-1}$. Then $(C_i/C_{i-1})IJ = (C_i/C_{i-1})J \neq 0$ and so $IJ \nleq Q_i$, proving that Q_i is prime.

Now assume that R/Q_i is right Goldie, let D be the torsion submodule of C_i/C_{i-1} as a right (R/Q_i)–module, and note that D is a sub–bimodule of C_i/C_{i-1}. If $D \neq 0$, then $D = C_i/C_{i-1}$ and Lemma 7.1 shows that $(R/Q_i)_R$ embeds in a finite direct sum of copies of D_R, which is impossible (because R/Q_i is not torsion as a right module over itself). Therefore $D = 0$ and C_i/C_{i-1} is torsionfree over R/Q_i.

(b) This is immediate from the Jordan–Hölder Theorem (Theorem 3.11), which says that $n = m$ and there is a permutation π of $\{1,2,...,m\}$ such that $D_i/D_{i-1} \cong C_{\pi(i)}/C_{\pi(i)-1}$ (as $(T^{op} \otimes_{\mathbb{Z}} R)$–modules) for $i = 1,...,m$. □

We intend to refer to the primes Q_i occurring in Proposition 7.19 as "right affiliated primes" of C, which *a priori* conflicts with our previous usage. As long as R is right noetherian, however, the next lemma shows that we may talk about right affiliated primes without ambiguity.

LEMMA 7.20. *Let ${}_TC_R$ be a bimodule such that ${}_TC$ has finite length and R is right noetherian. Let*

$$0 = C_0 < C_1 < ... < C_m = C$$

be a bimodule composition series for C, and let

$$0 = B_0 < B_1 < ... < B_n = C$$

be a right affiliated series for C. Let $Q_i = r.ann_R(C_i/C_{i-1})$ for $i = 1,...,m$ and $P_j = r.ann_R(B_j/B_{j-1})$ for $j = 1,...,n$. Then

$$\{Q_1,...,Q_m\} = \{P_1,...,P_n\}.$$

Proof. Since by Proposition 7.19 the set $\{Q_1,...,Q_m\}$ is independent of the choice of a bimodule composition series for C, we may assume that our bimodule composition series is a refinement of the given right affiliated series. Thus there exist integers

$$i(0) = 0 < i(1) < ... < i(n) = m$$

such that $C_{i(j)} = B_j$ for $j = 0,...,n$. To prove the lemma, it suffices to show that $Q_i = P_j$ for $j = 1,...,n$ and $i = i(j-1)+1,...,i(j)$, that is, we need only prove the lemma for each of the bimodules B_j/B_{j-1}.

Hence, we may assume that the right affiliated series for C is just

$$0 = B_0 < B_1 = C$$

and we must prove that all $Q_i = P_1$. Since $P_1 = r.ann_R(C) \leq Q_i$ for $i = 1,...,m$, we may replace R by R/P_1. Thus we may assume that R is prime and C_R is faithful, and it remains to show that all $Q_i = 0$.

By Proposition 7.5, C is torsionfree as a right R–module. As each C_i is a bimodule which has finite length on the left, Lemma 7.8 then says that C_i is divisible as a right R–module. Then $C_i x = C_i \nsubseteq C_{i-1}$ for all regular elements $x \in R$, whence the ideal Q_i contains no regular elements. Therefore $Q_i = 0$, as desired. □

DEFINITION. In the situation of Proposition 7.19, the prime ideals $Q_1,\dots,Q_m$ are called the *right affiliated primes of* C, and if a prime Q_j appears in the list $Q_1,\dots,Q_m$ exactly m_j times, we say that Q_j is *affiliated with multiplicity* m_j. It is clear from part (b) of Proposition 7.19 that the set $\{Q_1,\dots,Q_m\}$ and the multiplicities are independent of the choice of a bimodule composition series for C.

If instead of a bimodule composition series for C we take a right affiliated series, then as long as R is right noetherian the right affiliated primes will be the same (Lemma 7.20), but the multiplicities may differ. For instance, if $R = \mathbb{Q}$ and $T = C = M_2(\mathbb{Q})$, then a right affiliated series for ${}_TC_R$ is $0 < C$, whereas a bimodule composition series has length 2. The only right affiliated prime, namely 0, appears with multiplicity 1 with respect to the affiliated series but with multiplicity 2 with respect to the bimodule composition series.

PROPOSITION 7.21. *Let ${}_TC_R$ be a bimodule such that ${}_TC$ has finite length, and let $A > B$ be sub–bimodules of C. Then any prime ideal of R minimal over $r.ann_R(A/B)$ is right affiliated to C. In particular, $R/r.ann_R(A/B)$ has only finitely many minimal prime ideals.*

Proof. By stringing together bimodule composition series for B, A/B, and C/A, we may choose a bimodule composition series for C that passes through B and A, from which it is clear that any prime of R right affiliated to A/B is also right affiliated to C. Hence, we may replace A/B by C.

If $Q_1,\dots,Q_m$ is the list of right affiliated primes corresponding to a bimodule composition series

$$C_0 = 0 < C_1 < C_2 < \dots < C_m = C,$$

it is clear that $CQ_mQ_{m-1}\cdots Q_1 = 0$, whence $Q_mQ_{m-1}\cdots Q_1 \leq r.ann_R(C)$. On the other hand, $r.ann_R(C)$ annihilates each C_i/C_{i-1} and so is contained in each Q_i. Therefore any prime ideal Q of R containing $r.ann_R(C)$ must contain some Q_j, and if Q is minimal over $r.ann_R(C)$ then $Q = Q_j$. □

♦ ADDITIVITY PRINCIPLES

Given a bimodule ${}_TC_R$ with ${}_TC$ having finite length, we wish to compare the length of ${}_TC$ with the ranks of R modulo the right affiliated primes. For this it suffices to assume that these prime factors of R are all right Goldie; however, to avoid excessive hypotheses we shall assume that R is right noetherian. Also, for any prime ideal Q of R we write rank(R/Q) for the rank of R/Q as a right module over itself; note that this equals the rank of R/Q as a right R–module.

LEMMA 7.22. *Let ${}_TC_R$ be a bimodule such that ${}_TC$ has finite length, R is prime right noetherian, and C_R is torsionfree. Then the rank of R divides the length of ${}_TC$.*

Proof. If Q is the right Goldie quotient ring of R, then by Lemma 7.8 we may view C as a (T,Q)–bimodule. As rank(R) = rank(Q) (Exercise 5G), there is no loss of generality in replacing R by Q. Hence, we may assume that R is a simple artinian ring.

Now R is isomorphic to the $m \times m$ matrix ring over a division ring, where $m = \text{rank}(R)$, and so there exists a set of $m \times m$ matrix units $e_{ij} \in R$ (that is, $e_{ij}e_{kl} = 0$ for $j \neq k$ and $e_{ij}e_{jk} = e_{ik}$, while also $e_{11} + ... + e_{mm} = 1$). We now note that as a left T–module, C is the direct sum of the submodules Ce_{ii}, for $i = 1,...,m$. Also, right multiplication by e_{ji} and e_{ij} provide inverse isomorphisms between Ce_{ii} and Ce_{jj}. Hence, if Ce_{11} has length t, so does each Ce_{ii}, and therefore $\text{length}({}_TC) = mt$. □

THEOREM 7.23. (The Additivity Principle) *Let ${}_TC_R$ be a bimodule such that ${}_TC$ has finite length and R is right noetherian, and let X be the set of primes in R right affiliated to C. Then there are positive integers z(Q), for $Q \in X$, such that*

$$\text{length}({}_TC) = \sum_{Q \in X} z(Q)\,\text{rank}(R/Q).$$

Moreover, if $Q \in X$ is affiliated with multiplicity m(Q), then $z(Q) \geq m(Q)$.

Proof. Let $C_0 = 0 < C_1 < ... < C_n = C$ be a bimodule composition series for C, and for $i = i,...,n$ set $B_i = C_i/C_{i-1}$ and $Q_i = \text{r.ann}_R(B_i)$. Then $X = \{Q_1,...,Q_n\}$, each B_i is a torsionfree right (R/Q_i)–module (Proposition 7.19), and it follows from Lemma 7.22 that there are positive integers z_i such that $z_i\text{rank}(R/Q_i) = \text{length}({}_TB_i)$. We add up these equations to obtain

$$\text{length}({}_TC) = \sum_{i=1}^{n} \text{length}({}_TB_i) = \sum_{i=1}^{n} z_i\text{rank}(R/Q_i),$$

from which the conclusion of the theorem is clear. □

COROLLARY 7.24. *Let* $_TC_R$ *be a bimodule such that* $_TC$ *has finite length and* R *is right noetherian, and let* Q *be a prime ideal of* R *minimal over* $r.ann_R(C)$. *Then* $rank(R/Q) \leq length(_TC)$, *and if equality holds, then* $r.ann_R(C) = Q$.

Proof. If X is the set of primes in R right affiliated to C, then by Proposition 7.21 $Q \in X$. As the coefficient $z(Q)$ in Theorem 7.23 is positive, it is immediate that $rank(R/Q) \leq length(_TC)$. Now if $rank(R/Q) = length(_TC)$, we infer from Theorem 7.23 that $X = \{Q\}$ and that Q is affiliated to C with multiplicity 1. Consequently, C has a bimodule composition series of length 1 (that is, $0 < C$), and $Q = r.ann_R(C)$. □

Note that if $r.ann_R(C) = Q$ in the situation of Corollary 7.24, we cannot conclude that $rank(R/Q) = length(_TC)$ – for example, let $T = C = M_2(\mathbb{Q})$ while $R = \mathbb{Q}$ and $Q = 0$.

The Additivity Principle given in Theorem 7.23 is more general than the original, which applies to primes in ring extensions – see Theorem 7.25.

DEFINITION. Let R be a subring of a right noetherian ring S, and let P be a prime ideal of S. If C is the right Goldie quotient ring of S/P, we may view C as a (C,R)–bimodule, and of course $_CC$ has finite length. In this context the right affiliated primes of C are called the *primes of* R *right affiliated to* P. If we had regarded C as an (R,C)–bimodule, we would have obtained primes of R *left affiliated to* P. That the primes left affiliated to P are the same as those right affiliated to P, with the same multiplicities, is immediate from the following exercise.

EXERCISE 7H. Let R be a subring of a simple artinian ring C, let

$$C_0 = 0 < C_1 < \ldots < C_m = C$$

be a bimodule composition series for $_CC_R$, and set $Q_i = r.ann_R(C_i/C_{i-1})$ for $i = 1,\ldots,m$. If $D_i = r.ann_C(C_i)$ for $i = 0,1,\ldots,m$, show that $C_i = l.ann_C(D_i)$ and that

$$D_m = 0 < D_{m-1} < \ldots < D_0 = C$$

is a bimodule composition series for $_RC_C$. Then show that $Q_i = l.ann_R(D_{i-1}/D_i)$ for $i = 1,\ldots,m$. □

If R is a subring of a right noetherian ring S, and P is a prime ideal of S, then by Proposition 7.21 all prime ideals of R minimal over $P \cap R$ are right affiliated to P. However, the primes affiliated to P need not all be minimal over $P \cap R$, as the following example shows.

EXERCISE 7I. Let k be a field and x an indeterminate, let $S = M_2(k[x])$, and let P be the prime ideal 0 in S. Let R be the subring of S consisting of all matrices $\begin{pmatrix} f & 0 \\ 0 & f(0) \end{pmatrix}$ for $f \in k[x]$. Show that the primes in R right affiliated to P are 0 and $\begin{pmatrix} xk[x] & 0 \\ 0 & 0 \end{pmatrix}$. In particular, the latter is not minimal over $P \cap R$. □

THEOREM 7.25. [Joseph–Small, Borho, Warfield] *Let R be a right noetherian subring of a right noetherian ring S, let P be a prime ideal of S, and let X be the set of primes in R right affiliated to P. Then there are positive integers $z(Q)$, for $Q \in X$, such that*

$$\mathrm{rank}(S/P) = \sum_{Q \in X} z(Q)\mathrm{rank}(R/Q).$$

Moreover, if $Q \in X$ is affiliated with multiplicity $m(Q)$, then $z(Q) \geq m(Q)$.

Proof. This is immediate from Theorem 7.23 by letting T and C equal the right Goldie quotient ring of S/P, and noting that $\mathrm{length}({}_TC) = \mathrm{rank}(C) = \mathrm{rank}(S/P)$. □

COROLLARY 7.26. *Let R be a right noetherian subring of a right noetherian ring S, let P be a prime ideal of S, and let Q be a prime ideal of R minimal over $P \cap R$. Then $\mathrm{rank}(R/Q) \leq \mathrm{rank}(S/P)$, and if equality holds, then $P \cap R = Q$.* □

EXERCISE 7J. Let R be a right noetherian subring of a left artinian ring S, let N be the prime radical of R, and assume that every element of $\mathscr{C}_R(N)$ is invertible in S. Choose a bimodule composition series $S_0 = 0 < S_1 < \dots < S_m = S$ for ${}_SS_R$.

(a) For $i = 1,\dots,m$, show that S_i/S_{i-1} is a torsionfree divisible right (R/N)–module. [Hint: If this has been proved for $i = 1,\dots,j$, show that $S_jc = S_j$ for all $c \in \mathscr{C}_R(N)$.]

(b) Show that the primes in R right affiliated to ${}_SS_R$ are precisely the minimal primes. □

♦ NORMALIZING EXTENSIONS

To conclude the chapter, we briefly consider the behavior of primes affiliated to a prime ideal in an extension ring generated by elements that "normalize" the subring in the following sense.

DEFINITION. If R is a subring of a ring S, then an element $x \in S$ is said to *normalize* R if $xR = Rx$. (Note that we do not require that x commute with the elements of R, although that would certainly suffice.) We say that S is a *normalizing extension of* R if S is generated as a right (equivalently, left) R–module by a (possibly infinite) set of elements that normalize R. The reader should be warned that in much of the literature, the term "normalizing extension" is reserved for a *finite* normalizing extension, i.e., a ring extension $S \supseteq R$ in which S is generated by a finite set of normalizing elements.

For instance, if α is an automorphism of a ring R, then $R[\theta; \alpha]$ and $R[\theta, \theta^{-1}; \alpha]$ are normalizing extensions of R. For another example, if k is a field and H is a normal subgroup of a group G, the group algebra k[G] is a normalizing extension of the group algebra k[H]. (Here the normalizing generators can be taken to be a set of coset representatives for H in G, and k[G] is actually a free k[H]–module on this set of generators.)

EXERCISE 7K. Let R be a subring of a ring S, let x be an element of S that normalizes R, and let A be an R–submodule of a right S–module B. Observe that Ax is an R–submodule of B. Set $Q_1 = \mathrm{ann}_R(A)$ and $Q_2 = \mathrm{ann}_R(Ax)$. If $\mathrm{ann}_A(x) = 0$, show that $R/Q_1 \cong R/Q_2$. [Hint: Given $r \in R$, there exists $r' \in R$ such that $rx = xr'$; map $r + Q_1$ to $r' + Q_2$.] □

THEOREM 7.27. [Warfield] *Let R be a subring of a right noetherian ring S such that S is generated as an R–module by a (possibly infinite) set X of elements normalizing R, and let P be a prime ideal of S.*

(a) $P \cap R$ is a semiprime ideal of R.

(b) R has only finitely many prime ideals minimal over $P \cap R$, say $Q_1, \ldots, Q_m$.

(c) The factor rings R/Q_i are all isomorphic.

(d) If X consists of units of S, then $\{Q_1, \ldots, Q_m\} = \{x^{-1}Q_1x \mid x \in X\}$.

(e) If R is right noetherian, then $\mathrm{rank}(R/Q_1)$ is a divisor of $\mathrm{rank}(S/P)$.

Proof. As before, we let C be the right Goldie quotient ring of S/P. First regard C as a (C,S)–bimodule. If C' is any (C,S)–sub–bimodule of C, then since it has finite length on the left, by Lemma 7.8 it is divisible as a right (S/P)–module. Hence, C' is an ideal of C, and since C is a simple ring, either $C' = C$ or $C' = 0$. Thus C is a simple (C,S)–bimodule.

Now regard C as a (C,R)–bimodule, and choose a simple sub–bimodule $A \leq C$. Note that if $x \in X$, then while right multiplication by x is not necessarily a bimodule map, it comes quite close: because $Rx = xR$ the sets Ax and $\{a \in A \mid ax \in B\}$ (for any sub–bimodule $B \leq C$) are sub–bimodules of C. Hence, either $Ax = 0$ or Ax is a simple sub–bimodule of C.

We next look at the sub–bimodule $\sum_{x \in X} Ax = AS$. As this is a nonzero (C,S)–sub–

bimodule of C, we obtain $C = \sum_{x \in X} Ax$, and thus C is a sum of simple (C,R)–sub–bimodules. Therefore C is a semisimple $(C^{op} \otimes_{\mathbb{Z}} R)$–module, and so by Proposition 3.1 we can write it as a direct sum of simple sub–bimodules, in fact a finite direct sum because ${}_C C$ has finite length. Therefore $C = C_1 \oplus ... \oplus C_n$ for some simple sub–bimodules C_j, and each $C_j \cong Ax_j$ for some $x_j \in X$ (since $\sum_{x \in X} Ax$ is not contained in $\bigoplus_{i \neq j} C_i$). Observe that any simple sub–bimodule of C must be isomorphic to some C_j, and hence to Ax_j.

Since A was arbitrary, the same conclusion applies to any pair of simple sub–bimodules $B, B' \leq C$: there exists $x \in X$ such that $B' \cong Bx$. Note in this situation that $\text{l.ann}_B(x)$ is a proper sub–bimodule of B, whence $\text{l.ann}_B(x) = 0$.

(a) The ideals $Q_j' = \text{r.ann}_R(C_j)$ (for $j = 1,...,n$) are the primes in R right affiliated to P, and clearly $P \cap R = \text{r.ann}_R(C) = Q_1' \cap ... \cap Q_n'$. Thus statement (a) is proved.

(b) By Proposition 7.21, the primes of R minimal over $P \cap R$ are all contained in the set $\{Q_1',...,Q_n'\}$.

(c) If $i,j \in \{1,...,n\}$, then as observed above, there exists $x_{ij} \in X$ such that $C_j \cong C_i x_{ij}$ and $\text{l.ann}_{C_i}(x_{ij}) = 0$. Thus $R/Q_i' \cong R/Q_j'$ by Exercise 7K. (We will use this fact in the proof of part (e).) Statement (c) now follows using the proof of (b).

(d) As just observed, there are simple sub–bimodules $B_1,...,B_m \leq C$ such that each $Q_i = \text{r.ann}_R(B_i)$, and each $B_i \cong B_1 y_i$ for some $y_i \in X$. Since $y_i R = R y_i$ and y_i is a unit of S, we have $y_i R y_i^{-1} = R$. For $r \in R$, observe that $r \in Q_i$ if and only if $B_1 y_i r y_i^{-1} = 0$, if and only if $y_i r y_i^{-1} \in Q_1$. Therefore $Q_i = y_i^{-1} Q_1 y_i$.

Conversely, if $x \in X$ then $x^{-1}Rx = R$ and the rule $r \mapsto x^{-1}rx$ defines an automorphism of R. Hence, $x^{-1}Q_1x$ must be a prime of R minimal over $x^{-1}(P \cap R)x$. However, $x^{-1}Px = P$ (because P is an ideal of S), whence $x^{-1}(P \cap R)x = P \cap R$, and thus $x^{-1}Q_1x$ is minimal over $P \cap R$. Therefore $x^{-1}Q_1x \in \{Q_1,...,Q_m\}$.

(e) As observed in the proof of (c), the rings R/Q_j' (for $j = 1,...,n$) are pairwise isomorphic, and so they all have the same rank. Statement (e) now follows from Theorem 7.25. □

COROLLARY 7.28. *Let R be a noetherian ring and α an automorphism of R. If $S = R[\theta,\theta^{-1}; \alpha]$ and P is a prime ideal of S, then there exist a prime ideal Q of R and a positive integer m such that $P \cap R = Q \cap \alpha(Q) \cap ... \cap \alpha^{m-1}(Q)$ and $\alpha^m(Q) = Q$.* □

EXERCISE 7L. Give a more direct proof of Corollary 7.28 by first observing that an ideal I of R is of the form $J \cap R$ for some ideal J of S if and only if $\alpha(I) = I$. □

EXERCISE 7M. Let k be a field and H a normal subgroup of a finite group G.

(a) If P is any maximal ideal of k[G], show that $P \cap k[H] = M_1 \cap \ldots \cap M_n$ for some maximal ideals M_i of k[H], and that each $M_i = g_i^{-1}M_1g_i$ for some $g_i \in G$.

(b) If V is any simple k[G]–module, show that V is semisimple as a k[H]–module. (This is sometimes known as *Clifford's Theorem.*) □

EXERCISE 7N. Let $T = \mathbb{Q}[x_1,x_2,\ldots]$ where the x_i are independent commuting indeterminates, and let ϕ be the $\mathbb{Q}$–algebra endomorphism of T such that $\phi(x_1) = 0$ and $\phi(x_n) = x_{n-1}$ for all $n > 1$. Let $S = M_2(T)$, and let R be the subring of S consisting of all matrices of the form $\begin{pmatrix} t & 0 \\ 0 & \phi(t) \end{pmatrix}$, where $t \in T$. Show that S is a (finite) normalizing extension of R, but that one of the primes in R right affiliated to the prime 0 in S is not a minimal prime of R. □

♦ NOTES

Lenagan's Theorem. Lenagan proved that if an ideal in a right noetherian ring has finite length on the left, then it also has finite length on the right [1975, Proposition].

Noetherian Rings with Essential Socles. That a noetherian ring with essential socle must be artinian is due to Ginn and Moss [1975, Theorem].

Transfer of (Semi–) Primitivity across Bimodules. The semiprimitive part of Theorem 7.16 is due to Jategaonkar [1979, Theorem D; 1981, Theorem 6.1], the primitive part to Letzter [Pa, Lemma 1.3].

Jacobson Condition in Finite Ring Extensions. That a finite ring extension of a noetherian Jacobson ring must be Jacobson is due to Cortzen and Small [1988, Theorem 1].

Additivity Principles. The additivity principle for ring extensions given in Theorem 7.25 was first proved by Joseph and Small for certain factor rings of enveloping algebras, where X turns out to be the set of primes of R minimal over $P \cap R$ [1978, Theorem 3.9]. Borho then developed a version for noetherian rings with suitable symmetric dimension functions, where X again turns out to be the set of primes minimal over $P \cap R$ [1982, Theorem 7.2]. The general theorem was proved by Warfield, assuming just that all prime factors of R are right Goldie and that S/P is Goldie on one side [1983, Theorem 1].

Primes in Infinite Normalizing Extensions. Theorem 7.27 is due to Warfield, assuming only suitable Goldie conditions on prime factors of R and S [1983, Theorem 3].

8 FULLY BOUNDED RINGS

One major obstacle to adapting commutative noetherian ring theory to the noncommutative case in general is the lack of ideals. For example, the Weyl algebras over division rings of characteristic zero are simple noetherian domains, yet their module structure is quite complicated. Thus to derive much structure theory similar to the commutative theory we should work in a context where a large supply of ideals is guaranteed. One such context is introduced and investigated in this chapter.

♦ BOUNDEDNESS

DEFINITION. A ring R is *right bounded* if every essential right ideal of R contains an ideal which is essential as a right ideal.

For instance, every commutative ring is right bounded, as is every semisimple ring (since a semisimple ring has no proper essential right ideals). Note that a prime ring R is right bounded if and only if every essential right ideal of R contains a nonzero ideal (recall Exercise 3L).

DEFINITION. A ring R is *right fully bounded* provided every prime factor ring of R is right bounded.

A priori, a right fully bounded ring need not be right bounded. However, in a right fully bounded right noetherian ring, it can be shown that all factor rings are right bounded (Exercise 8F).

DEFINITION. A *right (left) FBN ring* is any right (left) fully bounded right (left) noetherian ring. An *FBN ring* is any right and left FBN ring.

In order to emphasize what ingredients are going into our proofs, we state many results for right FBN rings or for left noetherian right FBN rings. However, as there are no known examples of noetherian rings which are fully bounded on one side but not on the other, the extra generality may be gratuitous.

PROPOSITION 8.1. *(a) If R is a prime module–finite algebra over a commutative noetherian ring S, then every essential right or left ideal of R contains a nonzero central element. Consequently, R is right and left bounded.*

(b) More generally, if R is a module–finite algebra over a commutative ring S, then all factor rings of R are right and left bounded, and in particular R is right and left fully bounded.

Proof. (a) Note first that R is noetherian. If I is an essential right ideal of R, then by Proposition 5.9 there exists a regular element $x \in I$. Since R is a noetherian S–module, the S–submodule $\sum x^iS$ is finitely generated, say by $1,x,x^2,...,x^{n-1}$. Then

$$x^n = 1s_0 + xs_1 + x^2s_2 + ... + x^{n-1}s_{n-1}$$

for some $s_i \in S$. Since x is regular, $x^n \neq 0$, and so not all s_i can be zero. If j is the least index such that $s_j \neq 0$, then

$$x^{n-j} = 1s_j + xs_{j+1} + ... + x^{n-j-1}s_{n-1}$$

(because x is regular). Thus $1s_j \in I$, and $1s_j$ is a nonzero central element of R. In particular, it follows that s_jR is a nonzero ideal contained in I. Therefore R is right bounded. The left–hand versions of these properties follow by symmetry.

(b) Since all factor rings of R are module–finite S–algebras, it suffices to show that R is right and left bounded. Let $x_1,...,x_n$ be S–module generators for R.

Given $I \leq_e R_R$, set $I_j = \{r \in R \mid x_jr \in I\}$ for $j = 1,...,n$, and note that each $I_j \leq_e R_R$ (Proposition 3.21). Hence, the intersection $J = I_1 \cap ... \cap I_n$ is an essential right ideal of R. Then RJ is essential as a right ideal of R, and

$$RJ = (x_1S + ... + x_nS)J = (x_1J + ... + x_nJ)S \leq x_1I_1 + ... + x_nI_n \leq I.$$

Thus R is right bounded, and, by symmetry, left bounded. □

It also can be shown that Proposition 8.1 holds for P.I. rings.

In particular, Proposition 8.1 shows that any matrix ring over a commutative ring is fully bounded. It can be shown that any subring of a matrix ring over a commutative ring is fully bounded as well. As another example, Proposition 8.1 shows that the subring $\mathbb{Z} + \mathbb{Z}i + \mathbb{Z}j + \mathbb{Z}k$ of the quaternions is an FBN ring.

EXERCISE 8A. For any field k of characteristic $p > 0$, show that the Weyl algebras $A_n(k)$ are FBN rings. [Hint: The p–th. powers of the standard generators are central.] □

EXERCISE 8B. For any positive integer n, show that all proper factor rings of $A_n(\mathbb{Z})$ are FBN, but that $A_n(\mathbb{Z})$ itself is neither right nor left bounded. [Hint: Use the simplicity of $A_n(\mathbb{Q})$ to show that every nonzero ideal of $A_n(\mathbb{Z})$ contains a nonzero integer.] □

EXERCISE 8C. Let k be a field of characteristic $p > 0$, and let δ be a derivation on k. If δ is a nilpotent map (that is, $\delta^n = 0$ for some $n \in \mathbb{N}$), show that $k[\theta;\delta]$ is an FBN ring. [Hint: Show that k is finite–dimensional over its subfield of constants.] □

LEMMA 8.2. *Let R be a prime ring. Then R is right bounded if and only if R has no faithful finitely generated singular right modules.*

Proof. If R is not right bounded, then R has an essential right ideal I which contains no nonzero ideals. Since $\text{ann}(R/I)$ is an ideal contained in I, it must be zero. Thus R/I is a faithful finitely generated singular right R–module.

Conversely, assume that R is right bounded, and let A be any finitely generated singular right R–module. Choose generators $a_1,\ldots,a_n$ for A. Each a_j is annihilated by an essential right ideal $I_j \leq R$, and the intersection $I = I_1 \cap \ldots \cap I_n$ is essential. By assumption, there exists a nonzero ideal $J \leq I$, and $a_jJ = 0$ for each $j = 1,\ldots,n$. Since J is an ideal, $AJ = 0$, and hence A is not faithful. □

COROLLARY 8.3. *Let R be a right bounded prime right Goldie ring, and let A be a finitely generated uniform right R–module. If A is faithful, then A is torsionfree.*

Proof. Proposition 6.11 and Lemma 8.2. □

PROPOSITION 8.4. *Let R be a right FBN ring. If P is a right primitive ideal of R (in particular, if P is a maximal ideal), then R/P is a simple artinian ring.*

Proof. Without loss of generality, $P = 0$. Then there exists a faithful simple right R–module, which we may write as R/M for some maximal right ideal M of R. As R/M is faithful, M does not contain a nonzero ideal of R, and hence $M \not\leq_e R_R$. Thus there exists a nonzero right ideal $J \leq R$ for which $M \cap J = 0$. Then $J \cong R/M$, and so $J \leq \text{soc}(R_R)$. As $\text{soc}(R_R) \neq 0$, Corollary 6.16 shows that R is simple artinian. □

It follows from Proposition 8.4 that every right primitive ideal in a right FBN ring is maximal.

COROLLARY 8.5. *Let R be a right FBN ring, and let A be a simple right R–module. Then $\text{ann}(A)$ is a maximal ideal of R and $R/\text{ann}(A)$ is a simple artinian ring. Hence, A is isomorphic to a right ideal of $R/\text{ann}(A)$, and $R/\text{ann}(A)$ is isomorphic to a finite direct sum of copies of A.* □

♦ EMBEDDING MODULES INTO FACTOR RINGS

Corollary 8.5 shows that any nonzero artinian right module over a right FBN ring R has nonzero submodules isomorphic to right ideals in prime factor rings of R. We now prove this for all right R–modules.

THEOREM 8.6. *Let R be a right FBN ring, and let A be a finitely generated right R–module. Then there exist submodules*

$$A_0 = 0 < A_1 < A_2 < \ldots < A_n = A$$

such that for each $i = 1,\ldots,n$, *the ideal* $P_i = ann(A_i/A_{i-1})$ *is a prime ideal of R and* A_i/A_{i-1} *is isomorphic to a uniform right ideal of* R/P_i.

Proof. Since A is noetherian, it suffices to show that if A is nonzero, then A has a nonzero submodule B such that ann(B) is a prime ideal of R and B is isomorphic to a uniform right ideal of R/ann(B).

By Corollary 4.15, A has a uniform submodule C. Let P be the assassinator of C, and set $D = ann_C(P)$. Then P is a prime ideal of R, while D is a fully faithful right (R/P)–module and $P = ann_R(D)$. Since D is finitely generated and uniform, Corollary 8.3 shows that D is a torsionfree right (R/P)–module. By Lemma 6.17, D has a submodule B isomorphic to a uniform right ideal of R/P. As D is fully faithful, $ann_R(B) = P$. □

Theorem 8.6 should be compared with Propositions 2.13 and 7.5. In particular, Theorem 8.6 does for right modules over a right FBN ring what Proposition 7.5 does only for bimodules over arbitrary noetherian rings. Of course boundedness is necessary in Theorem 8.6, as Lemma 8.2 shows.

EXERCISE 8D. Let R be a right FBN ring, A a finitely generated right R–module, and $A_0 = 0 < A_1 < \ldots < A_n = A$ an affiliated series for A, with corresponding affiliated primes $P_1,\ldots,P_n$. Show that each A_i/A_{i-1} is a torsionfree (R/P_i)–module. □

Returning to Corollary 8.5, we observe that if R is a right FBN ring, M a maximal ideal of R, and A a finitely generated faithful right (R/M)–module, then R/M is a simple artinian ring and it embeds in some finite direct sum of copies of A. We shall prove a corresponding statement for faithful modules over arbitrary factor rings of R. As it is much easier for prime factor rings, we do that case first by way of illustration.

PROPOSITION 8.7. *Let R be a prime right Goldie ring, and let A be a finitely generated faithful right R–module. If R is right bounded, then* R_R *embeds in some finite direct sum of copies of A.*

Proof. By Lemma 8.2, A is not torsion. Then by Lemma 6.17, A has a uniform submodule U which is isomorphic to a right ideal of R. Now Corollary 6.25 shows that

R_R embeds in U^n for some $n \in \mathbb{N}$, whence R_R embeds in A^n. □

LEMMA 8.8. *Let R be a prime right noetherian ring, and let A be a finitely generated right R–module. If $A_0 > A_1 > ... > A_n$ is a chain of submodules of A such that each A_{i-1}/A_i is nonzero and torsionfree, then $n \leq rank(A)$.*

Proof. We proceed by induction on n, the case $n = 0$ being trivial. Now let $n > 0$ and assume the lemma holds for shorter chains. Then $n - 1 \leq \text{rank}(A_1)$.

Since A_0/A_1 is not a torsion R–module, $A_1 \not\leq_e A_0$, and so there is a nonzero submodule $B \leq A_0$ such that $B \cap A_1 = 0$. Thus

$$\text{rank}(A) \geq \text{rank}(B \oplus A_1) = \text{rank}(B) + \text{rank}(A_1) > \text{rank}(A_1) \geq n - 1,$$

whence $\text{rank}(A) \geq n$. □

THEOREM 8.9. [Cauchon] *Let R be a right FBN ring, and let A be a finitely generated right R–module. Then there exists a finite subset $X \subseteq A$ such that $ann_R(X) = ann_R(A)$. Consequently, $R/ann_R(A)$ is isomorphic to a submodule of some finite direct sum of copies of A.*

Proof. If $\text{ann}(A) = \text{ann}\{x_1,...,x_n\}$ for some $x_1,...,x_n \in A$, then the map $r \mapsto (x_1,...,x_n)r$ defines a homomorphism $R_R \to A^n$ with kernel $\text{ann}(A)$. Thus it suffices to prove the annihilator statement.

If the theorem fails, we work with a minimal criminal A/B, i.e., we choose a submodule B of A maximal with respect to the property that the theorem fails for A/B. Thus, after replacing A by A/B, we may assume that while $\text{ann}(A)$ does not equal the annihilator of any finite subset of A, the annihilator of any proper factor of A is equal to the annihilator of a finite subset. Also, after replacing R by $R/\text{ann}(A)$, we may assume that A is faithful.

By Theorem 8.6, A contains a uniform submodule U which is isomorphic to a right ideal of R/P for some prime ideal P in R. Because A is a minimal criminal, there exist cosets $x_1 + U,...,x_k + U$ in A/U such that

$$\text{ann}\{x_1 + U,...,x_n + U\} = \text{ann}(A/U).$$

Hence, if $I = \text{ann}\{x_1,...,x_k\}$, then $AI \leq U$, and consequently $AIP = 0$. Then $IP = 0$ (because A is faithful), and so I is a (finitely generated) right (R/P)–module.

Since the theorem fails for A, we have $I > \text{ann}(A)$, and so there exists $x_{k+1} \in A$ such that $x_{k+1}I \neq 0$. Set $I_1 = \text{ann}\{x_1,...,x_{k+1}\} = I \cap \text{ann}(x_{k+1})$, and note that $I/I_1 \cong x_{k+1}I \leq AI \leq U$. Thus I/I_1 is isomorphic to a nonzero right ideal of R/P, whence I/I_1 is torsionfree.

Similarly, there exists $x_{k+2} \in A$ such that if $I_2 = I_1 \cap \text{ann}(x_{k+2})$ then I_1/I_2 is nonzero and torsionfree. Continuing in this manner, we obtain an infinite descending chain $I > I_1 > I_2 > \ldots$ of right ideals such that each I_{j-1}/I_j is torsionfree. However, this contradicts Lemma 8.8.

Therefore the theorem holds. □

EXERCISE 8E. If A is a finitely generated right module over a right FBN ring R, show that R satisfies the DCC on right annihilators of subsets of A. □

EXERCISE 8F. If R is a right FBN ring, show that all factor rings of R are right bounded. □

EXERCISE 8G. Let R be a right FBN ring, P a prime ideal of R, and A a finitely generated faithful right R–module. Show that A has a subfactor isomorphic to a uniform right ideal of R/P. □

COROLLARY 8.10. *Let R be a right FBN ring, and let A be a finitely generated right R–module. If A is artinian, then $R/\text{ann}(A)$ is right artinian.* □

EXERCISE 8H. Show that a right noetherian ring R is right fully bounded if and only if every finitely generated right R–module A has a finite subset X such that $\text{ann}(X) = \text{ann}(A)$. [This condition is sometimes called *Condition (H)* in the literature.] □

♦ ARTINIAN MODULES

THEOREM 8.11. [Jategaonkar, Schelter] *Let R be a left noetherian right FBN ring, and let E be a finitely generated right R–module. If E has an essential artinian submodule, then E is artinian.*

Proof. Without loss of generality, we may assume that E is faithful. Then by Theorem 8.9 R_R embeds in E^n for some $n \in \mathbb{N}$. As E has an essential artinian submodule, so do E^n and R_R. As a result, every nonzero right ideal of R has a simple submodule, whence $\text{soc}(R_R)$ is an essential right ideal of R. By Theorem 7.14, R is artinian, and therefore E is artinian. □

Without the fully bounded hypothesis, Theorem 8.11 holds for right ideals of R (Exercise 7F), but not for arbitrary modules, as the following example shows.

EXERCISE 8I. Let $k[x]$ be a polynomial ring over a field k of characteristic zero, let $\delta = x\frac{d}{dx}$, and let $S = k[x][\theta;\delta]$. Set $A = \theta S/\theta(x-1)S$ and $E = S/\theta(x-1)S$, and observe from Exercise 2Q(b) that A is a simple right S–module. Show that $A \leq_e E$ but that E is not artinian. [Hints: (a) If $0 \neq f \in k[x]$ and $\deg(f) = n$, show that

$$f\theta(\theta+1)\cdots(\theta+n-1) \in kx^n + \theta S \qquad \text{and} \qquad f\theta(\theta+1)\cdots(\theta+n) \in \theta S.$$

(b) Observe that $S/\theta S \cong k[x]$ as $k[x]$–modules, and then show that $S/\theta S$ is not an artinian S–module.] □

EXERCISE 8J. Show that the ring $R = \begin{pmatrix} \mathbb{Z} & \mathbb{Q} \\ 0 & \mathbb{Q} \end{pmatrix}$ is right FBN, and that R has a right ideal which has an essential simple submodule but is not artinian. □

EXERCISE 8K. If R is a left noetherian right FBN ring and A is an artinian right R–module, show that $E(A) = \bigcup_{n=1}^{\infty} \operatorname{soc}^n(E(A))$. □

A special case of the Krull Intersection Theorem says that in any commutative noetherian ring R, the intersection of the powers of the Jacobson radical is zero, that is, $\bigcap_{n=1}^{\infty} J(R)^n = 0$. That this should hold for any noetherian ring R is known as *Jacobson's Conjecture*, and we now verify it for FBN rings. Jacobson's Conjecture fails for one–sided noetherian rings, as Exercise 8L shows.

THEOREM 8.12. [Cauchon, Jategaonkar, Schelter] *If R is a left noetherian right FBN ring, then*

$$\bigcap_{n=1}^{\infty} J(R)^n = 0.$$

Proof. Set $J = \bigcap_{n=1}^{\infty} J(R)^n$. If $J \neq 0$, then J_R has a maximal proper submodule, and so there is a simple right R–module A which is an epimorphic image of J_R. Since an epimorphism of J_R onto A extends to a homomorphism $R_R \to E(A)$, there exists $x \in E(A)$ such that $xJ = A$.

Note that $xR \cap A \neq 0$, whence $A \leq xR$. Thus xR has an essential simple submodule, and so by Theorem 8.11 xR is artinian. Consequently, xR has finite length, whence $(xR)J(R)^n = 0$ for some $n \in \mathbb{N}$ (because $J(R)$ annihilates each composition factor of xR). But then $xJ = 0$, contradicting the fact that $xJ = A$.

Therefore $J = 0$. □

EXERCISE 8L. Let $R = \begin{pmatrix} S & \mathbb{Q} \\ 0 & \mathbb{Q} \end{pmatrix}$ where S is the subring of $\mathbb{Q}$ consisting of all rational numbers with odd denominators. Show that R is right noetherian. Show that $J(R) = \begin{pmatrix} 2S & \mathbb{Q} \\ 0 & 0 \end{pmatrix}$ and that $\bigcap_{n=1}^{\infty} J(R)^n = \begin{pmatrix} 0 & \mathbb{Q} \\ 0 & 0 \end{pmatrix}$. □

♦ UNIFORM INJECTIVE MODULES

Consider trying to classify the uniform injective right modules over a right noetherian ring R by analogy with the commutative case (Proposition 4.24). Given a prime ideal P in R, the module $(R/P)_R$ need not be uniform (unless R/P is a domain), and so $E((R/P)_R)$ need not be uniform. To obtain a uniform injective module, we take a uniform right ideal $U \le R/P$ and use $E(U_R)$, which by Lemma 4.23 is a uniform injective with assassinator P. In case R is right fully bounded, a prime ideal of R cannot appear as the assassinator of two non–isomorphic uniform injective right R–modules, as follows.

PROPOSITION 8.13. *Let R be a right FBN ring. If E is a uniform injective right R–module and P is its assassinator, then $E \cong E(U_R)$ for every uniform right ideal U of R/P. Consequently, every uniform injective right R–module with assassinator P is isomorphic to E.*

Proof. Choose a nonzero finitely generated submodule $A \le \text{ann}_E(P)$. Then A is a faithful uniform right (R/P)–module. By Corollary 8.3, A is a torsionfree right (R/P)–module. If U is any uniform right ideal of R/P, then by Lemma 6.22 U is isomorphic to a submodule of A, whence $E(U_R)$ is isomorphic to a direct summand of E. As E is uniform, $E(U_R) \cong E$. □

Proposition 8.13 may fail if R has fewer prime ideals than uniform injective modules, as in Exercise 4U. In fact, the only right noetherian rings for which Proposition 8.13 holds are the right fully bounded ones, as follows.

THEOREM 8.14. [Gordon–Robson, Krause, Lambek–Michler] *Let R be a right noetherian ring. Then R is right fully bounded if and only if the rule*

$$E \mapsto (\text{assassinator of } E)$$

yields a bijection between the isomorphism classes of uniform injective right R–modules and the prime ideals of R.

Proof. If R is right fully bounded, Lemma 4.23 and Proposition 8.13 show that this rule yields a bijection.

If R is not right fully bounded, there is a prime ideal P of R such that R/P is not right bounded. Then R has a right ideal $I > P$ such that I/P is an essential right ideal of R/P but I/P contains no nonzero ideals of R/P. By noetherian induction, we may assume that I is maximal with respect to the property that I/P contains no nonzero ideals of R/P. Since $\mathrm{ann}_{R/P}(R/I)$ is an ideal of R/P contained in I/P, it must be zero. Thus R/I is a faithful right (R/P)–module.

We claim that R/I is actually a fully faithful (R/P)–module, and that R/I is uniform.

Any nonzero submodule of R/I has the form J/I for some right ideal J of R which properly contains I. By maximality of I, the factor J/P must contain a nonzero ideal of R/P, whence $\mathrm{ann}_R(R/J) > P$. As

$$\mathrm{ann}_R(R/J)\mathrm{ann}_R(J/I) \leq \mathrm{ann}_R(R/I) = P,$$

we obtain $\mathrm{ann}_R(J/I) = P$, and so J/I is a faithful (R/P)–module, as claimed.

If J and K are right ideals of R properly containing I, then as above, $\mathrm{ann}_R(R/J) > P$ and $\mathrm{ann}_R(R/K) > P$. Then

$$P < \mathrm{ann}_R(R/J) \cap \mathrm{ann}_R(R/K) = \mathrm{ann}_R(R/(J \cap K)),$$

whence $J \cap K > I$. Thus $(J/I) \cap (K/I) \neq 0$, proving that R/I is uniform, as claimed.

Set $E = E((R/I)_R)$, which is a uniform injective right R–module. Since R/I is a fully faithful (R/P)–module, the assassinator of E must be P.

Choose a uniform right ideal U in R/P, and set $E' = E(U_R)$. By Lemma 4.23, E' is a uniform injective right R–module with assassinator P. Since R/I is a torsion right (R/P)–module while U is a torsionfree right (R/P)–module, R/I cannot be isomorphic to a submodule of E'. Therefore $E \not\cong E'$. □

EXERCISE 8M. Let $R = A_n(k) = k[x_1,\ldots,x_n][\theta_1,\ldots,\theta_n; \partial/\partial x_1,\ldots,\partial/\partial x_n]$ for some $n \in \mathbb{N}$ and some field k of characteristic zero. Set $B_j = R/(x_1R + \ldots + x_jR)$ for each $j = 0,1,\ldots,n$. Show that $E(B_0),\ldots,E(B_n)$ are pairwise non–isomorphic uniform injective right R–modules with the same assassinator. [Hint: Show that B_j is torsionfree as a module over $k[x_{j+1},\ldots,x_n]$.] □

EXERCISE 8N. Let $R = A_1(k) = k[x][\theta; d/dx]$ for some field k of characteristic zero. For each $\alpha \in k$, set $B_\alpha = R/(x - \alpha)R$. Show that the modules $E(B_\alpha)$ are pairwise non–isomorphic uniform injective right R–modules with the same assassinator. [Hint: Show that the B_α are pairwise non–isomorphic simple R–modules.] □

♦ NOTES

Finite Annihilation for Modules over FBN Rings. Theorem 8.9 was first proved by

Cauchon for a right "T–ring", meaning a right noetherian ring over which non–isomorphic uniform injective right modules have distinct assassinators [1973, Proposition 2; 1976, Théorème II 8, p. 25]. (By Theorem 8.14 the right T–rings are exactly the right FBN rings.)

Essential Extensions of Artinian Modules over FBN Rings. Theorem 8.11 was proved first by Jategaonkar for a right and left FBN ring [1973, Theorem 8; 1974b, Corollary 3.6], and then by Schelter for a left noetherian right FBN ring in which all left primitive factor rings are artinian [1975, Theorem].

Jacobson's Conjecture for FBN Rings. This was proved by Cauchon for a right and left T–ring [1974, Théorème 5; 1976, Théorème I 2, p. 36], by Jategaonkar for a right and left FBN ring [1973, Theorem 8; 1974b, Theorem 3.7], and by Schelter for a left noetherian right FBN ring in which all left primitive factor rings are artinian [1975, Corollary].

Assassinators of Injectives over FBN Rings. Theorem 8.14 was proved independently by Gordon and Robson [1973, Theorem 8.6 and Corollary 8.11] and Krause [1972, Theorem 3.5]. That assassinators yield a bijection between uniform injectives and primes over a right FBN ring was also proved by Lambek and Michler [1973, Corollary 3.12].

9 RINGS OF FRACTIONS

One of the first constructions that an undergraduate student of algebra meets is the *quotient field* of a commutative integral domain, constructed as a set of fractions a/b subject to an obvious equivalence relation. Later on, there is a somewhat more general construction. Suppose that R is a commutative ring and X a subset of R. We want to find a "larger" ring in which the elements of X become units. First of all, since all products of elements of X would necessarily become units in our new ring, we may enlarge X and assume that X is multiplicatively closed, and that $1 \in X$. We then build a new ring RX^{-1} (often written R_X in the commutative literature, but this notation can cause confusion later) as a set of fractions r/x where $r \in R$ and $x \in X$. There must again be an equivalence relation on these fractions, and the situation is made slightly more complex by the fact that X may contain zero–divisors, in which case the map $R \to RX^{-1}$ taking r to $r/1$ is not injective. The correct equivalence relation turns out to be the following: we say that r/x and r'/x' define the same element of RX^{-1} if and only if $(rx' - r'x)y = 0$ for some $y \in X$. Some easy calculations show that we can define a ring RX^{-1} in this way. If we do, and if $\phi : R \to RX^{-1}$ is the ring homomorphism taking r to $r/1$, then we have the following conclusions:

(i) For each $x \in X$, the element $\phi(x)$ is a unit of RX^{-1};

(ii) The kernel of ϕ is $\{r \in R \mid rx = 0$ for some $x \in X\}$;

(iii) If $\psi : R \to T$ is any ring homomorphism such that $\psi(x)$ is a unit of T for each $x \in X$, then ψ factors uniquely through ϕ, that is, there is a unique ring homomorphism $\eta : RX^{-1} \to T$ such that $\psi = \eta\phi$.

Extending this simple idea to the noncommutative case has turned out not to be so simple. The general question was first considered by Ore, who investigated whether a noncommutative integral domain R can be embedded in a division ring D such that every element of D is of the form rx^{-1} for some $r,x \in R$ with $x \neq 0$. (In the terminology of Chapter 5, the question is whether R is a right order in a division ring.) This turns out to be false in general, but Ore succeeded in finding a necessary and sufficient condition for it to hold (Theorem 5.17): a division ring D as described exists if and only if any two nonzero elements of R have a nonzero common right multiple. The necessity of this common multiple condition is immediate from the special form of elements of D, for if r,s are nonzero elements of R, then D contains the element $s^{-1}r$, which must be of the form xy^{-1} for some nonzero $x,y \in R$, and the equation $s^{-1}r = xy^{-1}$ in D translates to the equation $ry = sx$ in R. This right common multiple condition is known as *Ore's condition* (although we shall use the term in a more general setting below).

We close these introductory remarks by reminding the reader that a right noetherian domain always satisfies Ore's condition (Corollary 5.16). (This fact was not noticed for many years, presumably because the importance of the noetherian condition had not become clear.) That arbitrary domains (even principal left ideal domains!) need not satisfy Ore's condition (on the right) was observed in the remarks following Corollary 5.16.

♦ NECESSARY CONDITIONS

Given any set X of elements in a ring R, one can look for a ring homomorphism $\phi : R \to S$ such that $\phi(x)$ is a unit of S for each $x \in X$, and such that any such ring homomorphism factors through ϕ. It is not hard to show that one can always do this, but we will not, because the resulting ring S is often quite unreasonable, as the following example shows.

EXERCISE 9A. Let V be an infinite–dimensional vector space over a field, let R be the ring of all linear transformations on V, and let X be the set of all surjective linear transformations in R. (We regard linear transformations as acting on the left of vectors.) First note that the elements of X are already right invertible in R. If $\phi : R \to S$ is a ring homomorphism such that $\phi(x)$ is a unit of S for all $x \in X$, show that $S = 0$. □

We are thus led to look for conditions on X which will lead to a useful construction of a ring in which the elements of X can be inverted. We start by specifying that we would like our end product to resemble the rings of fractions constructed in the commutative case.

DEFINITION. A *multiplicative set* in a ring R is any subset $X \subseteq R$ such that $1 \in X$ and X is closed under multiplication. A *right ring of fractions* (or *right Ore quotient ring*, or *right Ore localization*) *for* R *with respect to* X is a ring homomorphism $\phi : R \to S$ such that

(a) $\phi(x)$ is a unit of S for all $x \in X$.

(b) Each element of S has the form $\phi(r)\phi(x)^{-1}$ for some $r \in R$, $x \in X$.

(c) $\ker(\phi) = \{r \in R \mid rx = 0 \text{ for some } x \in X\}$.

By abuse of notation, we usually refer to S as the right ring of fractions, and later we will write elements of S in the form rx^{-1} for $r \in R$, $x \in X$. A *left ring of fractions for* R *with respect to* X is defined symmetrically; in particular, in a left ring of fractions the denominators are all written on the left.

For instance, if R is a semiprime right Goldie ring and X is the set of regular elements of R, then the right Goldie quotient ring of R is a right ring of fractions for R with respect to X.

Just as we found when studying Ore domains, a right ring of fractions cannot exist without imposing some restrictions on the multiplicative set.

LEMMA 9.1. *Let X be a multiplicative set in a ring R, and assume that there exists a right ring of fractions $\phi : R \to S$ for R with respect to X.*

(a) $rX \cap xR$ is nonempty for all $r \in R$ and $x \in X$.

(b) If $r \in R$ and $x \in X$ such that $xr = 0$, then there exists $x' \in X$ such that $rx' = 0$.

Proof. (a) The element $\phi(x)^{-1}\phi(r)$ in S must have the form $\phi(s)\phi(y)^{-1}$ for some $s \in R$, $y \in X$. Then $\phi(r)\phi(y) = \phi(x)\phi(s)$ and so $\phi(ry - xs) = 0$, whence $(ry - xs)z = 0$ for some $z \in X$. Thus $ryz = xsz$ with $yz \in X$.

(b) Since $\phi(x)\phi(r) = 0$ and $\phi(x)$ is invertible, $\phi(r) = 0$. □

DEFINITION. Let X be a multiplicative set in a ring R. Then X satisfies the *right Ore condition* if and only if X satisfies condition (a) of Lemma 9.1, while X is *right reversible* if and only if X satisfies condition (b). A *right Ore set* is any multiplicative set satisfying the right Ore condition, while a *right denominator set* is any right reversible right Ore set. After some study of Ore sets, we shall prove that right rings of fractions exist precisely for right denominator sets (Theorem 9.7).

EXERCISE 9B. If R is a semiprime right Goldie ring, show that the set of regular elements of R is a right denominator set, without using the right Goldie quotient ring. □

EXERCISE 9C. Show that the multiplicative set X in Exercise 9A is right Ore but not right reversible. □

EXERCISE 9D. In a ring with no nonzero nilpotent elements, show that every right Ore set is right reversible. □

We show later (Proposition 9.9) that in a right noetherian ring, every right Ore set is right reversible.

LEMMA 9.2. *Let X be a right Ore set in a ring R.*

(a) Given any $x_1,\dots,x_n \in X$, the set $x_1R \cap \dots \cap x_nR \cap X$ is nonempty.

Now suppose that $\phi : R \to T$ is a ring homomorphism such that $\phi(x)$ is a unit of T for all $x \in X$.

(b) Given any $r_1,\ldots,r_n \in R$ *and* $x_1,\ldots,x_n \in X$, *there exist* $s_1,\ldots,s_n \in R$ *and* $y \in X$ *such that* $\phi(r_i)\phi(x_i)^{-1} = \phi(s_i)\phi(y)^{-1}$ *for all* i. *(I.e., the "fractions"* $\phi(r_i)\phi(x_i)^{-1}$ *have a "common denominator"* $\phi(y)$.*)*

(c) The set $S = \{\phi(r)\phi(x)^{-1} \mid r \in R$ *and* $x \in X\}$ *is a subring of* T.

Proof. (a) By induction, it is enough to establish the case $n = 2$. In this case, the right Ore condition gives $x_1y = x_2s$ for some $y \in X$ and $s \in R$, and $x_1y \in X$.

(b) By (a), there exist $y \in X$ and $a_1,\ldots,a_n \in R$ such that $y = x_ia_i$ for each i. Note from the equations $\phi(y) = \phi(x_i)\phi(a_i)$ that each $\phi(a_i)$ is a unit in T. Then if $s_i = r_ia_i$ for each i, we obtain

$$\phi(s_i)\phi(y)^{-1} = \phi(r_i)\phi(a_i)\phi(a_i)^{-1}\phi(x_i)^{-1} = \phi(r_i)\phi(x_i)^{-1}.$$

(c) Obviously $1 = \phi(1)\phi(1)^{-1} \in S$, and it is clear from (b) that S is closed under addition and subtraction. Given $r_1,r_2 \in R$ and $x_1,x_2 \in X$, we may use the right Ore condition to obtain $r_2y = x_1s$ for some $y \in X$ and $s \in R$. Then $\phi(r_2)\phi(y) = \phi(x_1)\phi(s)$ and so $\phi(x_1)^{-1}\phi(r_2) = \phi(s)\phi(y)^{-1}$. Hence,

$$[\phi(r_1)\phi(x_1)^{-1}][\phi(r_2)\phi(x_2)^{-1}] = \phi(r_1)\phi(s)\phi(y)^{-1}\phi(x_2)^{-1} = \phi(r_1s)\phi(x_2y)^{-1},$$

which lies in S. Therefore S is closed under multiplication. □

LEMMA 9.3. *Let* X *be a right Ore set in a ring* R. *If* A *is any right* R*–module, the set*

$$B = \{a \in A \mid ax = 0 \text{ for some } x \in X\}$$

is a submodule of A.

Proof. Given $a_1,a_2 \in B$, there exist $x_1,x_2 \in X$ such that each $a_ix_i = 0$. By Lemma 9.2, there is some $y \in x_1R \cap x_2R \cap X$, and $(a_1 \pm a_2)y = 0$, whence $a_1 \pm a_2 \in B$. Given $r \in R$, the right Ore condition yields $ry = x_1s$ for some $y \in X$ and $s \in R$. Then $a_1ry = a_1x_1s = 0$ and hence $a_1r \in B$. □

DEFINITION. Let X be a right Ore set in a ring R. For any right R–module A, the submodule

$$t_X(A) = \{a \in A \mid ax = 0 \text{ for some } x \in X\}$$

is called the *X–torsion submodule* of A. The module A is *X–torsion* if and only if $t_X(A) = A$ and *X–torsionfree* if and only if $t_X(A) = 0$. Note that $t_X(R)$ is an ideal of R, and that if there exists a right ring of fractions $\phi : R \to S$ for R with respect to X, then $\ker(\phi) = t_X(R)$.

For instance, if R is a semiprime right Goldie ring and X is the set of regular elements of R, then $t_X(A)$ is just the singular submodule of A (Proposition 6.9).

EXERCISE 9E. Let X be a multiplicative set in a ring R. If, for all right R–modules A, the set $\{a \in A \mid ax = 0 \text{ for some } x \in X\}$ is a submodule of A, show that X is a right Ore set. □

PROPOSITION 9.4. *Let X be a right denominator set in a ring R, and suppose that there exists a right ring of fractions $\phi : R \to S$ for R with respect to X. If $\psi : R \to T$ is any ring homomorphism such that $\psi(x)$ is a unit in T for each $x \in X$, then there exists a unique ring homomorphism $\eta : S \to T$ such that $\psi = \eta\phi$.*

Proof. Certainly η must be unique if it exists, since it must satisfy

$$\eta(\phi(r)\phi(x)^{-1}) = \psi(r)\psi(x)^{-1}$$

for all $r \in R$ and $x \in X$.

If $r \in \ker(\phi)$, then $rx = 0$ for some $x \in X$, and so $\psi(r)\psi(x) = 0$. As $\psi(x)$ is invertible in T, it follows that $\psi(r) = 0$. Thus $\ker(\phi) \subseteq \ker(\psi)$, whence there exists a ring homomorphism $\eta_0 : \phi(R) \to T$ such that $\psi = \eta_0\phi$. We will first extend η_0 to S as a module homomorphism and then show that the extension is a ring homomorphism.

We may view S and T as right R–modules via the maps ϕ and ψ, that is, the right R–module multiplication in S is given by the rule $sr = s\phi(r)$, and similarly in T. Since the elements of $\psi(X)$ are units of T, we see that T_R is X–torsionfree. Then $T \cap t_X(E(T_R)) = 0$, whence $t_X(E(T_R)) = 0$. Thus $E(T_R)$ is X–torsionfree.

Observe that η_0 is a right R–module homomorphism, since for all $r,r' \in R$ we have

$$\eta_0(\phi(r)r') = \eta(\phi(r)\phi(r')) = \eta_0\phi(rr') = \psi(rr') = \psi(r)\psi(r') = \psi(r)r' = (\eta_0\phi(r))r'.$$

Hence, η_0 extends to a right R–module homomorphism $\eta : S \to E(T_R)$, and of course $\eta\phi = \eta_0\phi = \psi$. Given any $s \in S$, we may write $s = \phi(r)\phi(x)^{-1}$ for some $r \in R$, $x \in X$. Then

$$\eta(s)x = \eta(sx) = \eta\phi(r) = \psi(r) = \psi(r)\psi(x)^{-1}x.$$

As $E(T_R)$ is X–torsionfree, $\eta(s) = \psi(r)\psi(x)^{-1} \in T$.

Thus η actually maps S into T. It is already additive, and $\eta(1) = \eta\phi(1) = \psi(1) = 1$. Given $s_1,s_2 \in S$, write $s_2 = \phi(r)\phi(x)^{-1}$ for some $r \in R$, $x \in X$. Then

$$\eta(s_1s_2)x = \eta(s_1s_2x) = \eta(s_1\phi(r)) = \eta(s_1r) = \eta(s_1)r = \eta(s_1)\psi(r)$$
$$= \eta(s_1)\eta\phi(r) = \eta(s_1)\eta(s_2x) = \eta(s_1)\eta(s_2)x,$$

and consequently $\eta(s_1s_2) = \eta(s_1)\eta(s_2)$. Therefore η is a ring homomorphism. □

COROLLARY 9.5. *Let X be a right denominator set in a ring R, and suppose that $\phi_1 : R \to S_1$ and $\phi_2 : R \to S_2$ are right rings of fractions for R with respect to X. Then there is a (unique) ring isomorphism $\eta : S_1 \to S_2$ such that $\eta\phi_1 = \phi_2$.*

Proof. By Proposition 9.4, there exists a unique ring homomorphism $\eta : S_1 \to S_2$

such that $\eta\phi_1 = \phi_2$, as well as a unique ring homomorphism $\eta' : S_2 \to S_1$ such that $\eta'\phi_2 = \phi_1$. Then $\eta'\eta\phi_1 = \phi_1$, and so the uniqueness assertion in Proposition 9.4 implies that $\eta'\eta$ is the identity map on S_1. Similarly, $\eta\eta'$ is the identity map on S_2. □

Note that we have not yet established that rings of fractions exist – we do this in the next section.

EXERCISE 9F. (a) Let X be a nonempty subset of a ring R, and let Y be the multiplicative set generated by X (i.e., the set of all products of elements of X). If X satisfies condition (a) (condition (b)) of Lemma 9.1, prove that Y is right Ore (right reversible).

(b) Show that the multiplicative set generated by a nonempty family of right Ore (right reversible) sets is right Ore (right reversible).

(c) Conclude for a multiplicative set $Z \subseteq R$ that there exists a right denominator set contained in Z which contains all right denominator sets contained in Z. □

♦ EXISTENCE

We return briefly to our discussion of the construction of RX^{-1} for a commutative ring R and a multiplicative set X, in order to indicate a different procedure for the construction. Clearly the ideal $t_X(R)$ is the kernel of the map from R to RX^{-1}, and so RX^{-1} contains (a copy of) the ring $T = R/t_X(R)$. From the form of the fractions in RX^{-1}, for each element $s \in RX^{-1}$ there is an $x \in X$ with $sx \in T$, and so RX^{-1}/T is an X–torsion module. Hence, in retrospect, we could have obtained RX^{-1} by embedding T in its injective hull E(T) (either the R–module injective hull or the T–module injective hull would do) and taking

$$RX^{-1} = \{e \in E(T) \mid ex \in T \text{ for some } x \in X\},$$

or, in other terms, $RX^{-1}/T = t_X(E(T)/T)$. It should be clear to the reader that this is a correct description of RX^{-1}, at least as an R–module, although we have not provided all of the details. A difficulty with this approach is that it is not immediately apparent how to define multiplication of elements in this description of RX^{-1}. However, we may proceed as in our earlier approach to Goldie's Theorems (recall in particular Proposition 4.25) by showing that the module RX^{-1} is group–theoretically isomorphic to its own endomorphism ring, from whence we import the ring structure to RX^{-1}. This turns out to be a profitable approach in the noncommutative case, where it saves a mass of ugly calculations with an obstreporous equivalence relation (see Exercise 9H).

In order to simplify the notation, we make the following easy observation, which enables us to reduce to the case that R_R is X–torsionfree.

EXERCISE 9G. Let X be a right denominator set in a ring R, set $T = R/t_X(R)$, let $\pi : R \to T$ be the quotient map, and set $Y = \pi(X)$.

(a) Show that Y is a right denominator set in T, and that $t_Y(T) = 0$.

(b) If there exists a right ring of fractions $\psi : T \to S$ for T with respect to Y, show that $\psi\pi$ is a right ring of fractions for R with respect to X.

(c) If there exists a right ring of fractions $\phi : R \to S$ for R with respect to X, show that $\phi = \psi\pi$ for a ring homomorphism $\psi : T \to S$, and that ψ is a right ring of fractions for T with respect to Y. □

LEMMA 9.6. *Let X be a right Ore set in a ring R. Assume that $t_X(R) = 0$, and define a submodule A of $E(R_R)$ by $A/R = t_X(E(R_R)/R)$.*

(a) There is an abelian group isomorphism $\psi : End_R(A) \to A$ given by the rule $\psi(f) = f(1)$.

(b) There is a unique ring structure on A compatible with its right R–module structure, and when this ring structure is used R is a subring of A.

Proof. Note that $R \cap t_X(A) = t_X(R_R) = 0$ and so $t_X(A) = 0$. Thus A is X–torsionfree. We note also, from the definition of A, that $E(R_R)/A$ is X–torsionfree.

(a) Obviously there is a group homomorphism ψ as defined. If $f \in \ker(\psi)$, then from $f(1) = 0$ we have $f(R) = 0$, and there is an induced homomorphism $f^\# : A/R \to A$. Since A/R is X–torsion while A is X–torsionfree, $f^\# = 0$, whence $f = 0$. Thus ψ is injective.

Given $a \in A$, there is a homomorphism from R_R to A sending 1 to a, and this extends to some $f \in End_R(E(R_R))$ by injectivity. Since f sends R into A it induces a homomorphism $f^\# : A/R \to E(R_R)/A$. As $E(R_R)/A$ is X–torsionfree, $f^\# = 0$, and so $f(A) \subseteq A$. Hence, $f|_A$ is an endomorphism of A such that $\psi(f|_A) = f(1) = a$. Therefore ψ is surjective.

(b) In view of (a), there is a ring structure on A utilizing the given addition together with the multiplication rule

$$a * b = \psi(\psi^{-1}(a)\psi^{-1}(b)) = [\psi^{-1}(a)](b).$$

If $a \in A$ and $r \in R$, we have

$$a * r = [\psi^{-1}(a)](r) = \psi^{-1}(a)(1 \cdot r) = (\psi^{-1}(a)(1))r = ar.$$

Thus this ring structure on A is compatible with its right R–module structure.

The uniqueness of this ring structure follows from the X–torsionfreeness of A as in Proposition 4.26. It is clear that R is a subring of $(A,*)$. □

THEOREM 9.7. *Let X be a multiplicative set in a ring R. Then there exists a right ring of fractions for R with respect to X if and only if X is a right denominator set.*

Proof. The necessity is given by Lemma 9.1. Conversely, assume that X is a right denominator set. In view of Exercise 9G, we may assume that $t_X(R) = 0$.

Let $A/R = t_X(E(R_R)/R)$, and equip A with the ring structure given by Lemma 9.6. Then R is a subring of A.

We want to show that each $x \in X$ is a unit in A. First observe that $\text{r.ann}_R(x) = 0$: if $a \in R$ such that $xa = 0$, then $ay = 0$ for some $y \in X$ (right reversibility), and so $a = 0$. Consequently, there is a right R–module homomorphism from xR to R sending x to 1, and this extends to an endomorphism f of $E(R_R)$ sending x to 1. Since $x = 1x$, we infer that $f(x) = f(1)x$ and so $1 = ux$ for some $u \in E(R_R)$. Then $ux \in R$ and so $u \in A$ (by definition of A). Observe that $(xu - 1)x = xux - x = 0$, and since A is X–torsionfree we obtain $xu = 1$. Thus x is invertible in A.

For any $a \in A$ there exists $x \in X$ such that $ax \in R$, and then $a = (ax)x^{-1}$. Therefore the inclusion map $R \to A$ is a right ring of fractions for R with respect to X. □

EXERCISE 9H. Let X be a right denominator set in a ring R. Define a relation $\sim$ on $R \times X$ where $(r,x) \sim (r',x')$ if and only if there exist $s \in R$ and $y \in X$ such that $ry = r's$ and $xy = x's$. Show that $\sim$ is an equivalence relation on $R \times X$. Define addition and multiplication rules for the equivalence classes of $\sim$, and verify some of the ring axioms for these operations. □

DEFINITION. Given a right denominator set X in a ring R, we now know that there exists a right ring of fractions $\phi : R \to S$ for R with respect to X. Because of the uniqueness given by Corollary 9.5, we shall denote S by RX^{-1}, and we shall refer to ϕ as *the natural map* from R to RX^{-1}. Whenever $t_X(R) = 0$, so that ϕ is injective, we shall assume that R has been identified with $\phi(R)$ as a subring of RX^{-1}. In the case of a left ring of fractions for R with respect to a left denominator set Y, we denote the left ring of fractions by $Y^{-1}R$.

Given a right ring of fractions $\phi : R \to RX^{-1}$, it is cumbersome to use the precise notation $\phi(r)\phi(x)^{-1}$ for elements of RX^{-1}. Henceforth, we shall write elements of RX^{-1} in the form rx^{-1}, even when $t_X(R) \neq 0$, unless there is more than one ring of fractions under discussion. To avoid confusion in case $t_X(R) \neq 0$, the image in RX^{-1} of an element $r \in R$ should be denoted $r1^{-1}$, which emphasizes that it is being viewed as a fraction. There should be no confusion if for an element $x \in X$ the inverse of $x1^{-1}$ in RX^{-1} is denoted just x^{-1}, rather than $1x^{-1}$. Analogous notation ($x^{-1}r$, $1^{-1}r$, etc.) will of course be used in left rings of fractions.

PROPOSITION 9.8. *If X is a right and left denominator set in a ring R, then any right (left) ring of fractions for R with respect to X is also a left (right) ring of fractions.*

Proof. Use the universal mapping property given by Proposition 9.4, as in Corollary 9.5. □

Because of Proposition 9.8, we may refer to *the* ring of fractions of a ring R with respect to a right and left denominator set. Of course not all right denominator sets are also left denominator sets, for instance the set of nonzero elements in a right Ore domain which is not left Ore.

EXERCISE 9I. If α is an automorphism of a ring R, show that in the skew polynomial ring $R[\theta;\alpha]$ the set $X = \{1,\theta,\theta^2,...\}$ is a right and left denominator set, and that $R[\theta;\alpha]X^{-1} \cong R[\theta,\theta^{-1};\alpha]$. □

We close this section by showing that in noetherian rings only the Ore condition needs to be checked, since then reversibility comes for free.

PROPOSITION 9.9. *Let X be a right Ore set in a ring R. If R has ACC on right annihilators of elements, then X is right reversible.*

Proof. Consider $r \in R$ and $x \in X$ such that $xr = 0$. Using the ACC on right annihilators of elements, there exists a positive integer n such that $r.ann(x^n) = r.ann(x^{n+1})$. From the right Ore condition, there exist $s \in R$ and $y \in X$ such that $ry = x^n s$. As $x^{n+1}s = xry = 0$, we have

$$s \in r.ann(x^{n+1}) = r.ann(x^n),$$

and therefore $ry = x^n s = 0$. □

EXERCISE 9J. Let X be a right Ore set in a right noetherian ring R, and let $I \geq J$ be ideals of R. If $(I/J)_R$ is X–torsionfree, show that ${}_R(I/J)$ is X–torsionfree in the sense that no nonzero element of I/J can be annihilated on the left by any element of X. □

♦ MODULES OF FRACTIONS

Having constructed the right ring of fractions RX^{-1} given a right denominator set X in a ring R, we turn to the construction of an analogous module of fractions AX^{-1} for each right R–module A. Of course the elements of AX^{-1} should resemble fractions with numerators from A and denominators from X, the kernel of the map from A to AX^{-1} should be $t_X(A)$, and this map $A \to AX^{-1}$ should have a suitable universal property with respect to homomorphisms into RX^{-1}–modules. As with rings of fractions, we begin by specifying the desired properties of modules of fractions and proving uniqueness, then existence.

DEFINITION. Let X be a right denominator set in a ring R, and let A be a right R–module. A *module of fractions for* A *with respect to* X consists of a right RX^{-1}–module B together with an R–module homomorphism $f : A \to B$ such that

(a) Each element of B has the form $f(a)x^{-1}$ for some $a \in A$, $x \in X$.

(b) $\ker(f) = t_X(A)$.

With the usual abuse of notation, we often refer to B as the module of fractions.

PROPOSITION 9.10. *Let X be a right denominator set in a ring R, let A be a right R–module, and suppose that there exists a module of fractions $f : A \to B$ for A with respect to X. If C is a right RX^{-1}–module and $g : A \to C$ is an R–module homomorphism, there exists a unique RX^{-1}–module homomorphism $h : B \to C$ such that $g = hf$.*

Proof. Observe that C is X–torsionfree, and consequently $\ker(g) \geq t_X(A) = \ker(f)$. Hence, there is a unique R–module homomorphism $h_0 : f(A) \to C$ such that $h_0f = g$. Then h_0 extends to an R–module homomorphism $h : B \to E(C_R)$, and of course $hf = h_0f = g$.

Any $b \in B$ has the form $f(a)x^{-1}$ for some $a \in A$, $x \in X$. Then $bx = f(a)$, whence $h(b)x = hf(a) = g(a)$. As $g(a)$ lies in the RX^{-1}–module C, there exists $c \in C$ such that $g(a) = cx$, and since $E(C_R)$ is X–torsionfree (because C is), we conclude that $h(b) = c$. Therefore h actually maps B to C.

To show that h is an RX^{-1}–module homomorphism, it suffices to show that

$$h(bx^{-1}) = h(b)x^{-1}$$

for all $b \in B$ and $x \in X$. However, this is immediate from the observation that $h(bx^{-1})x = h(b)$. Finally, the uniqueness of h follows from the fact that it must satisfy $h(f(a)x^{-1}) = g(a)x^{-1}$ for all $a \in A$, $x \in X$. □

COROLLARY 9.11. *Let X be a right denominator set in a ring R, let A be a right R–module, and suppose that $f_1 : A \to B_1$ and $f_2 : A \to B_2$ are modules of fractions for A with respect to X. Then there is a (unique) RX^{-1}–module isomorphism $h : B_1 \to B_2$ such that $hf_1 = f_2$.* □

We now turn to the construction of modules of fractions, for which we use an approach similar to the construction of rings of fractions in the previous section.

DEFINITION. Let X be a multiplicative set in a ring R. We say that a right R–module A is *X–divisible* if and only if $Ax = A$ for all $x \in X$.

PROPOSITION 9.12. *Let X be a right denominator set in a ring R.*

(a) Every right or left RX^{-1}–module is X–torsionfree and X–divisible as an R–module.

(b) If A is an X–torsionfree X–divisible right R–module, there exists a unique right RX^{-1}–module structure on A compatible with its right R–module structure.

Proof. (a) This is clear from the fact that $X1^{-1}$ is contained in the set of units of RX^{-1}.

(b) Set $T = \mathrm{End}_{\mathbb{Z}}(A)^{op}$, and let $\psi : R \to T$ be the ring homomorphism given by the right R–module structure on A, that is, $\psi(r)(a) = ar$ for all $r \in R$, $a \in A$. A right RX^{-1}–module structure on A is just a ring homomorphism $\eta : RX^{-1} \to T$, and such a structure is compatible with the right R–module structure on A if and only if $\eta\phi = \psi$, where ϕ is the natural map from R to RX^{-1}. Thus we just need to show that there is a unique ring homomorphism $\eta : RX^{-1} \to T$ such that $\eta\phi = \psi$. However, since A is X–torsionfree and X–divisible, $\psi(x)$ is a unit of T for each $x \in X$. Therefore the existence and uniqueness of η follows from Proposition 9.4. □

THEOREM 9.13. *If X is a right denominator set in a ring R, then there exists a module of fractions for any right R–module A with respect to X.*

Proof. Set $C = A/t_X(A)$ and define a submodule B of $E(C)$ by $B/C = t_X(E(C)/C)$. Then the map $a \mapsto a + t_X(A)$ gives an R–module homomorphism $f : A \to B$ such that $\ker(f) = t_X(A)$.

Note that $C \cap t_X(B) = t_X(C) = 0$ and so $t_X(B) = 0$. Thus B is X–torsionfree. In particular, it follows that $Bt_X(R) = 0$. Now given any $b \in B$ and $x \in X$, we find using right reversibility that

$$r.\mathrm{ann}_R(x) \leq t_X(R) \leq \mathrm{ann}_R(b).$$

Hence, there is an R–module homomorphism from xR to B sending x to b, and so there exists $c \in E(C)$ such that $cx = b$. By definition of B, it follows that $c \in B$, whence $b \in Bx$. Therefore B is X–divisible.

By Proposition 9.12, there exists a unique right RX^{-1}–module structure on B compatible with its right R–module structure. Given any $b \in B$, there exists $x \in X$ such that $bx \in C$. Then $bx = f(a)$ for some $a \in A$, whence $b = f(a)x^{-1}$. Thus all elements of B have the required form. □

DEFINITION. Given a right denominator set X in a ring R and a right R–module A, we now know that there exists a module of fractions $f : A \to B$ for A with respect to X. Because of the uniqueness given by Corollary 9.11, we shall denote B by AX^{-1}, and we shall refer to f as *the natural map* from A to AX^{-1}. As with rings of fractions, we write elements of AX^{-1} in the form ax^{-1}, and elements of $f(A)$ in the form $a1^{-1}$.

Now suppose that C is another right R–module, and let $g \in \mathrm{Hom}_R(A,C)$. In view of Proposition 9.10, there exists a unique right RX^{-1}–module homomorphism $h : AX^{-1} \to CX^{-1}$ such that $h(ax^{-1}) = g(a)x^{-1}$ for all $a \in A$ and $x \in X$. We refer to h as *the map induced by* g, and if a notation for h is needed, we shall use gX^{-1}.

EXERCISE 9K. Let X be a right denominator set in a ring R.

(a) If A is any X–torsionfree X–divisible right R–module, show that the natural map $A \to AX^{-1}$ is an isomorphism.

(b) If A,B are right (left) RX^{-1}–modules, show that all R–module homomorphisms from A to B are also RX^{-1}–module homomorphisms. □

The following proposition shows that any module of fractions AX^{-1} over a ring of fractions RX^{-1} is naturally isomorphic to $A \otimes_R RX^{-1}$. Hence, an alternative approach to modules of fractions would be to define them as tensor products.

PROPOSITION 9.14. *Let X be a right denominator set in a ring R, and let A be a right R–module. Then the "multiplication map" $A \times RX^{-1} \to AX^{-1}$ given by the rule $(a,s) \mapsto (a1^{-1})s$ induces an RX^{-1}–module isomorphism of $A \otimes_R RX^{-1}$ onto AX^{-1}.*

Proof. It is clear that the multiplication map induces a group homomorphism

$$g : A \otimes_R RX^{-1} \to AX^{-1}$$

such that $g(a \otimes s) = (a1^{-1})s$ for all $a \in A$, $s \in RX^{-1}$, and that g is an RX^{-1}–module homomorphism. Any element of AX^{-1} has the form ax^{-1} for some $a \in A$, $x \in X$, and $ax^{-1} = g(a \otimes x^{-1})$. Thus g is surjective.

Any element $m \in A \otimes_R RX^{-1}$ has the form $(a_1 \otimes s_1) + \ldots + (a_n \otimes s_n)$ for some $a_i \in A$ and $s_i \in RX^{-1}$. Using Lemma 9.2, we can find $r_1,\ldots,r_n \in R$ and $x \in X$ such that each $s_i = r_ix^{-1}$, whence

$$m = (a_1r_1 + \ldots + a_nr_n) \otimes x^{-1}.$$

Thus any element $m \in A \otimes_R RX^{-1}$ has the form $a \otimes x^{-1}$ for some $a \in A$, $x \in X$, and $g(m) = ax^{-1}$. If $m \in \ker(g)$, we must have $a1^{-1} = 0$, and hence $ay = 0$ for some $y \in X$. Consequently,

$$m = a \otimes x^{-1} = a \otimes (yy^{-1})x^{-1} = ay \otimes (xy)^{-1} = 0.$$

Therefore g is injective. □

COROLLARY 9.15. *If X is a right denominator set in a ring R, then RX^{-1} is a flat left R–module. That is, if $f : A \to B$ is a monomorphism of right R–modules, then the induced map*

$$f \otimes 1 \; : \; A \otimes_R RX^{-1} \to B \otimes_R RX^{-1}$$

is injective.

Proof. Let $h : B \otimes_R RX^{-1} \to BX^{-1}$ be the isomorphism induced from the multiplication map as in Proposition 9.14.

Any element $m \in A \otimes_R RX^{-1}$ has the form $a \otimes x^{-1}$ for some $a \in A$, $x \in X$. If $m \in \ker(f \otimes 1)$, then

$$(f(a)1^{-1})x^{-1} = h(f(a) \otimes x^{-1}) = h(f \otimes 1)(m) = 0$$

and so $f(a)1^{-1} = 0$. Hence, $f(a)y = 0$ for some $y \in X$. Since f is injective, we obtain $ay = 0$, and thus, as in the proof of the previous proposition, $m = 0$. Therefore $f \otimes 1$ is injective. □

COROLLARY 9.16. *Let X be a right denominator set in a ring R.*

(a) Every injective right RX^{-1}–module is X–torsionfree and injective as a right R–module.

(b) If A is an X–torsionfree injective right R–module, the natural map $A \to AX^{-1}$ is an R–module isomorphism, and AX^{-1} is an injective right RX^{-1}–module.

Proof. (a) The X–torsionfreeness follows from Proposition 9.12 and the injectivity from Corollary 9.15. (The latter implication is a standard fact: if $\phi : R \to S$ is a ring homomorphism under which S becomes a flat left R–module, then all injective right S–modules are also injective as R–modules. Readers who have not seen this should treat it as an exercise.)

(b) We first observe that A is X–divisible. For if $a \in A$ and $x \in X$, then

$$r.\mathrm{ann}_R(x) \leq t_X(R) \leq \mathrm{ann}_R(a)$$

and so there is a homomorphism from xR to A sending x to a; by injectivity we obtain $cx = a$ for some $c \in A$. Thus the natural map $A \to AX^{-1}$ is an isomorphism by Exercise 9K.

Now $B = AX^{-1}$ is a right RX^{-1}–module which is injective as an R–module. If $C \leq D$ are right RX^{-1}–modules and $g : C \to B$ is an RX^{-1}–module homomorphism, then g extends to an R–module homomorphism $h : D \to B$, and h is an RX^{-1}–module homomorphism by Exercise 9K. Therefore B is injective as an RX^{-1}–module. □

♦ SUBMODULES OF MODULES OF FRACTIONS

As in the commutative case, the submodules of a module of fractions AX^{-1} are closely related to the submodules of A. In order to discuss this relationship without the notation obscuring the details, we use the following abbreviations.

DEFINITION. Let X be a right denominator set in a ring R, and let A be a right R–module. If B is a submodule of AX^{-1}, the set $\{a \in A \mid a1^{-1} \in B\}$ is called *the contraction of* B *to* A and is denoted B^c. If C is a submodule of A, the set $\{cx^{-1} \mid c \in C,\ x \in X\}$ in AX^{-1} is called *the extension of* C *to* AX^{-1} and is denoted C^e.

THEOREM 9.17. *Let X be a right denominator set in a ring R, and let A be a right R–module.*

(a) If B is any RX^{-1}–submodule of AX^{-1}, then B^c is an R–submodule of A, the factor A/B^c is X–torsionfree, and $B = B^{ce}$.

(b) If C is any R–submodule of A, then C^e is an RX^{-1}–submodule of AX^{-1} and $C \leq C^{ec}$. Moreover, $C^{ec}/C = t_X(A/C)$, and so $C = C^{ec}$ if and only if A/C is X–torsionfree.

(c) Contraction and extension provide inverse lattice isomorphisms between the lattice of RX^{-1}–submodules of AX^{-1} and the lattice of those R–submodules C of A such that A/C is X–torsionfree.

Proof. Let $f : A \to AX^{-1}$ be the natural map. Let us say that a submodule $C \leq A$ is *X–closed* in A if and only if A/C is X–torsionfree. Observe that any intersection of X–closed submodules of A is X–closed, whence the X–closed submodules of A do form a lattice, with intersections for infima. (The supremum of a family $\{C_i\}$ of X–closed submodules in this lattice is the intersection of all X–closed submodules containing $\sum C_i$.)

(a) Obviously $B^c = f^{-1}(B)$ and is thus an R–submodule of A. Since f induces an R–module embedding of A/B^c into the RX^{-1}–module AX^{-1}/B, it is clear that A/B^c is X–torsionfree.

As $f(B^c) \leq B$, we certainly have $B^{ce} \leq B$. Given any $b \in B$, write $b = ax^{-1}$ for some $a \in A$, $x \in X$. Then $f(a) = bx \in B$ and so $a \in B^c$, whence $b \in B^{ce}$. Thus $B^{ce} = B$.

(b) Since any pair of elements of RX^{-1} has a common denominator, we see that C^e is the RX^{-1}–submodule of AX^{-1} generated by f(C). Clearly $C \leq C^{ec}$. Given $a \in C^{ec}$, we have $a1^{-1} = cx^{-1}$ for some $c \in C$, $x \in X$. Then $ax1^{-1} = c1^{-1}$ and so $axy = cy$ for some $y \in X$, whence $axy \in C$. This shows that C^{ec}/C is X–torsion. On the other hand, A/C^{ec} is X–torsionfree by (a), and thus $C^{ec}/C = t_X(A/C)$.

(c) From (a) and (b) we see that the rules $B \mapsto B^c$ and $C \mapsto C^e$ provide inverse bijections between the lattice $\mathscr{L}$ of RX^{-1}–submodules of AX^{-1} and the lattice $\mathscr{L}'$ of X–closed R–submodules of A. As these maps clearly preserve inclusions, they are order–isomorphisms (i.e., isomorphisms of partially ordered sets). Since infima and suprema are defined in terms of the order relations in $\mathscr{L}$ and $\mathscr{L}'$, we conclude that $(-)^c$ and $(-)^e$ are lattice isomorphisms. □

COROLLARY 9.18. *Let X be a right denominator set in a ring R, and let A be a right R–module.*

(a) If A is noetherian, or artinian, or of finite length, then so is AX^{-1} (as an RX^{-1}–module).

(b) If A is simple, then AX^{-1} is either zero or a simple RX^{-1}–module.

Proof. (a) The lattice isomorphisms in Theorem 9.17 guarantee that any strictly ascending (descending) chain of RX^{-1}–submodules of AX^{-1} contracts to a strictly ascending (descending) chain of R–submodules of A.

(b) By Theorem 9.17, the only RX^{-1}–submodules of AX^{-1} are 0 and AX^{-1}. □

EXERCISE 9L. Let X be a right denominator set in a ring R, and let $C \leq B \leq A$ be right R–modules. In the notation of Theorem 9.17, show that $B^e/C^e \cong (B/C)X^{-1}$. □

EXERCISE 9M. Let X be a right denominator set in a ring R, and let A be an X–torsionfree right R–module. Show that the rank of AX^{-1} (as either an R–module or an RX^{-1}–module) is the same as the rank of A. □

♦ IDEALS IN RINGS OF FRACTIONS

The results of the previous section (especially Theorem 9.17) give us a good grasp of the submodule structure of a module of fractions. In particular, this applies to the right ideal structure of a right ring of fractions RX^{-1}. Thus, for instance, if R is right noetherian, then so is RX^{-1}. However, the connections between two–sided ideals of R and two–sided ideals of RX^{-1} are not quite so nice in general, as the following examples indicate.

EXERCISE 9N. Let $S = k[x_n \mid n \in \mathbb{Z}]$, where k is a field and the x_n are independent commuting indeterminates, let α be the k–algebra automorphism of S such that $\alpha(x_n) = x_{n+1}$ for all n, and let $R = S[\theta;\alpha]$. By Exercise 9I, the multiplicative set $X = \{1,\theta,\theta^2,\ldots\}$ is a right and left denominator set in R. Show that $I = x_1R + x_2R + \ldots$ is an ideal of R, but that I^e is not an ideal of RX^{-1}. □

EXERCISE 9O. Let k be a field, let S be the ring of all functions from $\mathbb{Z}$ to k, and let α be the automorphism of S such that $\alpha(f)(n) = f(n-1)$ for all $f \in S$, $n \in \mathbb{Z}$. Set $R = S[\theta;\alpha]$. By Exercise 9I, the multiplicative set $X = \{1,\theta,\theta^2,\ldots\}$ is a right and left denominator set in R. Show that RX^{-1} is a prime ring while R is not even semiprime. Thus 0 is a prime ideal of RX^{-1} but 0^c is not a semiprime ideal of R. □

PROPOSITION 9.19. *Let X be a right denominator set in a ring R.*

(a) If J is an ideal of RX^{-1}, then J^c is an ideal of R and $J = J^{ce}$. Hence, all ideals of RX^{-1} are extended from ideals of R. However, if I is an ideal of R then I^e need not be an ideal of RX^{-1}.

(b) If J is an ideal of RX^{-1} and J^c is prime (semiprime), then J is prime (semiprime). However, if J is prime then J^c need not even be semiprime.

Proof. (a) Let $\phi : R \to RX^{-1}$ be the natural map. Obviously $J^c = \phi^{-1}(J)$ is an ideal of R, and $J = J^{ce}$ by Theorem 9.17. That extensions of ideals need not be ideals is shown by Exercise 9N.

(b) First suppose that J^c is prime, and note that since J^c is proper, J is proper. If A,B are ideals of RX^{-1} such that $AB \le J$, then $A^cB^c \le (AB)^c \le J^c$, and so either $A^c \le J^c$ or $B^c \le J^c$, whence either $A = A^{ce} \le J^{ce} = J$ or $B \le J$. Thus J is prime in this case. The proof for semiprimeness is analogous, and the final statement is shown by Exercise 9O. □

In the noetherian case the anomalies described in Proposition 9.19 do not occur. In fact, it suffices to assume that RX^{-1} is right noetherian, as follows.

THEOREM 9.20. *Let X be a right denominator set in a ring R, and assume that RX^{-1} is right noetherian. (For instance, R could be right noetherian.)*

(a) If I is any ideal of R, then I^e is an ideal of RX^{-1}.

(b) Let I be an ideal of R such that $(R/I)_R$ is X–torsionfree. Then I is prime (semiprime) if and only if I^e is prime (semiprime).

(c) An ideal J of RX^{-1} is prime (semiprime) if and only if J^c is prime (semiprime).

(d) Let P be a prime (semiprime) ideal of R. Then $P = Q^c$ for some prime (semiprime) ideal Q of RX^{-1} if and only if $X \subseteq \mathscr{C}(P)$.

Proof. Let $\phi : R \to RX^{-1}$ be the natural map.

(a) We first show that the image of X in R/I is right reversible. Thus given $x \in X$ and $a \in R$ with $xa \in I$, we must show that $aw \in I$ for some $w \in X$.

Set $J_n = \{r \in R \mid x^n r \in I\}$ for $n = 0,1,\ldots$. As RX^{-1} is right noetherian, $J_n{}^e = J_{n+1}{}^e$ for some n. From the right Ore condition, there exist $y \in X$, $b \in R$ such that $ay = x^n b$. Then $x^{n+1}b = xay \in I$, whence $b \in J_{n+1}$, and so $b1^{-1} \in J_{n+1}{}^e = J_n{}^e$. Thus $b \in J_n{}^{ec}$. Since $J_n{}^{ec}/J_n$ is X–torsion (Theorem 9.17), $bz \in J_n$ for some $z \in X$. Now $ayz = x^n bz \in I$, and as $yz \in X$ the claim is proved.

To prove that I^e is an ideal of RX^{-1} it suffices to show that $x^{-1}\phi(I) \le I^e$ for each $x \in X$. Given $a \in I$, we have $ay = xb$ for some $y \in X$, $b \in R$, by the right Ore condition. Since then $xb \in I$, the reversibility proved above implies that $bz \in I$ for some

$z \in X$. Therefore
$$x^{-1}\phi(a) = by^{-1} = (bz)(yz)^{-1} \in I^e,$$
as desired.

(b) By Theorem 9.17, $I^{ec} = I$. Hence if I is prime (semiprime), Proposition 9.19 shows that I^e is prime (semiprime). Conversely, assume that I^e is prime, note that I must be proper, and consider any ideals A and B in R such that $AB \leq I$. Since B^e is an ideal of RX^{-1}, we find that
$$A^eB^e = \phi(A)RX^{-1}B^e = \phi(A)B^e = \phi(A)\phi(B)RX^{-1} = \phi(AB)RX^{-1} = (AB)^e \leq I^e.$$
Thus either $A^e \leq I^e$ or $B^e \leq I^e$, and so either $A \leq A^{ec} \leq I^{ec} = I$ or $B \leq I$. Thus I is prime in this case, and the proof for semiprimeness is analogous.

(c) This follows immediately from (b), since $(R/J^c)_R$ is X–torsionfree and $J^{ce} = J$, by Theorem 9.17.

(d) We prove only the prime case, the semiprime case being analogous. If $X \subseteq \mathscr{C}(P)$, then clearly $(R/P)_R$ is X–torsionfree. In this case, P^e is a prime ideal of RX^{-1} by (b), and $P^{ec} = P$ by Theorem 9.17.

Conversely, assume that $P = Q^c$ for some (prime) ideal Q of RX^{-1}. As R/P embeds in the right noetherian ring RX^{-1}/Q, it must have ACC on right annihilators. Thus, by Proposition 9.9, the image of X in R/P is right reversible. Since $(R/P)_R$ is X–torsionfree (Theorem 9.17), we conclude that for any $x \in X$, the right and left annihilators of x in R/P are both zero. Therefore $X \subseteq \mathscr{C}(P)$. □

EXERCISE 9P. Let X be a right denominator set in a ring R, and let I be an ideal of R. If the image of X in R/I is right reversible, the proof of Theorem 9.20(a) shows that I^e is an ideal of RX^{-1}. Prove the converse. □

EXERCISE 9Q. Let X be a right denominator set in a ring R, and assume that RX^{-1} is right noetherian. If Q is a semiprime ideal of RX^{-1}, show that R/Q^c is right Goldie, and that the right Goldie quotient rings of R/Q^c and RX^{-1}/Q are isomorphic. □

LEMMA 9.21. *Let X be a right Ore set in a prime right Goldie ring R. If $0 \notin X$, then R_R is X–torsionfree. If also X is a right denominator set, then X consists of regular elements.*

Proof. If the ideal $t_X(R)$ is nonzero, $t_X(R) \leq_e R_R$ by Exercise 3L, and hence by Proposition 5.9 there exists a regular element $c \in t_X(R)$. But then $cx = 0$ for some $x \in X$ and so $x = 0$, contradicting our hypotheses. Thus $t_X(R) = 0$.

Now if $x \in X$ and $rx = 0$ for some $r \in R$, then $r \in t_X(R)$ and so $r = 0$. Assuming that X is right reversible, if $xs = 0$ for some $s \in R$, then $sy = 0$ for some $y \in X$, whence $s = 0$. Therefore x is a regular element. □

THEOREM 9.22. *Let X be a right denominator set in a right noetherian ring R. Then contraction and extension provide inverse bijections between the set of prime ideals of RX^{-1} and the set of those prime ideals of R that are disjoint from X.*

Proof. If Q is a prime ideal of RX^{-1}, then Q^c is a prime ideal of R by Theorem 9.20, and $Q^{ce} = Q$ by Theorem 9.17. Since Q is a proper ideal of RX^{-1}, it is clear that Q^c is disjoint from X.

Conversely, let P be a prime ideal of R disjoint from X. If Y denotes the image of X in R/P, then $0 \notin Y$, and so $(R/P)_{R/P}$ is Y–torsionfree by Lemma 9.21. Thus $(R/P)_R$ is X–torsionfree. Therefore P^e is a prime ideal of RX^{-1} by Theorem 9.20, and $P^{ec} = P$ by Theorem 9.17. □

♦ PRIME IDEALS IN ITERATED DIFFERENTIAL OPERATOR RINGS

In this section we give an application of rings of fractions by showing that for an iterated differential operator ring $S = R[\theta_1;\delta_1]\cdots[\theta_n;\delta_n]$ in which R is a commutative noetherian algebra over the rational numbers, every prime factor ring S/P is a domain. (We proved the case $n = 1$ of this in Theorem 2.22.) The most important example of an iterated differential operator ring is the enveloping algebra of a solvable Lie algebra over the field of complex numbers, in which case this result includes the theorem of Lie that every finite–dimensional irreducible representation is one–dimensional. We will first need two more facts about differential operator rings.

LEMMA 9.23. *Let D be a division ring of characteristic zero and δ a derivation of D, and let $S = D[\theta;\delta]$. Then either S is a simple ring or S is isomorphic to a polynomial ring $D[y]$.*

Proof. It follows from Proposition 1.14 that S is simple unless δ is an inner derivation. If δ is inner, then there is an element $a \in D$ such that for all $d \in D$ we have $\delta(d) = ad - da$. If we let $y = \theta - a$, then an easy computation shows that y is a central element in S, that the powers of y are linearly independent over D, and that $S = D[y]$. □

EXERCISE 9R. Let R be a subring of a ring S, let $W \subseteq S$, and assume that S is generated as a ring by $R \cup W$. Let X be a right Ore set in R, and assume that for all $w \in W$, $x \in X$ there exist $s \in S$, $y \in X$, $r \in R$ such that $xs = wy + r$. Prove that X is a right Ore set in S. □

EXERCISE 9S. Let X be a right denominator set in a ring R, and let δ be a derivation on R.

(a) Show that X is a right denominator set in $R[\theta;\delta]$.

(b) If $\phi : R \to RX^{-1}$ is the natural map, show that there is a unique derivation ∂ on RX^{-1} such that $\partial\phi = \phi\delta$. [Hint: Exercise 1F]. Show that $R[\theta;\delta]X^{-1} \cong RX^{-1}[\theta;\partial]$. □

DEFINITION. A *completely prime ideal* in a ring R is any (prime) ideal P such that R/P is a domain.

EXERCISE 9T. Let X be a right denominator set in a ring R.

(a) If P is a completely prime ideal of R and P is disjoint from X, show that P^e is a completely prime ideal of RX^{-1}.

(b) If Q is a completely prime ideal of RX^{-1}, show that Q^c is a completely prime ideal of R. □

THEOREM 9.24. [Lie, Dixmier, Gabriel, Lorenz, Sigurdsson] *If R is a commutative noetherian $\mathbb{Q}$–algebra and $S = R[\theta_1;\delta_1]\cdots[\theta_n;\delta_n]$ is an iterated differential operator ring, then every prime ideal of S is completely prime.*

Proof. The proof is by induction on n, where the case $n = 0$ is clear. Now let $n > 0$, let $S = S_{n-1}[\theta_n;\delta_n]$ where $S_{n-1} = R[\theta_1;\delta_1]\cdots[\theta_{n-1};\delta_{n-1}]$, and note that S_{n-1} is noetherian. If P is a prime of S, then $P \cap S_{n-1}$ is a prime of S_{n-1} by Lemma 2.21. Thus if $T = S_{n-1}/(P \cap S_{n-1})$, then by induction T is a noetherian domain. According to Lemma 2.18 and Exercise 2X, $P \cap S_{n-1}$ is a δ_n–ideal of S_{n-1}, and if we also use the name δ_n for the induced derivation on T, then $S/(P \cap S_{n-1})S \cong T[\theta_n;\delta_n]$. Under this isomorphism, $P/(P \cap S_{n-1})S$ corresponds to a prime Q of $T[\theta_n;\delta_n]$ such that $Q \cap T = 0$ and $T[\theta_n;\delta_n]/Q \cong S/P$. Thus we need to show that Q is completely prime.

We now let X be the set of nonzero elements in T, and we let D be the Ore quotient ring of T. Then X is a denominator set in T and $TX^{-1} \cong D$. Using Exercise 9S, we find that X is also a denominator set in $T[\theta_n;\delta_n]$, that δ_n extends uniquely to a derivation ∂ on D, and that $T[\theta_n;\delta_n]X^{-1} \cong D[\theta;\partial]$. Since $Q \cap T = 0$, the prime Q is disjoint from X. According to Theorem 9.22, Q^e is a prime of $T[\theta_n;\delta_n]X^{-1}$ which contracts to Q. Hence, by Exercise 9T it will suffice to show that Q^e is completely prime. Thus it is enough to show that all primes of $D[\theta;\partial]$ are completely prime.

If $D[\theta;\partial]$ is simple, then this is immediate from the fact that $D[\theta;\partial]$ is a domain. Otherwise, according to Lemma 9.23 $D[\theta;\partial]$ is isomorphic to a polynomial ring $D[y]$. It follows that X is a denominator set in $T[y]$ and that $T[y]X^{-1} \cong D[\theta;\partial]$. Consequently, in view of Theorem 9.22 and Exercise 9T we now see that it will suffice to show that all primes of $T[y]$ are completely prime. However, $T[y]$ is a factor ring of a polynomial ring $S_{n-1}[y]$, and we finish the proof by showing that all primes of $S_{n-1}[y]$ are completely prime. To do this, we show that $S_{n-1}[y]$ may be viewed as an $(n-1)$–fold iterated differential operator ring over $R[y]$.

In case $n > 1$, we have $S_{n-1} = S_{n-2}[\theta_{n-1};\delta_{n-1}]$ for a suitable coefficient ring S_{n-2}. Observe that the inner derivation $s \mapsto \theta_{n-1}s - s\theta_{n-1}$ on $S_{n-1}[y]$ restricts to a derivation ∂_{n-1} on $S_{n-2}[y]$. Since $S_{n-1}[y]$ is clearly a free left $S_{n-2}[y]$–module with basis $1, \theta_{n-1}, \theta_{n-1}^2, \ldots$, we may thus identify $S_{n-1}[y]$ with the differential operator ring $S_{n-2}[y][\theta_{n-1};\partial_{n-1}]$. Continuing in this fashion, we find that $S_{n-1}[y]$ may be identified with an iterated differential operator ring of the form $R[y][\theta_1;\partial_1]\cdots[\theta_{n-1};\partial_{n-1}]$. This iterated differential operator ring involves $n-1$ steps over the commutative noetherian $\mathbb{Q}$–algebra $R[y]$. Therefore by induction all primes of $S_{n-1}[y]$ are completely prime, and this completes the proof of the theorem. □

♦ ADDITIONAL EXERCISES

9U. If R is a semiprime right Goldie ring and P is a minimal prime ideal of R, show that $\mathscr{C}(P)$ is a right denominator set in R. [Hint: Show that there is some $c \in \mathscr{C}(P)$ satisfying $Pc = 0$.] Show that $R\mathscr{C}(P)^{-1}$ is isomorphic to the right Goldie quotient ring of R/P. □

9V. Let S be a prime right Goldie ring, and let ${}_SB_S$ be a bimodule. Let R be the subring of $\begin{pmatrix} S & B \\ 0 & S \end{pmatrix}$ consisting of all $\begin{pmatrix} s & b \\ 0 & s \end{pmatrix}$ where $s \in S$ and $b \in B$, and let $P = \begin{pmatrix} 0 & B \\ 0 & 0 \end{pmatrix}$. Observe that P is a prime ideal of R.

(a) Show that $\mathscr{C}_R(P)$ is a right Ore set in R if and only if $(B/cB)_S$ is torsion for all regular elements $c \in S$.

(b) If S is noetherian, B_S is finitely generated, and ${}_SB$ is torsionfree, show that $\mathscr{C}_R(P)$ is a right denominator set in R. □

9W. Let $X \subseteq Y$ be multiplicative sets in a ring R, and assume that X is a right denominator set. Show that Y is a right denominator set in R if and only if the set $Y_1 = \{y1^{-1} \mid y \in Y\}$ is a right denominator set in RX^{-1}, in which case $RY^{-1} \cong (RX^{-1})Y_1^{-1}$. □

9X. Let α be an endomorphism of a ring R, and let X be a right denominator set in R such that $X \subseteq \alpha(X)$.

(a) Show that X is a right denominator set in $R[\theta;\alpha]$.

(b) If $\alpha(X) = X$ and if $\phi : R \to RX^{-1}$ is the natural map, show that there is a unique ring endomorphism β of RX^{-1} such that $\beta\phi = \phi\alpha$. Show that $R[\theta;\alpha]X^{-1} \cong RX^{-1}[\theta;\beta]$. □

9Y. Let Y be the multiplicative set generated by a subset X of a ring R. Assume that for all $x \in X$ there exists $y \in Y$ such that $Ry \subseteq xR$. Show that Y is a right Ore set. □

♦ NOTES

Much of the subject of this chapter developed as folklore, and so we do not give precise attributions for many of the results.

Ore's Condition. The name honors Ore's result that a domain is a right order in a division ring if and only if its nonzero elements satisfy the right Ore condition [1931, Theorems 1, II].

Existence of Rings of Fractions. Asano proved that a ring has a right ring of fractions with respect to its regular elements if and only if its regular elements satisfy the right Ore condition [1939, Satz 1], and later that a ring has a right ring of fractions with respect to a multiplicative set X of regular elements if and only if X satisfies the right Ore condition [1949, Satz I]. For an arbitrary multiplicative set X in a ring R, Elizarov showed that a right ring of fractions exists if and only if the intersection of those ideals I such that $X \subseteq \mathscr{C}(I)$ and the image of X in R/I satisfies the right Ore condition yields an ideal with the same properties [1960, Theorem 2]. The general criterion using the Ore condition together with the reversibility condition was given by Gabriel [1962, Proposition 5, p. 415].

Two–Sided Rings of Fractions. Proposition 9.8 was first proved for multiplicative sets of regular elements by Asano [1949, Satz 6].

Reversibility of Ore Sets. Right reversibility of right Ore sets was first proved in the right noetherian case by Goldie [1964a, Proposition 2, p. 5–22].

Ideals in Rings of Fractions. Part (a) of Theorem 9.20 was first proved in a special case by Ludgate [1972, Theorem 5]. In the case that R is right noetherian and X consists of regular elements, parts (a) and (b) were proved by Jategaonkar [1974a, Proposition 1.4].

Complete Primeness in Iterated Differential Operator Rings. Lie's Theorem, one version of which appeared in [1893, § 131, Satz 7], implies that for a solvable finite–dimensional complex Lie algebra L, every finite–dimensional irreducible representation is one–dimensional. This is equivalent to saying that every finite–dimensional prime factor ring of $U(L)$ is a domain. Dixmier proved that for a solvable finite–dimensional Lie algebra L over an algebraically closed field of characteristic zero, every prime of $U(L)$ is completely prime [1966, Théorème 1.3], and then Gabriel removed the algebraically closed hypothesis [1971, Théorème 3.2]. Lorenz proved that if R is an algebra over a field k of characteristic zero, and if for each field extension K of k the algebra $R \otimes_k K$ is right noetherian with all primes being completely prime, the same holds for any differential operator ring $R[\theta;\delta]$

where δ is a k–linear derivation [1981, Theorem]. Sigurdsson proved that if R is a right noetherian $\mathbb{Q}$–algebra in which every prime is completely prime, the same holds for any iterated differential operator ring $R[\theta_1;\delta_1]\cdots[\theta_n;\delta_n]$ [1984, Corollary 2.6].

10 ARTINIAN QUOTIENT RINGS

Goldie's Theorem gives a characterization of those rings R such that R is an order in a ring which is semisimple, and, in particular, artinian. This naturally gives rise to the question: which rings can be orders in artinian rings? While this question has a certain abstract interest of its own, its significance turns out to be much greater than one might initially suspect. Rings arising in a natural way frequently are orders in artinian rings, and this may be an important fact in their study. In particular, as we shall see in Chapter 12, if R is a subring of a ring S and P is a prime ideal in S, then while $P \cap R$ need not be prime or semiprime, it is often possible to show that $R/(P \cap R)$ is an order in an artinian ring.

Discussion of orders in artinian rings is often phrased in the language of the previous chapter. If the set of regular elements in a ring R is a right (left) denominator set, the corresponding right (left) ring of fractions for R is called a *classical right (left) quotient ring of* R. Thus the problem discussed above may be rephrased as follows: under what conditions does a ring R have a classical right quotient ring which is right artinian, or, for short, when does R have a right artinian classical right quotient ring?

We first introduce a new notion of rank, known as "reduced rank" (different from the rank introduced in Chapter 4), which is useful in many arguments involving noetherian rings, and we give two naive examples of its use. We then use reduced rank to derive necessary and sufficient conditions for a noetherian ring R to be an order in an artinian ring. This basic criterion is very satisfactory in some ways – for instance, it is phrased entirely in terms of properties of individual elements of R – but not in others. We then turn to criteria involving ideals, particularly affiliated prime ideals (parallel to the theorem in the commutative case that a commutative noetherian ring is an order in an artinian ring if and only if its associated prime ideals are all minimal).

♦ REDUCED RANK

The notion of the rank of a module that was discussed in Chapter 4 goes back to the rank of a torsionfree abelian group. More generally, it has been found that a useful invariant of an arbitrary abelian group A is its "torsionfree rank", obtained by reducing A to the torsionfree group A/T(A) and taking the rank of this group. This idea we shall carry over directly to modules over a semiprime Goldie ring, and then we shall extend it to modules over noetherian rings.

DEFINITION. If M is a right module over a semiprime right Goldie ring R, the *reduced rank* (or *torsionfree rank*) of M is defined as

$$\rho_R(M) = \text{rank}(M/Z(M)),$$

which we may denote by $\rho(M)$ if the coefficient ring R is understood. (E.g., note that $\rho_{\mathbb{Z}}(\mathbb{Z}/2\mathbb{Z}) = 0$ whereas $\rho_{\mathbb{Z}/2\mathbb{Z}}(\mathbb{Z}/2\mathbb{Z}) = 1$.) In case $M/Z(M)$ does not have finite rank, we assign the value ∞ to $\rho_R(M)$. Other notations found in the literature for reduced rank include r.rank and red.rk.

Alternatively, if $Q(R)$ is the right Goldie quotient ring of R, we could define

$$\rho_R(M) = \text{length}(M \otimes_R Q(R))$$

where $M \otimes_R Q(R)$ is viewed as a (semisimple) right $Q(R)$–module, because of Exercise 6E. This description of $\rho_R(M)$ will play a key role in some of our calculations.

EXERCISE 10A. If M is a finitely generated right module over a semiprime right Goldie ring R, show that $\rho_R(M)$ is finite. □

If we wish to concoct a notion of reduced rank in more general circumstances, it is helpful to keep in mind as an analog the notion of length for artinian modules that are not semisimple. In particular, a module M of finite length has a chain of submodules

$$M_0 = 0 \le M_1 \le \ldots \le M_n = M$$

such that M_i/M_{i-1} is semisimple for $i = 1,\ldots,n$, and of course

$$\text{length}(M) = \sum_{i=1}^{n} \text{length}(M_i/M_{i-1}).$$

Thus length could have been first defined for semisimple modules and then extended to other modules by the formula above.

We follow exactly this route in defining reduced rank for modules over a right noetherian ring R. Since the prime radical N of R is nilpotent (Theorem 2.11), say $N^n = 0$, a right R–module M has a chain of submodules

$$M \ge MN \ge MN^2 \ge \ldots \ge MN^n = 0$$

such that each of the factors MN^i/MN^{i+1} is a right (R/N)–module. Thus each MN^i/MN^{i+1} has a reduced rank, and we could define a reduced rank for M to be

$$\sum_{i=0}^{n-1} \rho_{R/N}(MN^i/MN^{i+1}).$$

To make sure this behaves properly we need a lemma, and then we will give an alternate but slightly more useful definition.

LEMMA 10.1. *Let R be a right noetherian ring and N its prime radical, and let M be a right R–module. Consider two chains of submodules of M,*

$$M = L_0 \geq L_1 \geq \ldots \geq L_m = 0 \qquad \text{and} \qquad M = M_0 \geq M_1 \geq \ldots \geq M_n = 0,$$

such that $(L_i/L_{i+1})N = 0$ *and* $(M_j/M_{j+1})N = 0$ *for all indices i,j. Then*

$$\sum_{i=0}^{m-1} \rho_{R/N}(L_i/L_{i+1}) = \sum_{j=0}^{n-1} \rho_{R/N}(M_j/M_{j+1}).$$

Proof. We first show this in case $MN = 0$. If X is the set of regular elements in R/N, then $(R/N)X^{-1}$ is the right Goldie quotient ring of R/N, and for any right (R/N)–module A Proposition 9.14 shows that $A \otimes_R (R/N)X^{-1} \cong AX^{-1}$. Hence, $\rho_{R/N}(A) = \text{length}(AX^{-1})$. Returning to our given module M, and adopting the notation of Theorem 9.17, each L_i^e is an $(R/N)X^{-1}$–submodule of MX^{-1}, and $L_i^e/L_{i+1}^e \cong (L_i/L_{i+1})X^{-1}$ by Exercise 9L. Thus

$$\sum_{i=0}^{m-1} \rho_{R/N}(L_i/L_{i+1}) = \sum_{i=0}^{m-1} \text{length}(L_i^e/L_{i+1}^e) = \text{length}(MX^{-1}),$$

and similarly for the second summation, which proves the lemma in the case $MN = 0$.

It follows from this case that we may replace either of the given chains of submodules by a refinement without changing the value of the corresponding summation. (For instance, if the first chain is refined, the portion of the new summation dealing with the terms from L_i to L_{i+1} must add up to $\rho_{R/N}(L_i/L_{i+1})$.)

Finally, we cite the Schreier Refinement Theorem (Theorem 3.10) to say that the two chains have refinements in which the factors are isomorphic in pairs. That is, after replacing both chains by refinements, we may assume that $m = n$ and that there is a permutation π of $\{0,\ldots,m-1\}$ with $M_i/M_{i+1} \cong L_{\pi(i)}/L_{\pi(i)+1}$ for $i = 0,\ldots,m-1$. From this the desired result follows. □

With this lemma in hand, we can immediately see that the following definition makes sense, i.e., that the value given for the reduced rank of M is independent of the chain of submodules chosen.

DEFINITION. Let R be a right noetherian ring with prime radical N, and let M be a right R–module. Choose a chain of submodules

$$M = M_0 \geq M_1 \geq \ldots \geq M_n = 0$$

such that $(M_i/M_{i+1})N = 0$ for $i = 0,\ldots,n-1$. (Reminder: Such chains exist because N is nilpotent.) The *reduced rank of* M, denoted by $\rho_R(M)$ or by $\rho(M)$ if the coefficient ring is understood, is defined to be

$$\rho_R(M) = \sum_{i=0}^{n-1} \rho_{R/N}(M_i/M_{i+1}).$$

EXERCISE 10B. If M is a finite abelian group and $I = \mathrm{ann}_{\mathbb{Z}}(M)$, show that $\rho_{\mathbb{Z}}(M) = 0$ whereas $\rho_{\mathbb{Z}/I}(M) = \mathrm{length}(M)$. □

We record some easy but useful facts about reduced rank in the following lemmas.

LEMMA 10.2. *If R is a right noetherian ring and M a finitely generated right R–module, then $\rho(M)$ is finite.* □

LEMMA 10.3. *If R is a right noetherian ring and $M' \leq M$ are right R–modules, then*

$$\rho(M) = \rho(M') + \rho(M/M').$$

Proof. This follows from the fact that the definition of reduced rank is independent of the chain of submodules chosen, so that the defining chain for $\rho(M)$ may be chosen as a refinement of the chain $M \geq M' \geq 0$. □

The next lemma is a further special case.

LEMMA 10.4. *Let R be a right noetherian ring and $M' \leq M$ right R–modules, such that $\rho(M)$ is finite. Then $\rho(M') = \rho(M)$ if and only if $\rho(M/M') = 0$.* □

Lemma 10.4 will be useful because we can give a "torsion" interpretation of the condition "reduced rank 0", which is parallel to the usual condition when R is semiprime. Let N be the prime radical of R and $Q(R/N)$ the right Goldie quotient ring of R/N. If $MN = 0$, then $\rho(M) = 0$ if and only if $M \otimes_R Q(R/N) = 0$, if and only if M is a torsion (R/N)–module (recall Exercise 6E). In general, as the following lemma shows, R–modules with zero reduced rank can be thought of as "$\mathscr{C}(N)$–torsion" modules in an obvious manner (although care is required since $\mathscr{C}(N)$ is not necessarily a right Ore set).

LEMMA 10.5. *Let R be a right noetherian ring, N its prime radical, and M a right R–module. Then $\rho(M) = 0$ if and only if for each $x \in M$ there is an element $c \in \mathscr{C}(N)$ such* that $xc = 0$.

Proof. Choose submodules $M = M_0 \geq M_1 \geq \ldots \geq M_n = 0$ such that $(M_i/M_{i+1})N = 0$ for $i = 0,\ldots,n-1$. Because of Lemma 10.3, $\rho(M) = 0$ if and only if $\rho(M_i/M_{i+1}) = 0$ for each i, and by our previous remarks this occurs if and only if each M_i/M_{i+1} is torsion as a right (R/N)–module. The result is now clear. □

EXERCISE 10C. If Q is a prime ideal in a right noetherian ring R, show that $\rho_R(R/Q) > 0$ if and only if Q is a minimal prime. [Hint: Proposition 6.5.] □

♦ APPLICATIONS OF REDUCED RANK TO FINITE RING EXTENSIONS

Despite its innocent appearance, the notion of reduced rank has been useful in many contexts in recent work in ring theory. We give here two such applications. It should be emphasized that although the results are simple, they were unknown until this method was used to prove them in 1986 and 1987.

If R is a subring of a ring S, one can ask many questions concerning the relations between primes of R and those of S. In the commutative case, one says a prime P of S *lies over* a prime Q of R if $P \cap R = Q$. That is not a reasonable definition in the noncommutative setting since $P \cap R$ is not generally prime.

DEFINITION. If R is a subring of a ring S, and P and Q are primes of S and R respectively, then we say that P *lies over* Q if Q is minimal over $P \cap R$.

This definition has many advantages but has its dangers as well. For example, primes $P_1, P_2, P_3, \ldots$ could all lie over Q while the contractions $P_i \cap R$ could all be different, as in the following example.

EXERCISE 10D. Let k be a field and $S = M_2(k) \times \ldots \times M_n(k)$ for some integer $n > 2$. For $i = 2,\ldots,n$, let u_i be the matrix in $M_i(k)$ with $(j,j+1)$–entry equal to 1 for $j = 1,\ldots,i-1$ while all other entries are 0. If $u = (u_2,\ldots,u_n)$ and $R = k[u] \subseteq S$, show that all the minimal primes P_j of S lie over the prime uR, but that all the contractions $P_j \cap R$ are distinct. □

DEFINITION. Just as in the commutative case, we say *LO (lying over)* holds in a ring extension $S \supseteq R$ if for every prime Q of R there is a prime P of S lying over Q. We say that *INC (incomparability)* holds if for every prime Q of R, there do not exist primes P and P' of S lying over Q with $P > P'$.

THEOREM 10.6. [Letzter] *If R is a right noetherian ring and R is a subring of a ring S such that S is finitely generated as a right R–module, then INC holds.*

Proof. Suppose P and P' are primes in S lying over the prime Q in R, and $P > P'$. This means that Q is minimal over both $P \cap R$ and $P' \cap R$ and all of this will remain true if we factor out P'. Therefore, without loss of generality we may assume that $P' = 0$ and hence that S is prime. Since Q is minimal over $P' \cap R$, it follows that Q is a minimal prime of R.

We will now use reduced rank for right R–modules, noting that $\rho_R(R/Q) > 0$ since Q is a minimal prime of R (Exercise 10C). Since $P \cap R \leq Q$, we conclude that $\rho_R(R/(P \cap R)) > 0$. Moreover, $R/(P \cap R)$ is isomorphic to an R–submodule of S/P, and so $\rho_R(S/P) > 0$. Now P is a nonzero ideal of the prime right noetherian ring S and hence contains a regular element, say x. Since S and xS are isomorphic right R–modules, clearly $\rho_R(S) = \rho_R(xS)$, and since $\rho_R(xS) \leq \rho_R(P) \leq \rho_R(S)$, we conclude that $\rho_R(S) = \rho_R(P)$. From Lemma 10.4 it follows that $\rho_R(S/P) = 0$, a contradiction. □

THEOREM 10.7. [Letzter] *If R is a right FBN ring and R is a subring of a ring S such that S is finitely generated as a right R–module, then S is right FBN.*

Proof. Without loss of generality, we may assume that S is prime, and we need only prove that S is right bounded. Let I be an essential right ideal of S. To show that I contains a nonzero ideal of S, we again use reduced rank for S and I as right R–modules. Since I contains a regular element of S, it follows as in the previous proof that $\rho_R(S/I) = 0$. Lemma 10.5 than implies that S/I is $\mathscr{C}_R(N)$–torsion, where N is the prime radical of R.

According to Theorem 8.6 there is a chain

$$A_0 = 0 < A_1 < \ldots < A_n = S/I$$

of R–submodules of S/I such that for $i = 1,\ldots,n$ the factor A_i/A_{i-1} is isomorphic to a right ideal in R/P_i for some prime ideal P_i of R. Since S/I is $\mathscr{C}_R(N)$–torsion, it follows that none of the primes P_i can be minimal primes (see Proposition 6.5). Consequently, $P_nP_{n-1}\cdots P_1 \neq 0$, and hence S/I is not faithful as a right R–module. Since R is a subring of S, it follows that S/I is also not faithful as an S–module. Hence I contains a nonzero ideal of S (the annihilator of S/I) as required. □

♦ SMALL'S THEOREM

We now turn to Small's Theorem, which provides a criterion for a ring to be an order in an artinian ring. It is most often stated in the noetherian case, where the criterion is especially simple, and this is the case we shall do in the text. The original theorem did not have the noetherian condition but was very complicated to state. With more recently developed notions and methods, it is now possible to give a fairly simple statement (see the discussion at the end of the section), which emphasizes the analogy with Goldie's Theorem.

Rather than asking for conditions under which a ring R is a right order in a right artinian ring, one might consider the apparently weaker condition that R embeds in a right ring of fractions which is right artinian. However, this is not really weaker, as the following exercise shows.

EXERCISE 10E. Let R be a ring and X a right denominator set in R. If RX^{-1} is right artinian and the natural map $R \to RX^{-1}$ is an embedding, show that R is a right order in RX^{-1}. □

Note that whenever a ring R is a right order in a ring Q, then Q is a right ring of fractions for R with respect to $\mathcal{C}_R(0)$.

LEMMA 10.8. [Goldie] *Let R be a right noetherian ring and N its prime radical. Then $\mathcal{C}(0) \subseteq \mathcal{C}(N)$, and given any $c \in \mathcal{C}(0)$ and $a \in R$ there exists $d \in \mathcal{C}(N)$ such that $ad \in cR$.*

Proof. Let $c \in \mathcal{C}(0)$. Since $\text{r.ann}_R(c) = 0$ we obtain $cR \cong R_R$, whence $\rho(cR) = \rho(R_R)$ and so $\rho(R/cR) = 0$. By Lemma 10.5, this implies that for any $a \in R$ there is some $d \in \mathcal{C}(N)$ with $ad \in cR$. This verifies the second conclusion of the lemma (sometimes called the *pseudo–Ore condition*).

In particular, from the case $a = 1$ we find that cR contains an element of $\mathcal{C}(N)$, whence $c(R/N)$ contains a regular element of R/N. Then $c + N$ itself must be a regular element of R/N, by Proposition 5.9 and Lemma 5.7. Thus $c \in \mathcal{C}(N)$. □

THEOREM 10.9. [Small, Talintyre] *A right noetherian ring R with prime radical N is a right order in a right artinian ring if and only if $\mathcal{C}_R(0) = \mathcal{C}_R(N)$.*

Proof. Assuming $\mathcal{C}(0) = \mathcal{C}(N)$, Lemma 10.8 implies that $\mathcal{C}(0)$ is a right Ore set in R, and we can form the corresponding right quotient ring $Q = R\mathcal{C}(0)^{-1}$. We identify R with its image in Q, and then R is a right order in Q. Since $(R/N)_R$ is $\mathcal{C}(0)$–torsionfree (because $\mathcal{C}(0) = \mathcal{C}(N)$), we conclude from Theorems 9.20 and 9.17 that NQ is an ideal of Q satisfying $NQ \cap R = N$. Now the embedding $R/N \to Q/NQ$ makes R/N (isomorphic to) a right order in Q/NQ. Therefore Q/NQ is the right Goldie quotient ring of R/N (Exercise 5F), whence Q/NQ is semisimple, and in particular artinian.

As NQ is an ideal of Q, we have $QNQ = NQ$, and it follows that $(NQ)^i = N^iQ$ for all positive integers i. Thus NQ is nilpotent (because N is nilpotent). Each of the factors $N^iQ/N^{i+1}Q$ is a right module over Q/NQ and so is semisimple. Also, since Q is right noetherian (Corollary 9.18), $N^iQ/N^{i+1}Q$ is finitely generated. Hence, each $N^iQ/N^{i+1}Q$ is artinian as a right Q–module, and therefore Q is right artinian.

Conversely, if R is a right order in a right artinian ring Q, then $\mathcal{C}(0)$ is a right Ore set in R and Q is a right ring of fractions for R with respect to $\mathcal{C}(0)$. Since $\mathcal{C}(0) \subseteq \mathcal{C}(N)$ (Lemma 10.8), $(R/N)_R$ is $\mathcal{C}(0)$–torsionfree, and so NQ is an ideal of Q satisfying $NQ \cap R = N$ (Theorems 9.20 and 9.17). As in the previous paragraph NQ is nilpotent, and hence $NQ \leq J(Q)$.

Given $c \in \mathscr{C}_R(N)$ and $q \in Q$ with $cq \in NQ$, write $q = ad^{-1}$ with $a \in R$, $d \in \mathscr{C}(0)$, and observe that $ca = cqd \in NQ \cap R = N$, whence $a \in N$ and $q \in NQ$. It follows that $c(Q/NQ) \cong Q/NQ$ and so $c(Q/NQ)$ has the same length as Q/NQ, yielding $c(Q/NQ) = Q/NQ$. Thus $c + NQ$ is right invertible in Q/NQ. By Corollary 5.6, $c + NQ$ is invertible in Q/NQ. As $NQ \leq J(Q)$, we conclude from Exercise 2V that c is invertible in Q. Consequently, $c \in \mathscr{C}_R(0)$, and therefore $\mathscr{C}_R(0) = \mathscr{C}_R(N)$. □

Since the condition $\mathscr{C}(0) = \mathscr{C}(N)$ is left–right symmetric, Theorem 10.9 shows that a noetherian ring R is a right order in a right artinian ring if and only if R is a left order in a left artinian ring.

COROLLARY 10.10. *A noetherian ring R with prime radical N is a right and left order in an artinian ring if and only if $\mathscr{C}_R(0) = \mathscr{C}_R(N)$.*

Proof. Theorem 10.9 and Proposition 9.8. □

In the situation of Corollary 10.10, the classical right and left quotient rings of R coincide, and we say that R *has an artinian classical quotient ring*.

The general (non–noetherian) version of Small's Theorem requires extending the notion of reduced rank to certain non–noetherian modules, as in the following exercises.

EXERCISE 10F. Let R be a ring with prime radical N such that R/N is right Goldie.

(a) Prove Lemma 10.1 under these hypotheses (assuming the module M has chains of submodules as described). Hence, define $\rho_R(M)$ for any right R–module M satisfying $MN^n = 0$ for some positive integer n.

(b) Prove Lemmas 10.3–10.5 in the present context, assuming $MN^n = 0$ for some n.

(c) If N is nilpotent and $\rho(R_R)$ is finite, prove Lemma 10.8 in the present context. □

EXERCISE 10G. Let R be a ring with prime radical N such that (i) R/N is right Goldie, (ii) N is nilpotent, (iii) $\rho(R_R)$ is finite, and (iv) $\mathscr{C}_R(0) = \mathscr{C}_R(N)$.

(a) Show that R is a right order in a ring Q.

(b) Show that NQ is a nilpotent two–sided ideal of Q.

(c) Show that Q/NQ is a semisimple ring.

(d) Show that $\rho(Q_R)$ is finite, and use this to see that $(N^iQ/N^{i+1}Q)_Q$ is artinian for all i.

(e) Conclude that Q as right artinian. □

EXERCISE 10H. Let R be a ring with prime radical N, and assume that R is a right order in a right artinian ring Q.

(a) If N' is the prime radical of Q, show that $N' \cap R$ is a nilpotent semiprime ideal of R, whence $N' \cap R = N$. In particular, N is nilpotent.

(b) Observe that $NQ = N'$. In particular, $NQ \cap R = N$.

(c) Show that $\mathscr{C}_R(0) = \mathscr{C}_R(N)$.

(d) Show that R/N is a right order in Q/NQ. Thus R/N is right Goldie and Q/NQ is its right Goldie quotient ring.

(e) Show that $\rho(R_R)$ is finite. [Hint: Show that $(N^i/N^{i+1}) \otimes_R Q \cong N^iQ/N^{i+1}Q$ for all $i = 0,1,\ldots$.] □

To summarize the previous exercises: A ring R with prime radical N is a right order in a right artinian ring if and only if R/N is right Goldie, N is nilpotent, $\rho(R_R)$ is finite, and $\mathscr{C}_R(0) = \mathscr{C}_R(N)$.

♦ AFFILIATED PRIME IDEALS

In this section we develop ideal–theoretic criteria for a noetherian ring R to be an order in an artinian ring. In the commutative case, it is known that R has an artinian classical quotient ring if and only if the associated primes of R are all minimal. As we will see, this fails in the noncommutative case (Exercise 10L). The appropriate modification is to look at affiliated primes rather than just associated primes or annihilator primes.

LEMMA 10.11. *Let R be a right noetherian ring, X a right Ore set in R, and M a finitely generated X–torsionfree right R–module. If P is any annihilator prime for M, or any prime affiliated to M, then $(R/P)_R$ is X–torsionfree.*

Proof. If $P = \mathrm{ann}_R(Y)$ for some $Y \subseteq M$, then $(R/P)_R$ embeds in the direct product M^Y, whence $(R/P)_R$ is X–torsionfree.

Note that X is right reversible by Proposition 9.9, whence the localizations RX^{-1} and MX^{-1} exist. Now consider an affiliated series $0 = M_0 < M_1 < \ldots < M_n = M$ with affiliated primes $P_1,\ldots,P_n$. If $I_j = P_jP_{j-1}\cdots P_1$ for $j = 1,\ldots,n$, it is clear from the definition of an affiliated series that $M_j = \mathrm{ann}_M(I_j)$. Since M is X–torsionfree, we may identify it with a submodule of the localization $N = MX^{-1}$. If $N_j = \mathrm{ann}_N(I_j)$, then $M \cap N_j = M_j$ and M/M_j embeds in N/N_j. On the other hand, the extension I_j^e of I_j to RX^{-1} is an ideal of RX^{-1} (Theorem 9.20), and $N_j = \mathrm{ann}_N(I_j^e)$, whence N_j is an RX^{-1}–submodule of N. Thus N/N_j is X–torsionfree, and so is M/M_j.

Therefore $P_1,\ldots,P_n$ are annihilator primes of X–torsionfree modules, and hence $(R/P_j)_R$ is X–torsionfree for $j = 1,\ldots,n$, as shown in the first paragraph. □

LEMMA 10.12. *If R is a right noetherian ring which is a right order in a right artinian ring, then every right annihilator prime for R and every right affiliated prime of R is a minimal prime.*

Proof. By assumption, $\mathscr{C}(0)$ is a right Ore set in R, and of course R_R is $\mathscr{C}(0)$–torsionfree. Hence if P is an annihilator prime for R_R or an affiliated prime of R_R, Lemma 10.11 shows that $(R/P)_R$ is $\mathscr{C}(0)$–torsionfree. In particular, P is disjoint from $\mathscr{C}(0)$. If N is the prime radical of R, then $\mathscr{C}(0) = \mathscr{C}(N)$ by Theorem 10.9, and so P is disjoint from $\mathscr{C}(N)$. By Proposition 6.5, P therefore is a minimal prime of R. □

THEOREM 10.13. [Stafford] *For a noetherian ring R the following conditions are equivalent:*

(a) R is a right and left order in an artinian ring.

(b) Every prime ideal of R which is either a right or a left affiliated prime of R is minimal.

(c) There exist a right affiliated series and a left affiliated series for R such that the affiliated primes for both series are all minimal.

Proof. (a) ⇒ (b): Lemma 10.12.

(b) ⇒ (c) a priori.

(c) ⇒ (a): By Corollary 10.10 and Lemma 10.8 it suffices to show that $\mathscr{C}(0) \supseteq \mathscr{C}(N)$, where N is the prime radical of R.

There exists a right affiliated series $0 = M_0 < M_1 < \ldots < M_n = R$ such that the corresponding right affiliated primes $P_1,\ldots,P_n$ are all minimal. Each of the factors M_i/M_{i+1} is torsionfree as a right (R/P_i)–module by Proposition 7.5, and hence also torsionfree as a right (R/N)–module, by Proposition 6.8.

Now given $c \in \mathscr{C}(N)$ the left annihilator of c in M_i/M_{i+1} is zero for $i = 1,\ldots,n$, whence $M_i \cap \text{l.ann}_R(c) \leq M_{i-1}$. We conclude by induction that $\text{l.ann}_R(c) = 0$, and, since we have symmetric hypotheses, $\text{r.ann}_R(c) = 0$. Therefore $c \in \mathscr{C}(0)$, completing the proof. □

EXERCISE 10I. Show that in the ring $R = \begin{pmatrix} \mathbb{Z} & \mathbb{Z}/2\mathbb{Z} \\ 0 & \mathbb{Z}/2\mathbb{Z} \end{pmatrix}$, there exists a right affiliated series for which the corresponding right affiliated primes are all minimal, as well as a right affiliated series for which one of the corresponding right affiliated primes is not minimal. □

EXERCISE 10J. Show via Theorem 10.13 that a commutative noetherian ring R is an order in an artinian ring if and only if all annihilator primes of R are minimal. □

EXERCISE 10K. Show that in the ring $R = \begin{pmatrix} \mathbb{Z}/2\mathbb{Z} & \mathbb{Z}/2\mathbb{Z} & \mathbb{Z}/2\mathbb{Z} \\ 0 & \mathbb{Z} & \mathbb{Z}/2\mathbb{Z} \\ 0 & 0 & \mathbb{Z}/2\mathbb{Z} \end{pmatrix}$, all right annihilator primes are minimal and all left annihilator primes are minimal, while nonetheless R fails to be a right order in a right artinian ring. □

The following exercise isolates the role of the annihilator primes in the question of existence of an artinian classical quotient ring.

EXERCISE 10L. If R is a noetherian ring with prime radical N, show that R is a right order in a right artinian ring if and only if (i) $\mathscr{C}(N)$ is a right Ore set, and (ii) all right annihilator primes of R are minimal. □

EXERCISE 10M. Let $T = A_1(k)$ where k is a field of characteristic zero, let B be a simple right T–module, set $R = \begin{pmatrix} k & B \\ 0 & T \end{pmatrix}$, and let N be the prime radical of R. Show that $\mathscr{C}_R(N)$ is a right Ore set and that all prime ideals of R are minimal, but that R is not a right order in a right artinian ring. □

In using Stafford's criterion to check whether a noetherian ring R has an artinian classical quotient ring (Theorem 10.13), one first has to find a right affiliated series and a left affiliated series for R. Given a specific R, it is often easier to find a series of ideals in R such that the corresponding ideal factors are torsionfree bimodules over prime factor rings of R. The following exercise (together with Exercise 10L) shows that the primes arising in this fashion can be used in place of the affiliated primes.

EXERCISE 10N. Let R be a noetherian ring, and let N be its prime radical. By Corollary 7.6, there exists a chain of ideals $A_0 = 0 < A_1 < \ldots < A_n = R$ such that for $i = 1,\ldots,n$, the ideals $P_i = \text{l.ann}_R(A_i/A_{i-1})$ and $Q_i = \text{r.ann}_R(A_i/A_{i-1})$ are prime and A_i/A_{i-1} is torsionfree as a left (R/P_i)–module and as a right (R/Q_i)–module.

(a) Show that $\mathscr{C}(N)$ is a right Ore set if and only if the set $\{P_i \mid Q_i \text{ is minimal}\}$ contains only minimal primes. [Hints: If $\mathscr{C}(N)$ is right Ore, use Exercise 9J. For the converse, show that if $c \in \mathscr{C}(N)$ then $A_i/(cA_i + A_{i-1})$ is torsion as a right (R/N)–module for each $i = n, n-1, \ldots, 1$.]

(b) Show that the set $\{P_i \mid Q_i \text{ is minimal}\}$ is the same for any choice of the chain of ideals A_i. [Hint: Schreier Refinement.] □

♦ NOTES

Reduced Rank. This was introduced by Goldie [1964b, p. 274].

INC and FBN in Finite Ring Extensions. Theorems 10.6 and 10.7 are due to Letzter

[Pb, Corollary 2.4 and Proposition 4.9].

Pseudo–Ore Condition. A slightly more general version of Lemma 10.8 (using one–sided regularity) was proved by Goldie [1972a, Theorem 2.5].

Small's Theorem. This is due to Small [1966a, Theorems 2.10, 2.11, 2.12]. The necessity of the condition $\mathscr{C}_R(0) = \mathscr{C}_R(N)$ was shown independently by Talintyre [1966, Theorem 2.1], who had earlier proved that if $\mathscr{C}_R(0) = \bigcap_{r=1}^{\infty} \mathscr{C}_R(N^r)$ then R is a right order in a right artinian ring [1963, Theorems 2.2, 3.2]. Two different criteria for a non–noetherian ring to be a right order in a right artinian ring were obtained by Small [1966b, Theorem C] and Robson [1967, Theorem 2.10]; the criterion involving reduced rank discussed in Exercises 10F–H is due to Warfield [1979a, Theorem 3].

Affiliated Prime Criteria for Artinian Classical Quotient Rings. Theorem 10.13 is due to Stafford [1982a, Proposition 1.3].

11 LINKS BETWEEN PRIME IDEALS

The way in which affiliated series may be put together to form modules over a noetherian ring leads us to study certain connections, which we will call *links,* between the prime ideals of the ring. In this chapter we will introduce these links and see how they can give insight into the structure of modules. This will naturally lead us to a class of rings for which this connection between the links and the module theory is particularly satisfactory.

♦ LINKS

To investigate the structure of a finitely generated module M over a noetherian ring R, we may start by choosing an affiliated series

$$0 = M_0 < M_1 < ... < M_n = M$$

for M. This breaks M up into "layers" M_i/M_{i-1}, each of which is a fully faithful module over a prime factor ring of R. The study of M may thus be separated into two topics: first, the study of fully faithful modules over prime noetherian rings, and, second, the ways in which the layers M_i/M_{i-1} can be put together to form the module M. The first topic, particularly torsionfree modules, we have studied to a certain extent in previous chapters. Here we shall begin to study certain aspects of the second topic.

Since uniform modules are ubiquitous (e.g., our module M contains an essential direct sum of uniform submodules, and M is in turn essential in a direct sum of uniform modules), it is reasonable to begin our discussion with uniform modules. In the commutative case the situation is very nice, as the following exercise shows.

EXERCISE 11A. Let M be a finitely generated uniform module over a commutative noetherian ring R, and let P be the assassinator of M. Show first that M is annihilated by a power of P. [Hint: Exercise 4Z.] Moreover, show that M has a unique affiliated series

$$0 = M_0 < M_1 < ... < M_n = M,$$

that the affiliated primes all equal P, and that each M_i/M_{i-1} is a torsionfree (R/P)–module. [Hint: Set $X = R - P$ and consider the RX^{-1}–module MX^{-1}.] □

Thus in the commutative case, the prime affiliated to the first layer of a uniform module is also affiliated to the higher layers. In the noncommutative case, however, it is no longer true that the affiliated prime ideals of a uniform module are all the same, as follows.

EXERCISE 11B. Let $R = \begin{pmatrix} k & k \\ 0 & k \end{pmatrix}$ for some field k, and let $M = \begin{pmatrix} k & k \\ 0 & 0 \end{pmatrix}$, viewed as a right R–module. Show that M is uniform, that M has a unique affiliated series, that M has two distinct affiliated primes, and that M is not annihilated by any power of its assassinator. □

Thus the question arises, how are the affiliated primes of a uniform module over a noncommutative noetherian ring related to the assassinator of the module? More generally (since the assassinator of a uniform module is its unique associated prime), how are the affiliated primes of a finitely generated module M related to the associated primes? Since the associated primes of M are annihilators of submodules of M (rather than subfactor modules), we may view them as being more "accessible" than the affiliated primes. Then if we can find a way to relate the affiliated primes to the associated primes, we will have found a useful way to "approximate" the annihilator of M (because M is annihilated by a product of affiliated primes).

As a basic building block for relating prime ideals to each other, which in a large class of rings will provide a solution to the problems just discussed, we introduce a notion of "links" between prime ideals. We then study how links arise, and try to relate affiliated primes to assassinators and associated primes via chains of links.

Links are probably most easily introduced in the context of simple modules with co–artinian annihilators, as follows. Suppose that M is a uniform module over a noetherian ring R with an affiliated series $0 < U < M$ such that U and M/U are simple modules. Let P and Q be the corresponding affiliated primes of M, and assume that R/P and R/Q are artinian. Then $R/(P \cap Q)$ is a semisimple ring. On the other hand, U is not a direct summand of M, and so M is not a semisimple module. Since $MQP = 0$, the ring R/QP cannot be semisimple, and therefore $QP \neq P \cap Q$.

Conversely, if $QP \neq P \cap Q$ there must exist a module M of the above form, as follows.

EXERCISE 11C. Let P and Q be maximal ideals in a noetherian ring R, such that R/P and R/Q are artinian. If $QP \neq P \cap Q$, show that there exists a uniform right R–module M with an affiliated series $0 < U < M$ and affiliated primes P,Q such that U and M/U are simple modules. [Hint: Assume that $QP = 0$ and $\text{l.ann}_R(P \cap Q) = Q$, and then show that Q is essential as a right ideal of R.] □

In the context just discussed above, we say there is a "link from Q to P" if $QP \neq P \cap Q$. We will want to generalize this notion to arbitrary prime ideals P and Q and to generalize Exercise 11C as well. The general definition of a link is necessarily more complicated, since we do not just want $(P \cap Q)/QP$ to be nonzero – we need it to be large in a suitable sense. The definition we present will be justified by the results which follow.

DEFINITION. If P and Q are primes in a noetherian ring R then we say there is a *link from* Q *to* P, written $Q \rightsquigarrow P$, if there is an ideal A of R such that $Q \cap P > A \geq QP$ and $(Q \cap P)/A$ is nonzero and torsionfree as a right (R/P)–module and as a left (R/Q)–module. In this case, the bimodule $(Q \cap P)/A$ is called a *linking bimodule* between Q and P.

A link as just defined is sometimes called a *second layer link*, to distinguish it from other types of links which are sometimes used. One such is an *ideal link* (or *internal bond*) from Q to P, meaning that R contains ideals $I > J$ with $IP \leq J$ and $QI \leq J$ such that I/J is a torsionfree right (R/P)–module and a torsionfree left (R/Q)–module. More generally, there is a *bimodule link* (or *bond*) from Q to P if there exists a nonzero $(R/Q, R/P)$–bimodule which is finitely generated and torsionfree on each side.

Although the definition of a link involves only two–sided ideals, it is nonetheless asymmetric in that Q appears as the left annihilator of the linking bimodule $(Q \cap P)/A$ while P appears as the right annihilator. Hence, given any theorem in which a link $Q \rightsquigarrow P$ is related to right R–modules, in the corresponding left module theorem the roles of P and Q must be interchanged. Perhaps the easiest way to keep track of this is by using the opposite ring, R^{op}. Note first that, viewed as subsets of R^{op}, the primes P and Q are also prime ideals of R^{op}. Then observe that $Q \rightsquigarrow P$ in R if and only if $P \rightsquigarrow Q$ in R^{op}.

DEFINITION. The set of prime ideals in a ring R is called the *prime spectrum* of R, denoted $\mathrm{Spec}(R)$. For the present we just regard $\mathrm{Spec}(R)$ as a set, although in Chapter 14 we will turn it into a topological space.

DEFINITION. If R is a noetherian ring, the *graph of links* of R is the directed graph whose vertices are the elements of $\mathrm{Spec}(R)$ with an arrow from Q to P whenever $Q \rightsquigarrow P$. The connected components of this graph are called *cliques,* and if $P \in \mathrm{Spec}(R)$ then we denote by $\mathrm{Cl}(P)$ the (unique) clique containing P.

The following exercises exhibit the cliques in some specific rings. We will later find all the cliques in any artinian ring (Corollary 11.9), and in any ring which is module–finite over its center when the center is noetherian (Theorem 11.20).

EXERCISE 11D. In a commutative noetherian ring, show that all cliques are singletons. □

EXERCISE 11E. Show that in the ring $\begin{pmatrix} k & k \\ 0 & k \end{pmatrix}$ (where k is a field) there are two prime ideals, say P and Q, and only one link, $Q \rightsquigarrow P$. (Though this is the most trivial possible example, it is not hard to see that in any noetherian ring R, if P and Q are primes and $Q \rightsquigarrow P$ by way of the linking bimodule $(Q \cap P)/A$, then the ring R/A tends to resemble the triangular matrix ring above.) □

EXERCISE 11F. Show that in the ring $\begin{pmatrix} \mathbb{Z} & \mathbb{Z} \\ 2\mathbb{Z} & \mathbb{Z} \end{pmatrix}$ every prime ideal except 0 and the maximal ideals $P = \begin{pmatrix} 2\mathbb{Z} & \mathbb{Z} \\ 2\mathbb{Z} & \mathbb{Z} \end{pmatrix}$ and $Q = \begin{pmatrix} \mathbb{Z} & \mathbb{Z} \\ 2\mathbb{Z} & 2\mathbb{Z} \end{pmatrix}$ is linked to itself, that the only other links are between P and Q, and that between these there are links in both directions. Hence, one clique contains two elements and the others are all singletons. □

EXERCISE 11G. Consider the ring $R = \mathbb{C}[x][\theta; x\frac{d}{dx}]$, where x is an indeterminate. This ring has two prime ideals which are not maximal, namely 0, which has no links of any kind, and xR, which is linked only to itself. All other primes are maximal ideals of the form $M_\alpha = xR + (\theta - \alpha)R$, where $\alpha \in \mathbb{C}$ (Exercise 2Y). Show that each M_α is linked to itself, and that otherwise there is a link from M_α to M_β if and only if $\beta = \alpha - 1$. Hence two cliques in this ring are singletons, and the remaining cliques are all countably infinite. □

♦ LINKS AND SHORT AFFILIATED SERIES

We now turn to the central concern of this chapter, which is how links or chains of links arise naturally between the affiliated primes of a module M. As will be shown in the following section, the question can be reduced to the case in which M has an affiliated series of length two, say $0 < U < M$, where there are only two affiliated primes, say P (the assassinator of M) and Q (the annihilator of M/U). Although there need not be a link between Q and P, we will find a dichotomy saying that either $Q \rightsquigarrow P$ in a natural way or else a very different situation occurs. This is the context of the following result, often called *Jategaonkar's Main Lemma*, which will prove very useful to our study of links.

THEOREM 11.1. [Jategaonkar] *Let R be a noetherian ring, and let M be a right R–module with an affiliated series $0 < U < M$ and corresponding affiliated prime ideals P and Q, such that $U \leq_e M$. Let M′ be a submodule of M, properly containing U, such that the ideal $A = \mathrm{ann}_R(M')$ is maximal among annihilators of submodules of M properly containing U. Then exactly one of the following two alternatives occurs:*

(i) $Q < P$ and $M'Q = 0$. In this case, M′ and M′/U are faithful torsion (R/Q)–modules.

(ii) $Q \rightsquigarrow P$ and $(Q \cap P)/A$ is a linking bimodule between Q and P. In this case, if

U is torsionfree as a right (R/P)–module then M'/U is torsionfree as a right (R/Q)–module.

Proof. The existence of a submodule M' as described follows from the ACC on ideals in R. Our hypotheses are retained if M is replaced by M'. Thus we may assume that $M' = M$, that is, all submodules of M properly containing U have the same annihilator as M, namely A. Note that any submodule N of M that is not contained in U also has annihilator A. (Since $\text{ann}_R(N \cap U) = P = \text{ann}_R(U)$, we must have $\text{ann}_R(N) = \text{ann}_R(N + U) = A$.) Note also that statements (i) and (ii) are mutually exclusive, since in case (i), $A = Q = Q \cap P$, while in case (ii), $A < Q \cap P$.

We first consider the case that $M(P \cap Q) = 0$. Since $MP \neq 0$ (because $U = \text{ann}_M(P)$), it follows that $P \neq Q$. Also, $MPQ = 0$ because $M(P \cap Q) = 0$, and $MP \cap U \neq 0$ because $U \leq_e M$. Hence, as U is a fully faithful (R/P)–module, $Q \leq P$, and thus $Q < P$. In particular, $UQ = 0$. Since $Q < P$ and $Q = \text{ann}_R(M/U)$, we have $(M/U)P \neq 0$, and hence $MP \nleq U$. Our added hypothesis on M implies that $\text{ann}_R(MP) = \text{ann}_R(M)$, whence $MQ = 0$. Therefore case (i) occurs.

Now in case (i), M is an essential extension of U as an (R/Q)–module, and so M/U is a torsion (R/Q)–module. Moreover, U is torsion as an (R/Q)–module (because $UP = 0$), and hence M is a torsion (R/Q)–module. Finally, M/U is a faithful (R/Q)–module by hypothesis, and consequently M is a faithful (R/Q)–module.

We have shown that if $M(P \cap Q) = 0$ then statement (i) holds. We now assume that $M(P \cap Q) \neq 0$, and we will show from this that statement (ii) holds.

Clearly $Q \cap P > A \geq QP$; we must show that $(Q \cap P)/A$ is torsionfree as a right (R/P)–module and as a left (R/Q)–module. If B/A is the right torsion submodule of $(Q \cap P)/A$, then by Lemma 7.3, B/A must be annihilated on the right by an ideal I properly containing P. Hence, $Q \cap P \geq B \geq A$ and $BI \leq A$. Then $MBI = 0$ and $(MB \cap U)I = 0$. Since U is a fully faithful (R/P)–module, $MB \cap U = 0$, and thus $MB = 0$. This shows that $B = A$ and therefore $(Q \cap P)/A$ is torsionfree as a right (R/P)–module. Similarly, if C/A is the left torsion submodule of $(Q \cap P)/A$ then $JC \leq A$ for some ideal J properly containing Q, and $MJC = 0$. Since M/U is a faithful (R/Q)–module, $(M/U)J \neq 0$, and so $MJ \nleq U$. By the remarks in the first paragraph, $\text{ann}_R(MJ) = A$, whence $C = A$. Therefore $(Q \cap P)/A$ is torsionfree as a left (R/Q)–module, and $Q \rightsquigarrow P$ via $(Q \cap P)/A$. Thus case (ii) occurs.

Finally, we must show that in case (ii), if U is a torsionfree right (R/P)–module then M/U is a torsionfree right (R/Q)–module. Let D/U be the torsion submodule of M/U, and consider any $m \in D$. Then $mc \in U$ for some $c \in \mathscr{C}_R(Q)$. Since left multiplication by c gives an injective map of $(Q \cap P)/A$ into itself, we see that $(c(Q \cap P) + A)/A$ is an essential right submodule of $(Q \cap P)/A$ (Corollary 4.18), whence $(Q \cap P)/(c(Q \cap P) + A)$ is a torsion right (R/P)–module. As $mc \in U$ and $UP = 0$, we have $m(c(Q \cap P) + A) = 0$, and hence $m(Q \cap P)$ is torsion as a right (R/P)–module. However, $m(Q \cap P) \leq MQ \leq U$, and U is assumed to be torsionfree as a right (R/P)–module, so $m(Q \cap P) = 0$. Thus $D(Q \cap P) = 0$. Since $Q \cap P > A$, our hypothesis on annihilators of submodules of M implies that $D = U$. We have therefore shown that M/U is a torsionfree right (R/Q)–module, as required. □

It is an open question whether, given a module M as in Theorem 11.1, there can exist two different choices of submodule M' such that for one choice alternative (i) occurs while for the other choice alternative (ii) occurs. This would in particular entail both $Q < P$ and $Q \rightsquigarrow P$, and no examples of prime ideals with this behavior are known.

Examples of alternative (ii) in Theorem 11.1 are very easily obtained. For instance, the $\mathbb{Z}$–module $\mathbb{Z}/4\mathbb{Z}$ has an affiliated series $0 < 2\mathbb{Z}/4\mathbb{Z} < \mathbb{Z}/4\mathbb{Z}$ with both affiliated primes being $2\mathbb{Z}$, and $2\mathbb{Z}$ is linked to itself via the linking bimodule $2\mathbb{Z}/4\mathbb{Z}$. The following exercise gives an instance when alternative (i) occurs.

EXERCISE 11H. Let $S = A_1(k) = k[x][\theta; d/dx]$, where k is a field of characteristic zero, and let $R = k + \theta S$, which is a subring of S. By Exercises 2ZC and 2ZD, R is noetherian, $R/\theta S$ and S/R are simple right R–modules, and $(S/R)_R$ is faithful. Show that $(S/\theta S)_R$ is uniform, that $(S/\theta S)_R$ has an affiliated series $0 < R/\theta S < S/\theta S$ with affiliated primes θS and 0, and that 0 is not linked to θS. □

EXERCISE 11I. Show that if in Theorem 11.1 we only assume that R is right noetherian then the conclusions remain valid except that in (ii) we can only conclude that $(Q \cap P)/A$ is a fully faithful right (R/P)–module and a fully faithful left (R/Q)–module, and that if U is torsionfree as a right (R/P)–module then $(Q \cap P)/A$ is torsionfree as a right (R/P)–module. (Here in case (ii) when U is a torsionfree (R/P)–module, it is an open question whether M'/U must be a torsionfree (R/Q)–module.) □

We have shown that affiliated series of uniform modules sometimes give rise to links between primes. We will now show that all links between primes arise in this way. In fact, the existence of links is equivalent to the existence of uniform modules with affiliated series in which the corresponding factor modules are torsionfree.

For use in the upcoming proof, and in Proposition 11.3, we make the following observation: if $A, B_1, \dots, B_n$ are modules such that A is uniform and $A \leq B_1 \oplus \dots \oplus B_n$, then A embeds in B_j for some j. (To see this, note that the projections $p_j : A \to B_1 \oplus \dots \oplus B_n \to B_j$ satisfy $\ker(p_1) \cap \dots \cap \ker(p_n) = 0$. Since A is uniform, $\ker(p_j) = 0$ for some j.) In particular, if $A \leq B^n$, then A must embed in B.

THEOREM 11.2. [Jategaonkar, Brown] *Let R be a noetherian ring and P and Q prime ideals of R. Then $Q \rightsquigarrow P$ if and only if there exists a finitely generated uniform right R–module M with an affiliated series $0 < U < M$ such that U is isomorphic to a (uniform) right ideal of R/P and M/U is isomorphic to a uniform right ideal of R/Q.*

Proof. If such a module M exists, all submodules of M/U are torsionfree as (R/Q)–modules. Hence, case (i) of Theorem 11.1 cannot occur, and thus $Q \rightsquigarrow P$.

Conversely, assume that $Q \rightsquigarrow P$, and let $(Q \cap P)/A$ be a linking bimodule. Without loss of generality, we may assume that $A = 0$, so that $Q \cap P$ is a nonzero torsionfree right (R/P)–module and a torsionfree left (R/Q)–module. Since $\text{l.ann}_R(Q \cap P) = Q$, we conclude that $\text{l.ann}_R(Q) \leq Q$ and $\text{l.ann}_R(P) = Q$ (since $QP = 0$ already). If I is a nonzero right ideal of R, then either $IQ = 0$, and so $I \leq \text{l.ann}_R(Q) \leq Q$, or $IQ \neq 0$, whence $I \cap Q \neq 0$. Hence Q is essential as a right ideal of R.

Now $Q \cap P$ is a torsionfree right (R/P)–module, and $Q/(Q \cap P)$ is isomorphic to an ideal of R/P and hence is also torsionfree as a right (R/P)–module. Hence Q is a torsionfree right (R/P)–module, and so Q_R has an essential submodule isomorphic to a finite direct sum of uniform right ideals of R/P (Proposition 6.18). Since Q_R is essential in R_R, we know that $E(Q_R) = E(R_R)$, and thus $E(R_R) \cong E^n$ where E is the injective hull of a uniform right ideal of R/P and $n = \text{rank}(R_R)$. (Recall from Proposition 6.23 that E is independent of the choice of uniform right ideal.) Note that E has an essential submodule which is a torsionfree (R/P)–module, whence $\text{ann}_E(P)$ is torsionfree as an (R/P)–module.

Since $\text{l.ann}_R(P) = Q$, it follows that $Q = R \cap \text{ann}_{E(R_R)}(P)$. Hence $(R/Q)_R$ embeds in $(E/\text{ann}_E(P))^n$, and it follows that any uniform right ideal of R/Q embeds in $E/\text{ann}_E(P)$. Let K be a submodule of E such that $K > \text{ann}_E(P)$ and $K/\text{ann}_E(P)$ is isomorphic to a uniform right ideal of R/Q. Choose an element x in K not annihilated by P, and set $M = xR$ and $U = \text{ann}_M(P)$. Clearly M/U is isomorphic to a uniform right ideal of R/Q, and U is isomorphic to a uniform right ideal of R/P by Corollary 6.20. Since E is uniform so is M, and it is clear from the definition of U that $0 < U < M$ is an affiliated series for M. □

EXERCISE 11J. If we want to work with a right noetherian ring R, then a natural definition of a link from Q to P would require that the linking bimodule $(Q \cap P)/A$ be torsionfree as a right (R/P)–module and fully faithful as a left (R/Q)–module (cf. Exercise 11I). Show that the proof of Theorem 11.2 carries over to give a result in this case as well, where, however, we cannot conclude that U is isomorphic to a right ideal of R/P but only that it is uniform and torsionfree. □

♦ LINKS AND AFFILIATED PRIMES

In the situation of Theorem 11.1, the alternative (i) is regarded as the "bad" case, since no information is obtained linking or otherwise relating Q to P. Thus we shall mostly study alternative (ii), and in rings where this alternative always occurs we can develop a satisfactory solution to the annihilator problem discussed at the beginning of the first section. The terminology we shall use to denote this desirable situation is based on a rather technical notion of first and second layers of certain modules, due to Jategaonkar. A general notion of the first layer and the second layer of a module turns out to be cumbersome at best. However, as at the beginning of this chapter, we can think of a module M with an affiliated series $M_0 = 0 < M_1 < \ldots < M_n = M$ as being equipped with a set of layers M_i/M_{i-1} (depending on the affiliated series). Then the alternatives (i) and (ii) of Theorem 11.1 are statements about the kinds of modules that can occur as second layers of a module M, and we view the ring R as "good" if only second layers of type (ii) occur.

DEFINITION. If P is a prime ideal in a noetherian ring R, then P is said to satisfy the *right strong second layer condition* if, given the hypotheses of Theorem 11.1, the conclusion (i) never occurs. Similarly, P is said to satisfy the *right second layer condition* if, given the hypotheses of Theorem 11.1 and the additional hypothesis that U is torsionfree as an (R/P)–module, the conclusion (i) never occurs. We say the ring R satisfies the *right strong second layer condition*, or the *right second layer condition* if this holds for all prime ideals P of R. The *left strong second layer condition* and *left second layer condition* are defined similarly. Finally, we say that the ring itself satisfies the *strong second layer condition* (or the *second layer condition*) if it satisfies these conditions on both the left and the right.

For example, any artinian ring R satisfies the strong second layer condition (since R contains no primes P and Q satisfying $Q < P$). A more important example is that any right fully bounded noetherian ring R satisfies the right strong second layer condition, since no prime factors of R possess fully faithful torsion modules (Lemma 8.2). Some non–fully–bounded examples appear in Theorem 11.15. Exercise 11H provides an example of a ring that fails to satisfy the right second layer condition. No examples are known of noetherian rings satisfying the second layer condition but not the strong second layer condition.

As stated, the second layer condition and the strong second layer condition involve properties of modules which are not necessarily even finitely generated. We show next that it suffices to check these conditions in finitely generated uniform modules. For some alternative conditions equivalent to the second layer conditions, see Corollary 11.5 and Exercise 11L.

PROPOSITION 11.3. *Let P be a prime ideal in a noetherian ring R.*

(a) P satisfies the right strong second layer condition if and only if there does not exist a finitely generated uniform right R–module M with an affiliated series $0 < U < M$ and corresponding affiliated prime ideals P,Q such that M/U is uniform, $Q < P$, and $MQ = 0$.

(b) P satisfies the right second layer condition if and only if there does not exist a finitely generated uniform right R–module M with an affiliated series $0 < U < M$ and corresponding affiliated prime ideals P,Q such that U is a torsionfree (R/P)–module, M/U is uniform, $Q < P$, and $MQ = 0$.

Proof. (a) Clearly if P satisfies the right strong second layer condition, there cannot exist a module M as described. Conversely, if P fails to satisfy the condition, there exists a right R–module M_1 with an affiliated series $0 < U_1 < M_1$ and corresponding affiliated primes P,Q such that $U_1 \leq_e M_1$ while $Q < P$ and $M_1Q = 0$. (Here M_1 and U_1 have taken the roles of the modules M' and U in the statement of Theorem 11.1.) Next choose an element $x \in M_1 - U_1$, and set $M_2 = xR$ and $U_2 = M_2 \cap U_1$. Then M_2 has an affiliated series $0 < U_2 < M_2$ with affiliated primes P,Q, and $U_2 \leq_e M_2$ and $M_2Q = 0$.

Since M_2 is finitely generated, $E(M_2) = E_1 \oplus \dots \oplus E_n$ for some uniform injective modules $E_1,\dots,E_n$. Now $U_2 \leq_e E(M_2)$ and so $U_2 \cap E_i \neq 0$ for $i = 1,\dots,n$. As U_2 is a fully faithful (R/P)–module, so is $U_2 \cap E_i$, and hence so is the affiliated submodule $A_i = \text{ann}_{E_i}(P)$. Set

$$U_3 = A_1 \oplus \dots \oplus A_n = \text{ann}_{E(M_2)}(P),$$

and observe that U_3 is a fully faithful (R/P)–module. If $M_3 = M_2 + U_3$, then

$$M_3/U_3 \cong M_2/(M_2 \cap U_3) = M_2/U_2.$$

Thus M_3 has an affiliated series $0 < U_3 < M_3$ with affiliated primes P,Q, and $U_3 \leq_e M_3$. Since $Q < P$, we also have $U_3Q = 0$, and so $M_3Q = 0$. Therefore $M_3 \leq B_1 \oplus \dots \oplus B_n$, where $B_i = \text{ann}_{E_i}(Q)$ for $i = 1,\dots,n$.

Choose a uniform submodule $V \leq M_3/U_3$. Then V embeds in the direct sum $(B_1/A_1) \oplus \dots \oplus (B_n/A_n)$, whence V embeds in B_j/A_j for some j. Choose a submodule $M_4 \leq B_j$ such that $M_4 > A_j$ and $M_4/A_j \cong V$. Then M_4 is a uniform module with an affiliated series $0 < A_j < M_4$ with affiliated primes P,Q, while M_4/A_j is uniform and $M_4Q = 0$.

Finally, choose an element $y \in M_4 - A_j$, and set $M = yR$ and $U = M \cap A_j$. Then M is a finitely generated uniform right R–module with an affiliated series $0 < U < M$ and affiliated primes P,Q such that M/U is uniform and $MQ = 0$. Since we already have $Q < P$, part (a) is proved.

(b) The proof is similar to the proof of (a), and is left to the reader. □

Observe from Proposition 11.3 (or directly from the definitions) that a noetherian ring R satisfies the right (strong) second layer condition if and only if all prime factor rings R/Q satisfy the right (strong) second layer condition.

DEFINITION. Let R be a noetherian ring. A subset X of Spec(R) is said to be *right link closed* if whenever $P \in X$ and $Q \rightsquigarrow P$, it follows that $Q \in X$. The *right link closure* of a set X of primes is the smallest right link closed subset of Spec(R) containing X.

It may not be immediately clear why we call the above property "right link closed" rather than "left link closed", but the point is that right link closure is what arises in the study of right modules (as is already suggested by Theorems 11.1 and 11.2) and in the study of right Ore sets (see Chapter 12).

THEOREM 11.4. [Jategaonkar] *Let R be a noetherian ring satisfying the right strong second layer condition, and let M be a nonzero right R–module which has an affiliated series. Then all of the affiliated prime ideals of M are in the right link closure of the set Ass(M) of associated prime ideals. In particular, M is annihilated by some product of prime ideals from the right link closure of Ass(M).*

Proof. We use induction on the length of an affiliated series for M, say

$$0 = M_0 < M_1 < \ldots < M_n = M,$$

with corresponding affiliated primes $P_1,\ldots,P_n$. Since $P_1 \in \text{Ass}(M)$ the result is trivial when $n = 1$. Now assume that $n > 1$, and that, by induction, the primes $P_1,\ldots,P_{n-1}$ are in the right link closure of $\text{Ass}(M_{n-1})$. Since $\text{Ass}(M_{n-1}) \subseteq \text{Ass}(M)$, the primes $P_1,\ldots,P_{n-1}$ are also in the right link closure of Ass(M).

If M_{n-1}/M_{n-2} is not essential in M/M_{n-2}, choose a submodule M' of M such that $M' > M_{n-2}$ and $M' \cap M_{n-1} = M_{n-2}$, and observe that M'/M_{n-2} is a fully faithful (R/P_n)–module. Hence, an affiliated series for M' is

$$0 = M_0 < M_1 < \ldots < M_{n-2} < M',$$

with corresponding affiliated primes $P_1,\ldots,P_{n-2},P_n$. This affiliated series has length $n-1$, and so P_n is in the right link closure of Ass(M') by induction. Thus P_n is in the right link closure of Ass(M).

If, on the other hand, M_{n-1}/M_{n-2} is essential in M/M_{n-2}, then we apply Theorem 11.1 to the module M/M_{n-2}. Because R satisfies the right strong second layer condition, we conclude that $P_n \rightsquigarrow P_{n-1}$. Since P_{n-1} is in the right link closure of Ass(M), so is P_n.

The final statement of the theorem follows immediately, since $MP_nP_{n-1}\cdots P_1 = 0$. □

Suppose in particular, in the setup of Theorem 11.4, that M has an essential affiliated submodule U with corresponding affiliated prime P. (For instance, this occurs if M is uniform.) Then Ass(M) = {P}, and consequently all affiliated primes of M are in the right link closure of P.

COROLLARY 11.5. *A noetherian ring R satisfies the right strong second layer condition if and only if for each prime ideal Q of R, every finitely generated essential (R/Q)–module extension of an unfaithful right (R/Q)–module is unfaithful.*

Proof. The sufficiency of the given condition is clear from Proposition 11.3. Conversely, assume that R satisfies the right strong second layer condition, let $Q \in$ Spec(R), and let $U \leq_e M$ be an essential extension of finitely generated right (R/Q)–modules such that U is unfaithful; we must show that M is an unfaithful (R/Q)–module. Since R/Q satisfies the right strong second layer condition, there is no loss of generality in assuming that $Q = 0$; hence, R is a prime ring.

Every associated prime of M is also an associated prime of U (because $U \leq_e M$) and so is nonzero (because U is unfaithful). Since the prime 0 cannot be linked to any nonzero prime, it follows that all primes in the right link closure of Ass(M) are nonzero. By Theorem 11.4, M is annihilated by a product of nonzero primes, and therefore M is unfaithful, as desired. □

EXERCISE 11K. Show that a noetherian ring R satisfies the right strong second layer condition if and only if for every finitely generated right R–module A, all annihilator primes of A are associated primes. □

THEOREM 11.6. [Jategaonkar] *Let R be a noetherian ring satisfying the right second layer condition, let M be a nonzero right R–module, and assume that M has an affiliated series*

$$0 = M_0 < M_1 < \ldots < M_n = M,$$

with corresponding affiliated prime ideals $P_1, \ldots, P_n$. For all P in Ass(M), assume that $\mathrm{ann}_M(P)$ is torsionfree as an (R/P)–module. Then for $i = 1, \ldots, n$, the prime ideal P_i is in the right link closure of Ass(M), and M_i/M_{i-1} is torsionfree as a right (R/P_i)–module.

Proof. The proof is analogous to that of Theorem 11.4, where at each stage we apply the last conclusion of Theorem 11.1, part (ii) to conclude that the factors are torsionfree. □

EXERCISE 11L. This is an analog of Corollary 11.5. Let R be a noetherian ring, and for each prime ideal P of R consider the following condition:

(†) Whenever M is a finitely generated submodule of $E((R/P)_R)$ containing R/P such that $\mathrm{ann}_R(M)$ is a prime ideal, then $\mathrm{ann}_R(M) = P$.

Show that if R satisfies the right second layer condition then all prime ideals of R satisfy (†). Conversely, show that any prime ideal P of R which satisfies (†) must satisfy the right second layer condition, and hence if all prime ideals of R satisfy (†) then R satisfies the right second layer condition. □

Occasionally in the literature, condition (†) of Exercise 11L is taken as the definition of the right second layer condition for a prime P. However, the definition we have given is the most common, and it is strictly weaker than (†), as the following example shows.

EXERCISE 11M. Let $R = k + \theta S$ as in Exercise 11H, and set $T = \begin{pmatrix} R & R \\ \theta S & R \end{pmatrix}$, $P = \begin{pmatrix} R & R \\ \theta S & \theta S \end{pmatrix}$, $Q = \begin{pmatrix} \theta S & R \\ \theta S & R \end{pmatrix}$, $A = \begin{pmatrix} R & R \\ S & S \end{pmatrix}$.

(a) Show that T is a prime noetherian ring, and that P and Q are prime ideals of T.

(b) Show that A is a right T–submodule of $M_2(S)$, and that $(T/P)_T \leq_e (A/P)_T$. Show also that $\text{ann}_T(A/P) = 0$, and conclude that the prime P in T does not satisfy condition (†) of Exercise 11L.

(c) Let $E = E((R/\theta S)_R)$ and make the row $E' = (E \;\; E)$ into a right $M_2(R)$–module in the obvious way. Show that E' is an injective right $M_2(R)$–module. [Hint: Given a right $M_2(R)$–module homomorphism $f : A \to E'$, consider the restriction of f to $A\begin{pmatrix} 1 & 0 \\ 0 & 0 \end{pmatrix}$.] Observe that $M_2(R)$ is a projective left T–module [Hint: As a left T–module $M_2(R)$ is a direct sum of two copies of a left ideal of T], and conclude that E' is also injective as a right T–module.

(d) Observe that the row $B = (0 \;\; R/\theta S)$ is a right T–submodule of E', and that $B \cong (T/P)_T$. If M is any submodule of E' such that $M > B$, show that M contains the row $C = (R/\theta S \;\; R/\theta S)$. Conclude that if M has an affiliated series $0 = M_0 < M_1 < \dots < M_n = M$, the first two corresponding affiliated primes must be P and Q, and M_1 and M_2/M_1 must be torsionfree modules over T/P and T/Q respectively.

(e) Show that P satisfies the right strong second layer condition. [Hint: Proposition 11.3.] □

♦ ARTINIAN RINGS

Any noetherian or artinian ring can be written in a unique way as a direct product of indecomposable rings. In this section we show as an application of Theorem 11.4 that the cliques in an artinian ring (more generally, in a noetherian ring in which all primes are maximal) are in one–to–one correspondence with these indecomposable factors.

DEFINITION. A ring R is *indecomposable* (as a ring) provided R is not isomorphic to a direct product of two nonzero rings. Note that R is indecomposable if and only if R cannot be expressed as a direct sum of two nonzero ideals.

If a ring R is isomorphic to a direct product $R_1 \times ... \times R_n$ of rings R_i, it is often convenient to identify R with $R_1 \times ... \times R_n$, and to identify each R_i with a non–unital subring of R, as follows. For $i = 1,...,n$ let e_i denote the n–tuple in R with 1 in the i–th. position and 0 elsewhere. These elements e_i form a *complete set of orthogonal central idempotents* in R, meaning that (a) they are in the center of R, (b) $e_i^2 = e_i$ for all i, (c) $e_ie_j = 0$ for $i \neq j$, and (d) $e_1 + ... + e_n = 1$. Each R_i is identified with the subset e_iR of R, which is a ring with identity e_i and also an ideal of R. Observe that if I is any ideal of R, then

$$I = e_1I + ... + e_nI = (I \cap R_1) + ... + (I \cap R_n).$$

In particular, if P is a prime ideal of R then since $R_iR_j = 0$ for $i \neq j$ we see that P must contain all but one of the R_i. Thus if $P \supseteq R_i$ for all indices $i \neq k$, then

$1 - e_k = \sum_{i \neq k} e_i \in P$ and $P = (1 - e_k)R + (P \cap R_k)$, where $P \cap R_k$ is a prime ideal of R_k. Hence, P is in an obvious sense associated with the factor R_k. (Note also that P equals the inverse image of the prime $P \cap R_k$ under the natural projection map $R \to R_k$.)

Our next observation, applied in the special case in which $c = 1 - e_k$, shows in particular that primes associated with different factors of a ring decomposition cannot be linked.

LEMMA 11.7. *Let R be a noetherian ring, P a prime ideal of R, and c a central element in R. If $c \in P$, then all prime ideals in $Cl(P)$ contain c.*

Proof. It suffices to show that if Q is a prime satisfying either $Q \rightsquigarrow P$ or $P \rightsquigarrow Q$, then $c \in Q$. We prove only the first case, the second being symmetric. Let $(Q \cap P)/A$ be a linking bimodule where $A \geq QP$. Then since $(Q \cap P)P \leq A$ it follows that $(Q \cap P)c \subseteq A$. Since c is central, we have $c(Q \cap P) \subseteq A$, and so $c \in \text{l.ann}_R((Q \cap P)/A) = Q$. □

PROPOSITION 11.8. *If R is an indecomposable noetherian ring in which all prime ideals are maximal, then $Spec(R)$ consists of a single clique.*

Proof. Observe that R satisfies the strong second layer condition. Let X be a clique in $Spec(R)$, and let $Y = Spec(R) - X$. Set

$$I = \{a \in R \mid aP_1P_2 \cdots P_m = 0 \text{ for some } P_1,...,P_m \in X\}$$
$$J = \{a \in R \mid aQ_1Q_2 \cdots Q_n = 0 \text{ for some } Q_1,...,Q_n \in Y\},$$

and observe that I and J are ideals of R. Since all primes of R are maximal, no maximal ideal of R can contain both a prime from X and a prime from Y. Consequently,

$$(P_1P_2 \cdots P_m) + (Q_1Q_2 \cdots Q_n) = R$$

for any $P_1,\ldots,P_m \in X$ and $Q_1,\ldots,Q_n \in Y$. It follows that $I \cap J = 0$.

If K/I is a nonzero right submodule of R/I, then K/I cannot be annihilated by a prime in X, for then K would be annihilated by a product of primes in X and so $K \leq I$. Thus all the associated primes of $(R/I)_R$ are in Y. Since Y is right link closed, Theorem 11.4 shows that $(R/I)_R$ is annihilated by some product of primes from Y.

Now $R/(I + J)$ is annihilated (on the right) by a product of primes from Y. Similarly, $R/(I + J)$ is also annihilated by a product of primes from X, and we conclude that $I + J = R$. Hence, as R is indecomposable, either $I = R$ or $J = R$.

If $J = R$, some product of primes from Y equals zero, and so $\mathrm{Spec}(R) = Y$, which is impossible. Therefore $I = R$, and consequently $\mathrm{Spec}(R) = X$. □

COROLLARY 11.9. *If R is a noetherian ring in which all prime ideals are maximal, there is a direct product decomposition $R = R_1 \times \ldots \times R_n$ such that each R_i is an indecomposable ring and $Spec(R_i)$ consists of a single clique. If $e_1,\ldots,e_n$ are the corresponding central idempotents in R (so that $e_1 + \ldots + e_n = 1$ and each $R_i = e_iR$), then the cliques in Spec(R) are the sets*

$$X_i = \{P \in Spec(R) \mid 1 - e_i \in P\}$$

for $i = 1,\ldots,n$.

Proof. The existence of a ring decomposition $R = R_1 \times \ldots \times R_n$ with each R_i indecomposable follows from the chain conditions on R, and then by Proposition 11.8 each R_i has only one clique of primes. It follows that each of the sets X_i is contained in a clique of Spec(R). Then by Lemma 11.7 each X_i is a clique, and since any prime of R must contain one of the idempotents $1 - e_i$ there are no other cliques. □

If R is an artinian ring, the decomposition given in Corollary 11.9 is called the *block decomposition* of R.

COROLLARY 11.10. *Let R be a noetherian ring and A a finitely generated right R–module such that $R/ann_R(A)$ is artinian. Then there is a direct sum decomposition $A = A_1 \oplus \ldots \oplus A_n$ such that for $i = 1,\ldots,n$ the annihilators of the composition factors of A_i all lie in the same clique of Spec(R).*

Proof. Set $I = \mathrm{ann}_R(A)$. By Corollary 11.9, there exist orthogonal central idempotents $e_1,\ldots,e_n$ in R/I such that $e_1 + \ldots + e_n = 1$ and each of the sets

$$X_i = \{P \in \mathrm{Spec}(R/I) \mid 1 - e_i \in P\}$$

is a clique of Spec(R/I). Then $A = A_1 \oplus ... \oplus A_n$ where each $A_i = Ae_i$.

If Y_i is the set of primes of R occurring as annihilators of composition factors of A_i, then $Q/I \in X_i$ for all $Q \in Y_i$ and so $\{Q/I \mid Q \in Y_i\}$ is contained in a clique of Spec(R/I). It follows that Y_i is contained in a clique of Spec(R). □

♦ THE ARTIN–REES PROPERTY

In this section we introduce a new method for showing that a ring satisfies the strong second layer condition. We have already noted that fully bounded noetherian rings (and, in particular, rings finitely generated as modules over their centers, when the centers are noetherian) satisfy the right and left second layer conditions. In fact, many results such as Theorem 11.4 and the unfaithfulness property of Corollary 11.5 were initially proved by Jategaonkar for fully bounded rings (by the methods indicated here) before the more general class of rings was discovered. Many rings which are not fully bounded still satisfy the second layer condition. In the known examples, an important role is played by the AR–property (named after Artin and Rees) which we now discuss.

DEFINITION. An ideal I in a ring R has the *right AR–property* if for every right ideal K of R, there is a positive integer n such that $K \cap I^n \leq KI$. The *left AR–property* is defined similarly, and I has the *AR–property* if it has both the right and left AR–properties.

EXERCISE 11N. Show that of the two prime ideals in the ring $\begin{pmatrix} k & k \\ 0 & k \end{pmatrix}$, where k is a field, one has the right AR–property but not the left AR–property, while the other has the left AR–property but not the right AR–property. □

We should remark that no example is known of an ideal I in a *prime* noetherian ring such that I has the right AR–property but not the left AR–property.

EXERCISE 11O. Show that in the ring $\begin{pmatrix} \mathbb{Z} & \mathbb{Z} \\ 2\mathbb{Z} & \mathbb{Z} \end{pmatrix}$, the prime ideal $\begin{pmatrix} 2\mathbb{Z} & \mathbb{Z} \\ 2\mathbb{Z} & \mathbb{Z} \end{pmatrix}$ satisfies neither the right nor the left AR–property. □

LEMMA 11.11. *The right AR–property for an ideal I in a ring R is equivalent to each of the following properties:*

(a) For every finitely generated right R–module A and every submodule $B \leq A$, there is a positive integer n such that $B \cap AI^n \leq BI$.

(b) For every finitely generated right R–module M with an essential submodule N satisfying $NI = 0$, there is a positive integer n such that $MI^n = 0$.

Proof. Suppose first that I has the right AR–property, and consider modules M,N as in (b). Since M is finitely generated, to get $MI^n = 0$ we need only show that each element of a set of generators for M is annihilated by a power of I. Thus let $x \in M$, and let $K = \{r \in R \mid xr \in N\}$. Observe that $xKI \leq NI = 0$. By the right AR–property, there is a positive integer n such that $K \cap I^n \leq KI$, and we claim that $xI^n \cap N = 0$. Namely, if $r \in I^n$ and $xr \in N$, then $r \in K \cap I^n \leq KI$, whence $xr = 0$. Thus $xI^n \cap N = 0$, as claimed. Since $N \leq_e M$, we conclude that $xI^n = 0$, as desired. This proves (b).

(b) $\Rightarrow$ (a): Choose a submodule C of A maximal with respect to the property that $B \cap C = BI$. Then the embedding $B/BI \to A/C$ maps B/BI isomorphically onto an essential submodule of A/C (Proposition 3.22). Condition (b) now implies that $(A/C)I^n = 0$ for some positive integer n, or, in other words, $AI^n \leq C$. Thus $B \cap AI^n \leq B \cap C = BI$, which verifies (a).

Finally, it is clear that (a) implies the right AR–property (take $A = R$). □

We now present an important method for verifying the AR–property, originally introduced by Rees.

DEFINITION. If R is a ring and I an ideal of R then the *Rees ring* of I is the subring $\mathcal{R}(I)$ of the polynomial ring $R[x]$ defined by

$$\mathcal{R}(I) = R + Ix + I^2x^2 + \dots + I^ix^i + \dots .$$

LEMMA 11.12. *If I is an ideal in a ring R, and the Rees ring $\mathcal{R}(I)$ is right noetherian, then I satisfies the right AR–property.*

Proof. Let K be a right ideal of R and let

$$K^* = K[x] \cap \mathcal{R}(I) = K + (K \cap I)x + (K \cap I^2)x^2 + \dots + (K \cap I^i)x^i + \dots .$$

If $\mathcal{R}(I)$ is right noetherian then the right ideal K^* must be finitely generated, and therefore K^* is generated by $K + (K \cap I)x + \dots + (K \cap I^{n-1})x^{n-1}$ for some positive integer n. Since $(K \cap I^n)x^n \leq K^*$, we see by comparing coefficients that

$$K \cap I^n = \sum_{i=0}^{n-1} (K \cap I^i)I^{n-i} \leq KI. \ \square$$

THEOREM 11.13. [Artin, Rees] *If R is a noetherian ring and I is an ideal of R generated by central elements, then $\mathcal{R}(I)$ is noetherian and hence I satisfies the AR–property.*

Proof. The noetherian assumption ensures that I can be generated by finitely many central elements. If I is generated by n central elements, say $a_1,\dots,a_n$, then $\mathcal{R}(I)$ is generated as an R–algebra by the central elements $a_1x,...,a_nx$. Then $\mathcal{R}(I)$ is a homomorphic image of a polynomial ring $R[x_1,...,x_n]$, and hence is noetherian by the Hilbert Basis Theorem (or Theorem 1.12). □

The AR–property for ideals in commutative noetherian rings can also be obtained from Exercise 4Z and Lemma 11.11. We have concentrated on the Rees ring approach because that provides the method we shall use to produce ideals with the AR–property in differential operator rings (Theorem 11.15).

DEFINITION. A ring R is *right AR–separated* if for every pair of prime ideals P and Q in R such that $Q < P$, there is an ideal I such that $Q < I \leq P$ and I/Q satisfies the right AR–property in R/Q. *Left AR–separated* is defined symmetrically, and R is *AR–separated* if R is both left and right AR–separated.

LEMMA 11.14. *If R is a noetherian ring which is right AR–separated, then R satisfies the right strong second layer condition.*

Proof. If R does not satisfy the right strong second layer condition, there are prime ideals P and Q with $Q < P$, and a finitely generated right R–module M with an essential submodule U such that $UP = 0$ and $\text{ann}_R(M) = Q$. Since R is right AR–separated, there is an ideal I such that $Q < I \leq P$ and such that I/Q has the right AR–property. Since $UI = 0$ and $MQ = 0$, we can regard M as an (R/Q)–module and conclude that $MI^n = 0$ for some positive integer n (Lemma 11.11). Hence, $I^n \leq Q$, which is impossible since Q is prime and $Q < I$. □

THEOREM 11.15. [Bell, Sigurdsson] *Let R be a commutative noetherian ring with a derivation δ, and let $T = R[\theta;\delta]$. Then any ideal of T of the form IT, where I is a δ–ideal of R, has the AR–property. If, moreover, R is a $\mathbb{Q}$–algebra, then T is AR–separated and thus satisfies the strong second layer condition.*

Proof. First let I be a δ–ideal of R, and consider the Rees rings $\mathcal{R}(I)$ and $\mathcal{R}(IT)$. We can extend δ to a derivation δ^* on R[x] such that $\delta^*(x) = 0$, and since I is a δ–ideal of R, we see that δ^* restricts to a derivation on $\mathcal{R}(I)$. As $\mathcal{R}(I)$ is a noetherian ring (Theorem 11.13), we conclude from Theorem 1.12 that $\mathcal{R}(I)[\theta;\delta^*]$ is noetherian. On the other hand, a comparison of coefficients shows that there is a natural isomorphism $\mathcal{R}(I)[\theta;\delta^*] \cong \mathcal{R}(IT)$ (restricted from the natural isomorphism of $R[x][\theta;\delta^*]$ onto $R[\theta;\delta][x]$). Therefore $\mathcal{R}(IT)$ is noetherian, and hence IT has the right and left AR–properties.

Now assume that R is a $\mathbb{Q}$–algebra, and let P and Q be prime ideals of T with

$Q < P$. We must find an ideal J such that $Q < J \leq P$ and J/Q has the AR–property. If $Q = (Q \cap R)T$ and $P \cap R > Q \cap R$, the ideal $J = (P \cap R)T$ satisfies $Q < J \leq P$. Since $P \cap R$ is a δ–ideal of R (Lemma 2.18), J has the AR–property in T, and it follows that J/Q has the AR–property in T/Q.

Finally, if either $Q > (Q \cap R)T$ or $P \cap R = Q \cap R$, then there is a prime ideal $K > (Q \cap R)T$ such that $K \cap R = Q \cap R$ (namely, either $K = Q$ or $K = P$). Then according to Theorem 2.22, T/Q is commutative, and we can take $J = P$. Therefore T is AR–separated. □

Theorem 11.15 remains true even if R is not necessarily a $\mathbb{Q}$–algebra, but we shall not develop the methods needed to prove this.

EXERCISE 11P. Show that if R is a commutative noetherian ring and α an automorphism of R, then $R[\theta,\theta^{-1};\alpha]$ is AR–separated, and hence satisfies the strong second layer condition. [Hint: Use Proposition 8.1 in one step.] □

♦ LINK–FINITENESS

In this section we show how the presence of ideals with the AR–property places restrictions on possible links between prime ideals. Our first observation is a generalization of Lemma 11.7. We give a direct proof from the definitions; alternatively, Theorem 11.2 and Lemma 11.11 may be used.

PROPOSITION 11.16. *Let I be an ideal in a noetherian ring R, and let P and Q be prime ideals of R with $Q \rightsquigarrow P$. If $I \leq P$ and I satisfies the right AR–property, then $I \leq Q$. Similarly, if $I \leq Q$ and I satisfies the left AR–property, then $I \leq P$.*

Proof. Assume first that $I \leq P$ and that I satisfies the right AR–property. Let $(Q \cap P)/A$ be a linking bimodule for the link $Q \rightsquigarrow P$. In particular, $QP \leq A$ and $\text{l.ann}_R((Q \cap P)/A) = Q$. From the right AR–property, $Q \cap P \cap I^n \leq (Q \cap P)I$ for some positive integer n. Then

$$I^n(Q \cap P) \leq Q \cap P \cap I^n \leq (Q \cap P)I \leq QP \leq A,$$

whence $I^n \leq \text{l.ann}_R((Q \cap P)/A) = Q$, and therefore $I \leq Q$.

The final statement of the proposition is proved symmetrically. □

EXERCISE 11Q. Let $T = R[\theta;\delta]$ where R is a commutative noetherian ring, and let P and Q be prime ideals of T. If P and Q are in the same clique, show that $P \cap R = Q \cap R$. □

LEMMA 11.17. *If Q is a minimal prime ideal in a semiprime noetherian ring R, there are no prime ideals of R linked to or from Q.*

Proof. If $Q \rightsquigarrow P$ for some prime P, then by Theorem 11.2 there exists a uniform right R–module M with a nonzero submodule U such that M/U is isomorphic to a right ideal of R/Q. In particular, M/U is a torsionfree (R/Q)–module, and since Q is a minimal prime, M/U must also be torsionfree as an R–module (Proposition 6.8). However, since $U \leq_e M$ this is impossible. Therefore Q is not linked to any primes in R. By symmetry, no primes are linked to Q either. □

THEOREM 11.18. [Müller, Brown] *Let Q be a prime ideal in a noetherian ring R. If R is left AR–separated, Q is linked to at most finitely many prime ideals in R, while if R is right AR–separated, at most finitely many prime ideals are linked to Q.*

Proof. By symmetry, it suffices to prove the first statement. If the theorem fails, we may assume that R is a minimal criminal, i.e., that there is an infinite set $X \subseteq \mathrm{Spec}(R)$ such that Q is linked to each prime in X, but that if I is a nonzero ideal contained in Q, then Q/I is linked to at most finitely many primes of R/I.

Choose an ideal T of R maximal with respect to the property that T equals the intersection of some infinite subset of X. Given ideals H_1, H_2 properly containing T, it follows from the maximality of T that each H_i is contained in at most finitely many primes from X. Hence, H_1H_2 is contained in at most finitely many primes from X, and so $H_1H_2 \nleq T$. Thus T is a prime ideal. After replacing X by $\{P \in X \mid P \geq T\}$, we may assume that $T = \bigcap X$.

For each $P \in X$, there exists an ideal A_P such that $Q \cap P > A_P \geq QP$ and $(Q \cap P)/A_P$ is torsionfree as a left (R/Q)–module and as a right (R/P)–module. Note that if I is an ideal of R and $I \leq A_P$, then Q/I and P/I are primes of R/I satisfying $Q/I \rightsquigarrow P/I$. Thus, by our noetherian induction, any nonzero ideal of R can be contained in at most finitely many of the ideals A_P. In other words, $\bigcap \{A_P \mid P \in Y\} = 0$ for any infinite subset $Y \subseteq X$.

In particular, it follows that

$$QT \leq \bigcap \{QP \mid P \in X\} \leq \bigcap \{A_P \mid P \in X\} = 0,$$

and hence $Q \cap T$ is a left (R/Q)–module. Since $Q \cap T \leq Q \cap P$ for all $P \in X$, there is a natural left (R/Q)–module homomorphism

$$Q \cap T \to \prod_{P \in X} (Q \cap P)/A_P,$$

and this map is injective because $\bigcap \{A_P \mid P \in X\} = 0$. As each $(Q \cap P)/A_P$ is a torsionfree left (R/Q)–module, it follows that $Q \cap T$ must be torsionfree as a left (R/Q)–module.

We claim that $Q \cap T = 0$. If not, choose a nonzero ideal $B \leq Q \cap T$ with $\text{rank}({}_RB)$ as small as possible. Observe that if C is any nonzero ideal contained in B, then $\text{rank}({}_RC) = \text{rank}_R(B)$ (by minimality of $\text{rank}({}_RB)$), whence ${}_RC \leq_e {}_RB$ (Corollary 4.17), and consequently B/C is torsion as a left (R/Q)–module.

Now if $P \in X$ and $B \cap A_P \neq 0$, then $B/(B \cap A_P)$ is torsion as a left (R/Q)–module. However, $B/(B \cap A_P)$ embeds in the torsionfree left (R/Q)–module $(Q \cap P)/A_P$, whence $B/(B \cap A_P) = 0$ and so $B \leq A_P$. On the other hand, if $P \in X$ and $B \cap A_P = 0$, then B embeds in the torsionfree right (R/P)–module $(Q \cap P)/A_P$, and hence $\text{r.ann}_R(B) = P$. This is obviously possible for at most one prime $P \in X$. Thus $B \leq A_P$ for all but at most one $P \in X$. However, as $B \neq 0$, this contradicts the fact that $\bigcap \{A_P \mid P \in Y\} = 0$ for all infinite subsets $Y \subseteq X$. Therefore $Q \cap T = 0$, as claimed. In particular, R is a semiprime ring.

By Lemma 11.17, Q cannot be a minimal prime. If Q_0 is a minimal prime contained in Q, then as $QT = 0 \leq Q_0$ we must have $T \leq Q_0 < Q$, whence $T = Q \cap T = 0$. Thus R is a prime ring.

Since R is left AR–separated, there is a nonzero ideal $J \leq Q$ such that J satisfies the left AR–property. Proposition 11.16 then shows that $J \leq P$ for all $P \in X$. But now $J \leq \bigcap X = T = 0$, a contradiction.

Therefore the theorem holds. □

Theorem 11.18 also holds for FBN rings, as we shall see later (Theorem 14.22). In general, a prime ideal in a noetherian ring can be linked to or from at most countably many primes (Theorem 14.23). Stafford has constructed an example of a noetherian ring containing a prime linked to infinitely many primes [1985, Theorem 4.4], but this example is too complicated to reproduce here.

♦ MODULE–FINITE ALGEBRAS

Given a module–finite algebra R over a commutative noetherian ring S, we shall completely describe the cliques of prime ideals in R, in terms of their intersections with the center of R. Observe that the center of R is noetherian as an S–module and hence is a noetherian ring, and that R is a finitely generated module over its center. Hence, we may as well replace S by the center of R.

We proceed by using the AR–property, and then by reducing things to an artinian ring, where we apply Proposition 11.8.

EXERCISE 11R. Let R be a ring module–finite over its center S. If S is noetherian and I is an ideal of S, show that IR has the AR–property. □

DEFINITION. Given an ideal I in a ring R, we say that an element w in R/I *lifts* to an element x in R if $x+I=w$.

LEMMA 11.19. *Let R be a ring module–finite over its center S, and assume that S is noetherian. If I is any ideal of S, there exists a positive integer n such that every element of R/IR which can be lifted to the center of R/I^nR can also be lifted to S.*

Proof. Choose elements $x_1,\dots,x_t$ that generate R as an S–module, and set $A = R^t$, viewed as a left S–module. Since R is a finitely generated S–module, so is A. We may define an S–module homomorphism $f : R \to A$ according to the rule

$$f(r) = (rx_1 - x_1r,\ rx_2 - x_2r,\ \dots,\ rx_t - x_tr),$$

and we observe that $\ker(f) = S$. Since I has the AR–property, Lemma 11.11 shows that there exists a positive integer n such that $f(R) \cap I^nA \le If(R)$, that is, $f(R) \cap I^nA \le f(IR)$.

Any coset in R/IR which lifts to the center of R/I^nR has the form $r + IR$ where $r + I^nR$ is central in R/I^nR. Then $rx_i - x_ir \in I^nR$ for $i = 1,\dots,t$, from which we see that $f(r) \in I^nA$. As $f(R) \cap I^nA \le f(IR)$, we obtain $f(r) = f(x)$ for some $x \in IR$. Now $r-x \in \ker(f) = S$, and since $r-x+IR = r+IR$ we have lifted $r+IR$ to an element of S. □

EXERCISE 11S. Let X be a right denominator set in a noetherian ring R, and let P,Q be primes of R disjoint from X. By Theorem 9.22, the extended ideals P^e,Q^e are primes of RX^{-1}. Show that $Q^e \rightsquigarrow P^e$ if and only if $Q \rightsquigarrow P$. □

THEOREM 11.20. [Müller] *Let R be a ring module–finite over its center S, and assume that S is noetherian. Then for any prime ideal P of R, the contraction $P \cap S$ is a prime ideal of S, and*

$$Cl(P) = \{Q \in Spec(R) \mid Q \cap S = P \cap S\}.$$

Proof. If $x,y \in S$ and $xy \in P \cap S$, then $xRy = xyR \subseteq P$ and so either $x \in P$ or $y \in P$, whence either $x \in P \cap S$ or $y \in P \cap S$. Thus $P \cap S$ is a prime of S.

We next remark, as an immediate consequence of Lemma 11.7, that any two linked primes of R have the same contraction in S. Consequently, $Cl(P)$ is contained in the set

$$Y = \{Q \in Spec(R) \mid Q \cap S = P \cap S\},$$

and it remains to show that $Cl(P) \supseteq Y$.

Set $M = P \cap S$ and $X = S - M$, and observe that X is a right (and left) denominator set in R. Also (the natural image of) SX^{-1} equals the center of RX^{-1}. (Use the fact that any element of R which commutes with a set of generators for R_S will be in S.) By Theorem 9.22, each of the extended ideals Q^e, for $Q \in Y$, is a prime of RX^{-1}, and for such Q we note that

$$Q^e \cap SX^{-1} = P^e \cap SX^{-1} = M(SX^{-1}).$$

Moreover, if the primes in RX^{-1} contracting to $M(SX^{-1})$ all belong to the same clique in $\mathrm{Spec}(RX^{-1})$, it will follow from Exercise 11S that the primes in Y all belong to the same clique in $\mathrm{Spec}(R)$, and therefore $\mathrm{Cl}(P) = Y$.

Hence, without loss of generality, we may assume that M is a maximal ideal of S. Note that $MR \cap S \leq P \cap S$, whence $MR \cap S = M$.

According to Lemma 11.19, there exists a positive integer n such that every element of R/MR that lifts to the center of R/M^nR also lifts to S. We use this property to show that R/M^nR is indecomposable (as a ring). Let e be a central idempotent in R/M^nR, and write $e = a + M^nR$ for some $a \in R$. Then $a + MR$ is an element of R/MR that lifts to the center of R/M^nR, and so $a + MR$ lifts to S. Hence, $a + MR = b + MR$ for some $b \in S$. Since e is idempotent, so is $a + MR$, whence $b - b^2 \in MR \cap S = M$. As S/M is a field, either $b \in M$ or $1 - b \in M$. Then either $a \in MR$ or $1 - a \in MR$, whence either e or $1 - e$ lies in MR/M^nR, and so either e or $1 - e$ is nilpotent. We conclude that either $e = 0$ or $e = 1$. Therefore R/M^nR is indecomposable, as claimed.

Finally, S/M^n is an artinian ring, and R/M^nR is finitely generated as an (S/M^n)–module, whence R/M^nR is an artinian ring. By Proposition 11.8, all primes of R/M^nR lie in the same clique of $\mathrm{Spec}(R/M^nR)$. It follows that all primes of R containing M^nR lie in the same clique of $\mathrm{Spec}(R)$, and therefore $Y \subseteq \mathrm{Cl}(P)$. □

♦ NOTES

Links. Ideal links (not named, but with the notation $\rightsquigarrow$) appeared in work of Jategaonkar [1973, pp. 153–154]. Second layer links (called links, and with the notation $\rightsquigarrow$) between prime ideals in FBN rings appeared in work of Müller [1976b, p. 235].

Jategaonkar's Main Lemma. The first version was proved for FBN rings [1974b, Lemma 2.4], where alternative (i) does not occur. A version with U torsionfree (in the notation of Theorem 11.1) was given in [1981, Lemma 4.1], and the full version in [1982, Lemma 2.2]. Finally, a version for one–sided noetherian rings was proved in [1986, Lemma 6.1.3].

Characterization of Links via Uniform Modules. Theorem 11.2 was first proved by Jategaonkar under the hypothesis that Q is not properly contained in P [1986, Lemma 6.1.6]. Brown observed that this restriction is unnecessary [1985, Remark 2.3(ii)].

Second Layer Condition. The right second layer condition for a noetherian ring R (in the form given in Exercise 11L) was introduced by Jategaonkar in [1979, p. 167] under the name (*), and in [1981, p. 386] he defined the condition $(*)_r$ for single primes. He presented and named the right second layer condition for a right noetherian ring R (in a form similar to condition (b) of Proposition 11.3) in [1982, p. 47]. In the same paper, he introduced the right strong second layer condition for R (in the same form as Corollary 11.5) under the name $\binom{*}{*}_r$ [1982, pp. 23, 24]. Finally, he presented and named the right strong second layer condition for single primes in [1986, p. 220].

Links from Affiliated to Associated Primes. Jategaonkar first proved versions of Theorems 11.4 and 11.6 for noetherian bimodules (where the second layer condition hypotheses are not needed) in [1981, Theorem 4.2]. He then proved right module versions in [1982, Theorem 3.1] and [1986, Theorem 9.1.2].

Artin–Rees Property. The AR–property for ideals in commutative noetherian rings – in a stronger form than we have given – was proved independently by Artin (apparently unpublished; see Nagata [1962, p. 212]) and Rees [1956, Lemma 1]. Rees's proof used a version of what is now called the Rees ring, as in Lemma 11.12.

AR–Separation Implies Strong Second Layer Condition. Jategaonkar proved in [1979, Proposition 10] that AR–separated noetherian rings satisfy the second layer condition, and then improved the conclusion to the strong second layer condition in [1982, Proposition 4.1]. He used this to show that enveloping algebras of solvable Lie algebras satisfy the strong second layer condition [1982, p. 61; 1986, Theorem A.3.9], and that group rings of polycyclic–by–finite groups with commutative noetherian coefficient rings satisfy it [1979, Proposition 10; 1982, Theorem 4.5].

Second Layer Condition in Differential Operator Rings. Bell proved that in a large class of iterated differential operator rings with commutative noetherian coefficient rings, the second layer condition holds [1987, Theorem 7.3]. In the case $T = R[\theta;\delta]$ where R is a commutative noetherian $\mathbb{Q}$–algebra, Sigurdsson showed that T is AR–separated [1986, Proposition 2.4], from which it follows that T satisfies the strong second layer condition.

AR–Separation Implies Link–Finiteness. Müller proved that if Q is a prime ideal in a noetherian P.I. ring R, then at most finitely many primes of R can be linked to or from Q [1979, Theorem 7; 1980, Theorem 7]; his proof used central elements where the proof we have given of Theorem 11.18 uses AR–ideals. Brown proved the corresponding result for primes in the group ring of a polycyclic–by–finite group with a suitable commutative noetherian coefficient ring [1981b, Theorem 6.4], and he remarked that his proof could be modified to obtain the result for AR–separated noetherian rings satisfying certain Krull dimension symmetry hypotheses [1981b, p. 279]. The general theorem appeared in Jategaonkar [1986, Proposition 8.1.8].

Cliques in Module–Finite Algebras. Müller defined a finite set X of incomparable prime ideals in a noetherian ring R to be *classical* provided $\mathscr{C}(\cap X)$ is an Ore set and the Jacobson radical of the corresponding localization has the AR–property, and he defined a *cycle* (later: *clan*) to be a classical set of primes such that no proper subset is classical. (In the presence of the second layer condition, it follows from Theorem 12.21 that cycles/clans are the same as finite cliques.) Müller proved that if R is a noetherian ring which is module–finite over its center S, then the cycles/clans in R are precisely the sets $\{P \in \mathrm{Spec}(R) \mid P \cap S = Q\}$ for $Q \in \mathrm{Spec}(S)$ [1974b, Theorem 6.1; 1976a, Theorem 7].

12 RINGS SATISFYING THE SECOND LAYER CONDITION

The main theme of this chapter is to explore the ideal theory of noetherian rings satisfying the second layer condition. This is a very large class of rings (as we began to see in the previous chapter), including many iterated differential operator rings and iterated skew–Laurent extensions, as well as the group rings of polycyclic–by–finite groups and the enveloping algebras of finite–dimensional solvable Lie algebras. It turns out that these rings have many properties which are not shared by other noetherian rings, and which can be thought of as generalizations of well–known properties of commutative rings. We begin with a symmetry property of bimodules over these rings. This will give us immediate information about the graphs of links of these rings, and will also give us the key tool to prove two intersection theorems – a strong form of Jacobson's Conjecture, and an analogue of the Krull Intersection Theorem. Rings satisfying the second layer condition also behave well with respect to finite extensions. If R is a noetherian ring satisfying the second layer condition and R is a subring of a ring S such that S is finitely generated as both a left and as a right R–module, we prove that S also satisfies the second layer condition, and that "Lying Over" holds for the prime ideals in this setup. Finally, rings with the second layer condition turn out to be the natural class of rings in which to discuss localization at prime ideals and families of prime ideals. The results in this direction are still not complete, but we will present what is probably going to remain the definitive result on localization at semiprime ideals, and give some indications concerning the further development of the theory.

♦ CLASSICAL KRULL DIMENSION

Given a link $Q \rightsquigarrow P$ in a noetherian ring R, we would like to be able to say that the factor rings R/Q and R/P are similar in some fashion, for instance that there is some notion of "dimension" for rings which yields the same value for R/Q as for R/P. The dimension which has proved most useful in this regard, at least in the presence of the second layer condition, is "classical Krull dimension", which we introduce in this section. (A different "Krull dimension", better adapted to module theory, will be developed in the following chapter.) We should emphasize in advance that except in the finite case, the actual value of this dimension for a ring has not had great significance. The importance of the dimension is that it has provided an invariant with certain good features and with the property that it distinquishes between a prime ring R and a prime factor R/P. In particular, classical Krull dimension provides a basis for proofs via transfinite induction.

In this and the following chapters we will be using invariants with arbitrary ordinal values. In this way we obtain invariants which apply to all noetherian rings. The reader who is uncomfortable with infinite ordinals may be relieved to know that in most of the important examples which arise in applications, the invariants are actually finite. The reader may assume if desired that all of the invariants are finite in these chapters, but then, of course, the results will not be as general as those stated here.

Classical Krull dimension was first defined for commutative noetherian rings, by counting lengths of chains of prime ideals. As with composition series, it is the *gaps* between the primes that are counted, so that a single prime is viewed as a chain of length 0, and a chain $P_0 > P_1 > \dots > P_n$ has length n. The classical Krull dimension of a ring R was originally defined to be the supremum of the lengths of all chains of prime ideals in R. Then, in order to distinguish among rings with infinite classical Krull dimension, Krause introduced a refinement of the definition allowing infinite ordinal values.

The smallest classical Krull dimension that will occur for a nonzero ring is 0. In order to have a convenient value for the classical Krull dimension of the zero ring, and in order to conform with the standard usage for the "non–classical" Krull dimension to be introduced in the following chapter, we make the convention that -1 is to be considered an ordinal number.

DEFINITION. Let R be a ring. We define, by transfinite induction, sets X_α of prime ideals of R, for each ordinal α. To start with, let X_{-1} be the empty set. Next, consider an ordinal $\alpha \geq 0$; if X_β has been defined for all ordinals $\beta < \alpha$, let X_α be the set of those prime ideals P in R such that all prime ideals properly containing P belong to $\bigcup_{\beta<\alpha} X_\beta$. (In particular, X_0 is the set of maximal ideals of R.) If some X_γ contains all prime ideals of R, we say that *Cl.K.dim(R) exists*, and we set Cl.K.dim(R) – the *classical Krull dimension of* R – equal to the smallest such γ.

For example, any simple ring has classical Krull dimension 0, as does any artinian ring. The ring $\mathbb{Z}$ has classical Krull dimension 1. Some other examples will be given in the exercises.

PROPOSITION 12.1. *If R is a ring with the ACC on prime ideals, then Cl.K.dim(R) exists.*

Proof. Define the sets X_α of prime ideals as in the definition above. Since there is a bound on the cardinalities of these sets (e.g., $2^{\text{card}(R)}$), the transfinite chain $X_{-1} \subseteq X_0 \subseteq X_1 \subseteq \ldots$ cannot be properly increasing forever. Hence, there exists an ordinal γ such that $X_\gamma = X_{\gamma+1}$.

If Cl.K.dim(R) does not exist, then X_γ does not contain all the prime ideals of R. Using the ACC on prime ideals, there is a prime P in R maximal with respect to the property $P \notin X_\gamma$. Hence, all primes properly containing P lie in X_γ. But then $P \in X_{\gamma+1} = X_\gamma$, a contradiction.

Therefore Cl.K.dim(R) exists. □

EXERCISE 12A. Let R be a ring for which Cl.K.dim(R) exists.

(a) For any ideal I of R, show that Cl.K.dim(R/I) exists and is no larger than Cl.K.dim(R).

(b) If P and Q are prime ideals of R such that $P < Q$, show that Cl.K.dim(R/P) > Cl.K.dim(R/Q). Conclude that R satisfies the ACC on prime ideals.

(c) If R is right or left noetherian, P is a prime ideal of R, and I is an ideal of R with $I > P$, show that Cl.K.dim(R/I) < Cl.K.dim(R/P).

(d) If S is a polynomial ring in an infinite number of indeterminates over a nonzero ring, show that Cl.K.dim(S) does not exist. □

EXERCISE 12B. If $R = k[x_1,\ldots,x_n]$ is a polynomial ring over a field k in n independent indeterminates, show that Cl.K.dim(R) = n. [Hints: To see that Cl.K.dim(R) $\leq$ n, it is enough to show that if $P_0 > P_1 > \ldots > P_n$ is a chain of primes in R, then P_0 is a maximal ideal. By induction, $k(x_1)[x_2,\ldots,x_n]$ has classical Krull dimension $n-1$. Use this to show that $P_0 \cap k[x_1] \neq 0$, and similarly $P_0 \cap k[x_i] \neq 0$ for all i. Then show that R/P_0 is finite–dimensional.] □

LEMMA 12.2. *Let R be a ring for which Cl.K.dim(R) exists, and let Cl.K.dim(R) = γ. If α is any nonnegative ordinal strictly less than γ, then there is a prime ideal P of R such that Cl.K.dim(R/P) = α. If R is right or left noetherian, then there is a minimal prime P of R such that Cl.K.dim(R/P) = γ.*

Proof. Consulting the definition, we see for a prime ideal P that Cl.K.dim(R/P) = α if and only if $P \in X_\alpha$ while $P \notin X_\beta$ for all $\beta < \alpha$. If there is no prime ideal P such that Cl.K.dim(R/P) = α, then we must have $X_\alpha = X_{\alpha+1}$, from which it would follow that $X_\beta = X_\alpha$ for all $\beta > \alpha$. We would then have Cl.K.dim(R) $\leq \alpha$, contrary to hypothesis. Therefore there must be a prime P such that Cl.K.dim(R/P) = α. This argument does not apply when $\alpha = \gamma$. However, it does show that Cl.K.dim(R) is the supremum of the ordinals Cl.K.dim(R/P) as P ranges over the set of prime ideals. Clearly, since every prime contains a minimal prime, we may restrict this set of primes to be the minimal primes. When R is right or left noetherian there are only finitely many minimal primes, and therefore this supremum must actually be a maximum, so that for one of these minimal primes P we obtain $\text{Cl.K.dim}(R/P) = \gamma$. □

EXERCISE 12C. Let k be a field, let $x_1, x_2, \ldots$ be independent indeterminates, and set

$$R = k\cdot 1 + \bigoplus_{n=1}^{\infty} k[x_1,\ldots,x_n] \subseteq \prod_{n=1}^{\infty} k[x_1,\ldots,x_n].$$

Show that $\text{Cl.K.dim}(R) = \omega$ whereas $\text{Cl.K.dim}(R/P)$ is finite for all primes P of R. □

EXERCISE 12D. Let k be a field, let $X = \{x_{ij} \mid i,j \in \mathbb{N} \text{ and } j \leq i\}$ be a collection of independent indeterminates, and let $S = k[X]$. For $i \in \mathbb{N}$ set

$$P_i = x_{i1}S + x_{i2}S + \ldots + x_{ii}S.$$

Then set $U = \bigcup_{i \in \mathbb{N}} P_i$ and $R = \{ab^{-1} \mid a,b \in S \text{ and } b \notin U\} \subseteq k(X)$.

(a) Show that any ideal J of S contained in U must be contained in some P_i. [Hint: Apply Exercise 2ZI to $J \cap k[\{x_{ij} \mid i \leq n\}]$ for all $n \in \mathbb{N}$, and then to J itself.]

(b) Show that no P_i contains an infinite ascending chain of prime ideals of S. [Hint: First show that S localized at P_i is noetherian.]

(c) Show that any prime ideal P of S contained in U is finitely generated. [Hint: Consider $(P \cap k[\{x_{ij} \mid i \leq n\}])S$ for each $n \in \mathbb{N}$.] Conclude that all prime ideals of R are finitely generated, and therefore, by Exercise 2ZH, that R is noetherian.

(d) Show that $\text{Cl.K.dim}(R) = \omega$. □

♦ BIMODULE SYMMETRY AND INTERSECTION THEOREMS

LEMMA 12.3. *Let S and R be noetherian rings, such that R satisfies the right second layer condition, and let P be a prime ideal of R. Let ${}_SB_R$ be a bimodule, finitely generated on each side, such that B_R is faithful. Then B has sub–bimodules B′ and B″ with $B' > B''$ such that B'/B'' is a torsionfree right (R/P)–module as well as a torsionfree left (S/Q)–module for some prime Q of S.*

Proof. By noetherian induction, we may assume that the lemma holds in any case where the right–hand ring is a proper factor of R.

Choose a right affiliated series $0 = A_0 < A_1 < \ldots < A_m = B$ with corresponding right affiliated primes $P_1,\ldots,P_m$. By Proposition 7.5, each factor A_i/A_{i-1} is a torsionfree right (R/P_i)–module, and hence also a faithful right (R/P_i)–module. Since B_R is faithful, $P_mP_{m-1}\cdots P_1 = 0$, and so $P_j \leq P$ for some index j. Hence, we may replace R, P, B by

R/P_j, P/P_j, A_j/A_{j-1}. Thus, without loss of generality, we may assume that R is prime and that B is torsionfree as a right R–module.

If $P = 0$, just set $B'' = 0$ and let B' be any left affiliated sub–bimodule of B; by Proposition 7.5, B' is a torsionfree left module over $S/\text{l.ann}_S(B')$.

Now assume that $P \neq 0$, and note that all primes in the right link closure of $\{P\}$ are nonzero. Since B_R is faithful, R_R embeds in $(B_R)^m$ for some positive integer m (Lemma 7.1). Then $(R/P)_R$ is isomorphic to a subfactor of $(B_R)^m$, and so some uniform right ideal of R/P is isomorphic to B_1/B_2 where B_1 and B_2 are right R–submodules of B with $B_1 > B_2$.

Choose a right R–submodule C of B such that $C \geq B_2$ and C is maximal with respect to the property $C \cap B_1 = B_2$. Then $(B_1 + C)/C \cong B_1/B_2$ and $(B_1 + C)/C \leq_e B/C$. Hence, after replacing B_1 and B_2 by $B_1 + C$ and C, we may assume that $B_1/B_2 \leq_e B/B_2$.

Since R satisfies the right second layer condition, we see by Theorem 11.6 that all affiliated primes of B/B_2 are in the right link closure of $\{P\}$ and so are nonzero. Hence, there exists a nonzero ideal K such that $(B/B_2)K = 0$. Set $I = \text{r.ann}_R(B/BK)$, and observe that $0 \neq K \leq I$. On the other hand, since $BK \leq B_2$ and B_1/B_2 is a faithful right (R/P)–module, we have $I \leq P$. Applying the induction hypothesis to the bimodule B/BK, we obtain sub–bimodules B' and B'' of B such that $B' > B'' \geq BK$ and the desired properties hold. □

EXERCISE 12E. Let R be a prime noetherian ring satisfying the right second layer condition. If the intersection of all nonzero ideals of R is nonzero, show that R is a simple ring. □

THEOREM 12.4. [Jategaonkar] *Let S and R be noetherian rings, such that R satisfies the right second layer condition, and let ${}_SB_R$ be a bimodule which is finitely generated on each side. If B_R is faithful, then $\text{Cl.K.dim}(R) \leq \text{Cl.K.dim}(S)$.*

Proof. We use induction on the ordinal $\alpha = \text{Cl.K.dim}(R)$. Since the cases $\alpha = -1$ and $\alpha = 0$ are trivial, we may suppose that $\alpha > 0$.

By Lemma 12.2, there is a minimal prime P of R such that $\text{Cl.K.dim}(R/P) = \alpha$. Then by Lemma 12.3, there exist sub–bimodules B' and B'' of B with $B' > B''$ such that B'/B'' is a torsionfree right (R/P)–module and a torsionfree left (S/Q)–module for some prime Q of S. Since it is enough to show that $\text{Cl.K.dim}(S/Q) \geq \alpha$, we may reduce to the bimodule B'/B''. Thus there is no loss of generality in assuming that R and S are prime and that B is torsionfree on both sides.

Given any ordinal $\beta < \alpha$, there exists a prime P of R such that $\text{Cl.K.dim}(R/P) = \beta$

(Lemma 12.2). By Lemma 12.3, there exist sub–bimodules C' and C'' of B with $C' > C''$ such that C'/C'' is a torsionfree right (R/P)–module. If $I = \text{l.ann}_S(C'/C'')$, then $\text{Cl.K.dim}(S/I) \geq \beta$ by the induction hypothesis. Since C'/C'' is unfaithful as a right R–module, Proposition 7.4 implies that it is also unfaithful as a left S–module, whence $I \neq 0$. As $\text{Cl.K.dim}(S/I) \geq \beta$, we obtain $\text{Cl.K.dim}(S) \geq \beta + 1$ (Exercise 12A). Therefore $\text{Cl.K.dim}(S) \geq \alpha$. □

COROLLARY 12.5. [Jategaonkar] *Let S and R be noetherian rings satisfying the second layer condition, and ${}_SB_R$ a bimodule which is finitely generated and faithful on both sides. Then $\text{Cl.K.dim}(R) = \text{Cl.K.dim}(S)$.* □

COROLLARY 12.6. [Jategaonkar] *If R is a noetherian ring satisfying the second layer condition, and if P and Q are prime ideals of R in the same clique, then $\text{Cl.K.dim}(R/P) = \text{Cl.K.dim}(R/Q)$. Hence, if P and Q are distinct primes in the same clique, then P and Q are incomparable.* □

EXERCISE 12F. Let $S = A_1(k)$ and $R = k + \theta S$ as in Exercises 2ZC/ZD. Show that if we regard θS as an (R,S)–bimodule, then θS is finitely generated and faithful on both sides, but the conclusion of Corollary 12.5 does not hold (that is, R and S have different classical Krull dimensions.) □

EXERCISE 12G. Let R be a noetherian ring satisfying the second layer condition, and let P and Q be primes of R. Show that $Q \rightsquigarrow P$ if and only if the bimodule $(Q \cap P)/QP$ is faithful both as a left (R/Q)–module and as a right (R/P)–module. [Hint: Given the faithfulness assumption, choose an ideal A as large as possible such that $Q \cap P > A \geq QP$ and $(Q \cap P)/A$ is faithful as a left (R/Q)–module and as a right (R/P)–module.] □

We next verify Jacobson's Conjecture for any noetherian ring R satisfying the second layer condition. In fact, we prove that $\bigcap J^n = 0$ where J is the intersection of all the maximal ideals of R. Since the Jacobson radical is the intersection of all the primitive ideals of R, the ideal J is *a priori* larger than $J(R)$. Our intersection result is therefore a strong form of Jacobson's Conjecture.

LEMMA 12.7. *Let R be a noetherian ring. If I is a nonzero right ideal of R, there exists a maximal ideal M of R such that $\text{Hom}_R(I, E((R/M)_R)) \neq 0$.*

Proof. Set $J = \text{r.ann}_R(I)$. Since R is left noetherian, $(R/J)_R$ embeds in I^n for some positive integer n (Lemma 7.1). Choose a maximal ideal $M \geq J$, and observe that the quotient map $(R/J)_R \to (R/M)_R$ extends to a nonzero homomorphism $I^n \to E((R/M)_R)$. Therefore $\mathrm{Hom}_R(I, E((R/M)_R)) \neq 0$. □

THEOREM 12.8. [Jategaonkar] *Let R be a noetherian ring satisfying the second layer condition, and let J be the intersection of all the maximal ideals of R. Then $\bigcap J^n = 0$. In particular, $\bigcap J(R)^n = 0$.*

Proof. Let E be the direct sum of $E((R/M)_R)$ as M ranges over all maximal ideals of R. By Lemma 12.7, the intersection of the kernels of all homomorphisms from R_R to E is zero, and so R_R embeds in a direct product of copies of E. Hence, it suffices to show that $\bigcap J^n$ annihilates E. Thus it is enough to prove that any element x in any $E((R/M)_R)$ is killed by a power of J.

The right second layer condition implies that $xP_1P_2 \cdots P_n = 0$ for some primes P_i in the right link closure of $\{M\}$ (Theorem 11.6). From Corollary 12.6, we have

$$\mathrm{Cl.K.dim}(R/P_i) = \mathrm{Cl.K.dim}(R/M) = 0$$

for all i, and so each P_i is a maximal ideal. Therefore J is contained in each P_i, and hence $xJ^n = 0$, as desired. □

Another application of the Krull Intersection Theorem is that if R is a commutative noetherian domain and P a prime ideal in R, then $\bigcap P^n = 0$. This statement usually fails in the noncommutative case (Exercise 12H). There is, however, a version of this theorem for a prime noetherian ring with the second layer condition, as we now show. (Observe that Proposition 12.10 includes the commutative result just stated, since if R is commutative $\{P\}$ is right link closed.)

LEMMA 12.9. *Let R be a prime right noetherian ring and E a nonzero injective right R–module. If I is a nonzero right ideal of R, then $\mathrm{Hom}_R(I,E) \neq 0$.*

Proof. Since the ideal RI is nonzero, it contains a regular element, and hence R_R embeds in I^n for some positive integer n. There exist nonzero homomorphisms $R_R \to E$, and they all extend to nonzero homomorphisms $I^n \to E$. Thus $\mathrm{Hom}_R(I,E) \neq 0$. □

PROPOSITION 12.10. *Let R be a prime noetherian ring satisfying the right second layer condition, let P be a prime ideal of R, and let J be the intersection of the right link closure of $\{P\}$. Then $\bigcap J^n = 0$.*

Proof. Set $E = E((R/P)_R)$. By Lemma 12.9, R_R embeds in a direct product of copies of E. The right second layer condition implies that any element of E is killed by a product of primes in the right link closure of $\{P\}$ (Theorem 11.6). Hence, $\bigcap J^n$ annihilates E, and therefore $\bigcap J^n = 0$. □

EXERCISE 12H. Find an example of a prime fully bounded noetherian ring R with a prime ideal P such that $\bigcap P^n \neq 0$. □

EXERCISE 12I. For a ring R, let $\beta(R)$ be the intersection of the maximal ideals of R. This ideal of R is sometimes called the *Brown–McCoy radical* of R. We showed in Theorem 12.8 that if R is a noetherian ring satisfying the second layer condition then $\bigcap \beta(R)^n = 0$. Give an example to show that this may fail for noetherian rings not satisfying the second layer condition. □

♦ FINITE RING EXTENSIONS

In this section we return to the study of prime ideals in finite ring extensions, which we considered earlier in Chapters 7 and 10. Here we concentrate on the situation in which the smaller ring satisfies the second layer condition. For some results – in particular the first theorem – all that is needed from the second layer condition is the "bimodule symmetry" of classical Krull dimension obtained in Corollary 12.5. More precisely, the theorem can be proved for a noetherian ring with a suitable "symmetric dimension function", as indicated in Exercise 12J.

LEMMA 12.11. *Let R be a noetherian ring satisfying the second layer condition, and suppose that R is a subring of a ring S such that $_RS$ and S_R are finitely generated. If S is a prime ring, then the minimal prime ideals of R are exactly those primes Q for which Cl.K.dim(R/Q) = Cl.K.dim(R); moreover, the set of minimal primes of R is right and left link closed.*

Proof. By Lemma 12.2, R has at least one minimal prime Q' such that R/Q' has the same classical Krull dimension as R. If Q is any minimal prime of R, it follows from Lemma 7.15 and Corollary 12.5 that

$$\text{Cl.K.dim}(R/Q) = \text{Cl.K.dim}(R/Q') = \text{Cl.K.dim}(R).$$

On the other hand, if Q is a non–minimal prime of R, it properly contains a minimal prime Q'', and using Exercise 12A we conclude that

$$\text{Cl.K.dim}(R/Q) < \text{Cl.K.dim}(R/Q'') = \text{Cl.K.dim}(R).$$

The final conclusion of the lemma now follows from the above results and Corollary 12.6. □

THEOREM 12.12. [Joseph–Small, Borho, Warfield] *Let R be a noetherian ring satisfying the second layer condition, and suppose that R is a subring of a ring S such that ${}_RS$ and S_R are both finitely generated. If P is any prime ideal of S, the ring $R/(P \cap R)$ has an artinian classical quotient ring A. Moreover, if B is the Goldie quotient ring of S/P, the embedding $R/(P \cap R) \to S/P$ extends to an embedding $A \to B$ (in other words, $\mathcal{C}_R(P \cap R) \subseteq \mathcal{C}_S(P)$).*

Proof. Without loss of generality, we may assume that $P = 0$.

To show that R has an artinian classical quotient ring, it will suffice, by Theorem 10.13, to show that R has a left affiliated series and a right affiliated series such that the affiliated primes for both series are all minimal. By symmetry, it is enough to find such an affiliated series on the left.

We first choose a left affiliated series $0 = S_0 < S_1 < \ldots < S_n = S$ for the bimodule ${}_RS_S$ and show that the corresponding left affiliated primes $Q_1,\ldots,Q_n$ are all minimal primes of R. Then by Exercise 2M there will be a left affiliated series for R whose corresponding left affiliated primes are all from the set $\{Q_1,\ldots,Q_n\}$, and hence are minimal.

Let $T = S_i/S_{i-1}$ (for some $i \in \{1,\ldots,n\}$), and note that it follows from Propositions 7.5 and 7.7 that T is a (nonzero) torsionfree left (R/Q_i)–module as well as a torsionfree right S–module. In particular, T is faithful as a right S–module and hence also faithful as a right R–module. Next let $0 = T_0 < T_1 < \ldots < T_m = T$ be a right affiliated series for the bimodule ${}_RT_R$. Any choice of a minimal prime Q of R must be a right affiliated prime corresponding to some factor in this series (Proposition 2.14), say T_j/T_{j-1}. Then T_j/T_{j-1} is a nonzero $(R/Q_i,R/Q)$–bimodule which is finitely generated and torsionfree on each side (Propositions 7.5 and 7.7). Using Corollary 12.5 and Lemma 12.11, we conclude that

$$\text{Cl.K.dim}(R/Q_i) = \text{Cl.K.dim}(R/Q) = \text{Cl.K.dim}(R)$$

and hence that Q_i is a minimal prime of R, as desired.

Therefore R does have an artinian classical quotient ring A. Now by Corollary 10.10, $\mathcal{C}_R(0) = \mathcal{C}_R(N)$, where N is the prime radical of R.

To obtain the embedding of A into B, we must show that every regular element of R is a unit in B (see Proposition 9.4); hence, it is enough to show that every regular element of R is also regular as an element of S. Using other terminology, we would like S to be $\mathcal{C}_R(0)$–torsionfree both as a left R–module and as a right R–module. Since the primes Q_i corresponding to our left affiliated series for ${}_RS_S$ are all minimal prime ideals of R, we conclude that $\mathcal{C}_R(0) = \mathcal{C}_R(N) \subseteq \mathcal{C}_R(Q_i)$ for $i = 1,\ldots,n$ (Proposition 6.5). As the factors S_i/S_{i-1} are already $\mathcal{C}_R(Q_i)$–torsionfree on the left, they must also be $\mathcal{C}_R(0)$–torsionfree. Therefore S is $\mathcal{C}_R(0)$–torsionfree on the left, and, by symmetry, also on the right. □

EXERCISE 12J. Let R be a noetherian ring equipped with a *symmetric dimension function,* i.e., a function δ associating to each prime factor R/P of R an element of a partially ordered set (for example, an ordinal or a real number) such that the following two conditions are satisfied, for all prime ideals P and Q:

(a) If $P < Q$ then $\delta(R/P) > \delta(R/Q)$.

(b) If there exists a nonzero $(R/P, R/Q)$–bimodule which is finitely generated and torsionfree on each side, then $\delta(R/P) = \delta(R/Q)$.

(E.g., if R satisfies the second layer condition, δ can be taken to be the classical Krull dimension. There are other important examples, such as the *Gelfand–Kirillov dimension* for finitely generated algebras over a field, when this dimension is finite. For a discussion of this dimension, see Krause–Lenagan [1985] or McConnell–Robson [1987].) Show that the conclusions of Theorem 12.12 hold in this context, using δ in place of Cl.K.dim. □

The next two theorems relate the second layer condition and prime ideals in a noetherian ring R and a finite ring extension S. In both cases, we use the R–module structure of an S–module to produce a suitable ideal of S. We isolate the basic argument for this in a lemma (to avoid having to go through the argument three times).

LEMMA 12.13. *Let R be a noetherian ring, let R be a subring of a ring S such that ${}_RS$ and S_R are finitely generated, and assume that S is a prime ring. Let M be a finitely generated right S–module, let U,V be R–submodules of M such that $V < U$, and assume that none of the associated primes of U/V is a minimal prime of R. Suppose either that R satisfies the strong second layer condition, or that R satisfies the second layer condition and for all $T \in \mathrm{Ass}(U/V)$, the (R/T)–module $\mathrm{ann}_{U/V}(T)$ is torsionfree. Then there exists a nonzero ideal J in S such that $MJ \cap U \leq V$.*

Proof. In view of Theorems 12.12 and 10.9, $\mathscr{C}_R(N) = \mathscr{C}_R(0) \subseteq \mathscr{C}_S(0)$, where N is the prime radical of R.

Let W be an R–submodule of M such that $W \geq V$ and W is maximal with respect to the property that $W \cap U = V$. Then U/V is isomorphic to an essential R–submodule of M/W. Now if $T \in \mathrm{Ass}(M/W)$, then $T \in \mathrm{Ass}(U/V)$, and $\mathrm{ann}_{U/V}(T)$ is isomorphic to an essential submodule of $\mathrm{ann}_{M/W}(T)$. Hence, if $\mathrm{ann}_{U/V}(T)$ is torsionfree as an (R/T)–module, so is $\mathrm{ann}_{M/W}(T)$.

It now follows from either Theorem 11.4 or Theorem 11.6 that M/W is annihilated by a product of primes from the right link closure of $\mathrm{Ass}(U/V)$, say $MI_1I_2\cdots I_n \leq W$ for suitable $I_1,\ldots,I_n$ in the right link closure of $\mathrm{Ass}(U/V)$. Since, by assumption, there are no minimal primes in $\mathrm{Ass}(U/V)$, none of $I_1,\ldots,I_n$ can be a minimal prime (Lemma 12.11).

By Proposition 6.5, each I_j contains an element of $\mathscr{C}_R(N)$. Since $\mathscr{C}_R(N) \subseteq \mathscr{C}_S(0)$, the ideal $I = I_1I_2 \cdots I_n$ therefore contains a regular element of S. Hence, S/SI is torsion as a left S–module, and so, by Lemma 7.3, there is a nonzero ideal J of S such that $JS \leq SI$, that is, $J \leq SI$. Thus

$$MJ \leq MSI = MI = MI_1I_2 \cdots I_n \leq W,$$

and therefore $MJ \cap U \leq W \cap U = V$. □

THEOREM 12.14. [Letzter] *Let R be a noetherian ring and let R be a subring of a ring S such that ${}_RS$ and S_R are both finitely generated. If R satisfies the second layer condition (or the strong second layer condition), so does S.*

Proof. Suppose first that R satisfies the strong second layer condition, but that S does not, say on the right. Then by Proposition 11.3 there exists a finitely generated uniform right S–module M with an affiliated series $0 < U < M$ and corresponding affiliated primes P, Q such that $Q < P$ and $MQ = 0$. Without loss of generality, we may assume that $Q = 0$, and that M is faithful (because M/U is then faithful).

Let Y be the set of primes of R that contain $P \cap R$. If Y contains a minimal prime T, then T is minimal over both $0 \cap R$ and $P \cap R$, whence the primes 0 and P in S both lie over T. However, this contradicts Theorem 10.6 (INC). Therefore no element of Y is a minimal prime. Since $U(P \cap R) = 0$, it follows that there are no minimal primes in $\mathrm{Ass}(U_R)$.

Applying Lemma 12.13 (with $V = 0$), we obtain a nonzero ideal J in S such that $MJ \cap U = 0$. Since MJ is an S–submodule of M and M_S is uniform, it follows that $MJ = 0$. But as M is a faithful S–module and $J \neq 0$, this is a contradiction.

Therefore S satisfies the strong second layer condition on the right, and, by symmetry, on the left.

To prove that the second layer condition carries up from R to S, two changes are required in the above proof. First, we may assume that U is torsionfree as a right (S/P)–module, and second, in order to apply Lemma 12.13 we must show that for any $T \in \mathrm{Ass}(U_R)$, the (R/T)–module $\mathrm{ann}_U(T)$ is torsionfree.

Since $\mathrm{ann}_U(T) \neq 0$ and U is torsionfree as a right (S/P)–module, T must be disjoint from $\mathscr{C}_S(P)$. On the other hand, if $N/(P \cap R)$ is the prime radical of $R/(P \cap R)$, it follows from Theorems 12.12 and 10.9 that

$$\mathscr{C}_R(N) = \mathscr{C}_R(P \cap R) \subseteq \mathscr{C}_S(P).$$

Hence, T is disjoint from $\mathscr{C}_R(N)$. As $T \geq P \cap R$, Proposition 6.5 now shows that T is minimal over $P \cap R$. Finally, $\mathrm{ann}_U(T)$ is torsionfree as an (R/N)–module (because $\mathscr{C}_R(N) \subseteq \mathscr{C}_S(P)$), and therefore Proposition 6.8 shows that $\mathrm{ann}_U(T)$ is also torsionfree as an (R/T)–module, as desired. □

COROLLARY 12.15. *If R is a noetherian ring satisfying the second layer condition, and R is a subring of a ring S such that ${}_RS$ and S_R are both finitely generated, then R and S have the same classical Krull dimension.*

Proof. Theorem 12.14 and Corollary 12.5 (applied to the bimodule ${}_RS_S$). □

THEOREM 12.16. [Letzter] *If R is a noetherian ring satisfying the second layer condition and R is a subring of a ring S such that ${}_RS$ and S_R are both finitely generated, then Lying Over holds. That is, for every prime ideal Q of R there is a prime P of S such that Q is minimal over $P \cap R$.*

Proof. Let P be an ideal of S maximal with respect to the property that $P \cap R \leq Q$. It is easy to see that P is prime: if A,B are ideals of S properly containing P, then $A \cap R \nleq Q$ and $B \cap R \nleq Q$, whence $(A \cap R)(B \cap R) \nleq Q$, yielding $AB \cap R \nleq Q$ and so $AB \nleq P$. Without loss of generality, we may assume that $P = 0$, and, by way of contradiction, we assume that Q is not a minimal prime of R.

As a right R–module, R/Q has only one associated prime, namely Q, and of course R/Q is torsionfree as a right (R/Q)–module. We now apply Lemma 12.13, using $M = S_S$ while $U = R_R$ and $V = Q_R$. Thus there is a nonzero ideal J in S such that $SJ \cap R \leq Q$, that is, $J \cap R \leq Q$. However, this contradicts the fact that 0 is the largest ideal of S whose contraction to R is contained in Q. Therefore Q must be a minimal prime of R. □

♦ LOCALIZATION AT A SEMIPRIME IDEAL

If R is a commutative ring and P a prime ideal of R, then we can *localize* the ring R at the prime P by inverting the elements not in P. The resulting ring of fractions is denoted R_P and it is a local ring with maximal ideal PR_P. The factor R_P/PR_P is naturally isomorphic to the quotient field of R/P. It is natural to try to imitate this in the non–commutative setting, but a naive attempt to do so does not work. If R is noetherian and P is a prime ideal, then we would like to invert the set $\mathscr{C}(P)$. If $\mathscr{C}(P)$ is a right Ore set, then we again write the corresponding ring of fractions as R_P and it is easy to verify that PR_P is the unique maximal ideal of R_P, that it is also the only primitive ideal of R_P, and that R_P/PR_P can be naturally identified with the Goldie quotient ring of R/P. The trouble with all of this is that $\mathscr{C}(P)$ is usually not a right Ore set. The first obstruction to its being an Ore set is indicated in the following lemma.

LEMMA 12.17. *Let R be a noetherian ring and P and Q primes in R such that $Q \rightsquigarrow P$. Let $\mathscr{C}$ be a right Ore set in R and suppose that $\mathscr{C} \subseteq \mathscr{C}(P)$. Then $\mathscr{C} \subseteq \mathscr{C}(Q)$.*

Proof. If Q is linked to P via the linking bimodule $(Q \cap P)/A$, there is no loss of generality in assuming that $A = 0$. Now $Q \cap P$ is a torsionfree right (R/P)–module, and hence $Q \cap P$ is also $\mathscr{C}$–torsionfree on the right (because $\mathscr{C} \subseteq \mathscr{C}(P)$). By right reversibility (Proposition 9.9), $Q \cap P$ is also $\mathscr{C}$–torsionfree on the left.

As a left (R/Q)–module, $Q \cap P$ is nonzero and torsionfree, and hence faithful. It follows that R/Q embeds (as a left module) in some finite direct sum of copies of $Q \cap P$ (Lemma 7.1). Hence, R/Q is $\mathscr{C}$–torsionfree on the left, i.e., for each $c \in \mathscr{C}$ the coset $c + Q$ has zero right annihilator in R/Q. Therefore, by Lemma 5.7, these cosets are regular elements of R/Q, that is, $\mathscr{C} \subseteq \mathscr{C}(Q)$. □

It is clear from Lemma 12.17 that if we are given a prime P in a noetherian ring R and we want to find a right Ore set contained in $\mathscr{C}(P)$, then the largest set we can consider as a possible right Ore set is the intersection of the $\mathscr{C}(Q)$ for all primes Q in the right link closure of $\{P\}$. This set can certainly be infinite, as examples in Chapter 11 show. We will concentrate at first on the special case in which such a set is finite. Our main considerations will be restricted to rings satisfying the second layer condition, and so the primes in the right link closure of $\{P\}$ will be incomparable primes (Corollary 12.6). This puts us in the following situation: we have a finite set $\{P_1,\ldots,P_n\}$ of incomparable primes, such that $\{P_1,\ldots,P_n\}$ is right link closed, and we want to know whether $\mathscr{C}(P_1) \cap \ldots \cap \mathscr{C}(P_n)$ is a right Ore set.

In this situation we will want to look at the semiprime ideal $N = P_1 \cap \ldots \cap P_n$. Obviously any prime minimal over N is one of the P_i; conversely, since the P_i are incomparable they are all minimal over N. (Any P_i contains a prime Q minimal over N, and since $Q = P_j$ for some j, it follows from the incomparability of our set of primes that $P_i = P_j$, whence P_i is minimal over N.) Now by Proposition 6.5,

$$\mathscr{C}(N) = \mathscr{C}(P_1) \cap \ldots \cap \mathscr{C}(P_n).$$

Thus, to rephrase our problem: we have a semiprime ideal N such that the set of primes minimal over N is right link closed, and we want to know whether $\mathscr{C}(N)$ is a right Ore set. Assuming that R satisfies the right second layer condition, we shall prove that this is the case (Theorem 12.21). We first collect some elementary facts about the corresponding localization in case it exists.

DEFINITION. A semiprime ideal N in a right noetherian ring R is called *right localizable* provided $\mathscr{C}(N)$ is a right Ore set. When this occurs, we denote the corresponding localization by R_N. The ideal N is *classically right localizable* if, in addition, the injective hull of the right R_N–module $R_N/J(R_N)$ is the union of its socle series. Of course a semiprime ideal N in a noetherian ring R is called *(classically) localizable* if N is both right and left (classically) localizable.

If R is the right noetherian ring discussed in Exercise 8L, it is easily checked that $J(R)$

is right localizable but not classically right localizable. However, it is not known whether all right localizable semiprime ideals in a (right and left) noetherian ring are classically right localizable (cf. Exercise 12O).

LEMMA 12.18. *Let N be a right localizable semiprime ideal of a right noetherian ring R. Then $NR_N = J(R_N)$, the ring R_N/NR_N is semisimple, and the natural map $R/N \to R_N/NR_N$ extends to a ring isomorphism from the Goldie quotient ring of R/N onto R_N/NR_N.*

Proof. Since R/N is $\mathscr{C}(N)$–torsionfree, Theorems 9.17 and 9.20 show that $NR_N = N^e$ is an ideal of R_N that contracts to N. In particular, the natural map $R/N \to R_N/NR_N$ is a ring embedding. Now the map $R/N \to R_N/NR_N$ is a right ring of fractions for R/N with respect to the Ore set of regular elements of R/N, and therefore R_N/NR_N is the right Goldie quotient ring of the image of R/N.

In particular, R_N/NR_N is a semisimple ring, whence $J(R_N) \leq NR_N$. To prove the reverse inclusion, it suffices to show that any maximal right ideal I of R_N contains NR_N.

Now I^c is a proper right ideal of R such that $I^cR_N = I$ (Theorem 9.17). It follows that I^c must be disjoint from $\mathscr{C}(N)$, and so $R/(I^c + N)$ is not torsion as a right (R/N)–module. If $K/(I^c + N)$ is the torsion submodule of $R/(I^c + N)$, then K is a proper right ideal of R and R/K is $\mathscr{C}(N)$–torsionfree. By Theorem 9.17, $(KR_N)^c = K$. Since KR_N is a proper right ideal of R_N containing I^cR_N, which equals I, we must have $KR_N = I$. Therefore $NR_N \leq I$, as desired. □

EXERCISE 12K. Let N be a right localizable semiprime ideal in a right noetherian ring R, and let $E = E((R/N)_R)$. Show that N is classically right localizable if and only if $E = \bigcup_{i=0}^{\infty} \operatorname{ann}_E(N^i)$. [Hint: Show that the injective hull of the right R_N–module $R_N/J(R_N)$ is isomorphic to E.] □

If R is a noetherian ring and N is its prime radical, then we showed in Lemma 10.5 that an R–module M is $\mathscr{C}(N)$–torsion if and only if it has zero reduced rank. The following lemma can be thought of as a generalization of this result. To see the connnection, the reader may want to verify directly that a right R–module A has zero reduced rank if and only if $\operatorname{Hom}_R(A, E((R/N)_R)) = 0$.

LEMMA 12.19. *If R is a right noetherian ring and N a semiprime ideal of R, and $E = E((R/N)_R)$, then a right R–module A is $\mathscr{C}(N)$–torsion if and only if $\operatorname{Hom}_R(A,E) = 0$.*

Proof. Suppose first that there exist an element $f \in \mathrm{Hom}_R(A,E)$ and an element $a \in A$ such that $f(a) \neq 0$. There is an element $r \in R$ such that $f(a)r \in R/N$ and $f(a)r \neq 0$. If A were $\mathscr{C}(N)$–torsion then there would be an element $c \in \mathscr{C}(N)$ such that $arc = 0$, and hence $f(a)rc = 0$. Since R/N is $\mathscr{C}(N)$–torsionfree, this would imply that $f(a)r = 0$ which is false. Hence, $\mathrm{Hom}_R(A,E) \neq 0$ implies that A is not $\mathscr{C}(N)$–torsion.

Conversely, suppose that A is not $\mathscr{C}(N)$–torsion and choose $a \in A$ such that $ac \neq 0$ for all $c \in \mathscr{C}(N)$. We note first that $a + aN$ is not a $\mathscr{C}(N)$–torsion element of A/aN, since if $ac \in aN$ for some $c \in \mathscr{C}(N)$, then $ac = an$ for some $n \in N$, whence $a(c - n) = 0$, and clearly $c - n \in \mathscr{C}(N)$. Now aR/aN is an (R/N)–module which we have shown is not $\mathscr{C}(N)$–torsion, and so by Lemma 6.17, aR/aN has a submodule U which is isomorphic to a uniform right ideal of R/N. The embedding of U into R/N extends to a nonzero homomorphism $A/aN \rightarrow E$. (Of course, if R were also left noetherian, we could conclude from Proposition 6.19 that $\mathrm{Hom}_R(aR/aN, R/N) \neq 0$.) Thus we have shown that if A is not $\mathscr{C}(N)$–torsion then $\mathrm{Hom}_R(A,E) \neq 0$. □

Of course, if $\mathscr{C}(N)$ is a right Ore set, then Lemma 12.19 immediately implies that $E((R/N)_R)$ is $\mathscr{C}(N)$–torsionfree (since then the $\mathscr{C}(N)$–torsion elements of $E((R/N)_R)$ form a submodule). This turns out, in fact, to characterize the Ore condition for $\mathscr{C}(N)$.

THEOREM 12.20. [Jategaonkar] *If R is a right noetherian ring and N a semiprime ideal of R, and $E = E((R/N)_R)$, then N is right localizable if and only if E is $\mathscr{C}(N)$–torsionfree.*

Proof. It is immediate from the previous lemma that if N is right localizable then E is $\mathscr{C}(N)$–torsionfree. Conversely, suppose that E is $\mathscr{C}(N)$–torsionfree, and let $r \in R$ and $c \in \mathscr{C}(N)$. Since R/cR is generated by the $\mathscr{C}(N)$–torsion element $1 + cR$, and since E is $\mathscr{C}(N)$–torsionfree, we see that $\mathrm{Hom}_R(R/cR, E) = 0$, from which Lemma 12.19 allows us to conclude that R/cR is $\mathscr{C}(N)$–torsion. Hence there is an element $c' \in \mathscr{C}(N)$ such that $(r + cR)c' = 0$, or, in other words, $rc' \in cR$. Therefore $\mathscr{C}(N)$ is a right Ore set. □

THEOREM 12.21. [Müller, Jategaonkar] *Let R be a noetherian ring satisfying the right second layer condition and N a semiprime ideal of R. Then N is classically right localizable if and only if the set of prime ideals minimal over N is right link closed.*

Proof. Let $E = E((R/N)_R)$.

Assume first that N is classically right localizable, and consider a link $Q \rightsquigarrow P$ where P is a prime minimal over N. By Theorem 11.2, there exists a finitely generated uniform right R–module M with an affiliated series $0 < U < M$ such that U is isomorphic to a (uniform) right ideal of R/P and M/U is isomorphic to a uniform right ideal of R/Q. Since U is torsionfree as a right (R/P)–module, it is also torsionfree as a right (R/N)–module (Proposition 6.8), and so U is isomorphic to a right ideal of R/N (Corollary 6.20). Consequently, M can be embedded in E. Because of Exercise 12K, it follows that $MN^i = 0$ for some $i \in \mathbb{N}$. As M/U is a faithful right (R/Q)–module, we obtain $N^i \leq Q$, and hence $N \leq Q$.

We also have $\mathscr{C}(N) \subseteq \mathscr{C}(P)$ by Proposition 6.5, and then $\mathscr{C}(N) \subseteq \mathscr{C}(Q)$ by Lemma 12.17. Since Q contains N, we conclude from Proposition 6.5 that Q is minimal over N. Thus the set of primes minimal over N is right link closed.

Conversely, assume that the set of primes minimal over N is right link closed. Observe using Corollary 6.7 that if P is any associated prime of the module $(R/N)_R$, then P is minimal over N and $\text{l.ann}_{R/N}(P)$ is torsionfree as a right (R/P)–module. Hence, if M is a nonzero finitely generated submodule of E, and if $0 = M_0 < M_1 < \ldots < M_k = M$ is an affiliated series for M with corresponding affiliated primes $P_1,\ldots,P_k$, then according to Theorem 11.6 each P_i is in the right link closure of $\text{Ass}((R/N)_R)$, and each factor M_i/M_{i-1} is torsionfree as a right (R/P_i)–module. Our hypothesis on the primes minimal over N implies that each P_i contains N and is minimal over N. In particular, $MN^k = 0$. Since each P_i is minimal over N, each M_i/M_{i-1} is torsionfree as a right (R/N)–module (Proposition 6.8), that is, M_i/M_{i-1} is $\mathscr{C}(N)$–torsionfree. It follows that M is $\mathscr{C}(N)$–torsionfree.

We conclude from the previous paragraph that E is $\mathscr{C}(N)$–torsionfree and that $E = \bigcup_{i=0}^{\infty} \text{ann}_E(N^i)$. Therefore, by Theorem 12.20 and Exercise 12K, N is classically right localizable. □

COROLLARY 12.22. *Let R be a noetherian ring satisfying the second layer condition and N its prime radical. If all factors of R by minimal prime ideals have the same classical Krull dimension, then N is classically localizable.*

Proof. Corollary 12.6 and Theorem 12.21. □

EXERCISE 12L. Let R be a noetherian ring satisfying the second layer condition and P a prime of R. Assume that the right link closure of $\{P\}$ is a finite set, say $\{P_1,\ldots,P_n\}$.

(a) Show that the set $\mathscr{C} = \mathscr{C}(P_1) \cap \ldots \cap \mathscr{C}(P_n)$ is the largest right Ore set contained in $\mathscr{C}(P)$, and set $S = R\mathscr{C}^{-1}$.

(b) For $i = 1,\ldots,n$, show that P_iS is a maximal ideal of S, that S/P_iS is a simple artinian ring isomorphic to the Goldie quotient ring of R/P_i, and that $E((S/P_iS)_S)$ is the union of its socle series.

(c) Show that $P_1S,...,P_nS$ are the only right primitive ideals of S. □

EXERCISE 12M. Let N be a semiprime ideal in a noetherian ring R and let X be the set of primes minimal over N. Show that N is classically right localizable if and only if X is right link closed and every prime in X satisfies the right second layer condition. (Cf. Exercise 10N. This is an improvement of Theorem 12.21 above. However, the only cases in which it has been applied in rings not necessarily satisfying the right second layer condition for *all* primes is when N is the prime radical, in which case one can recover some of the results of Chapter 10.) □

The next two exercises address the question of whether the second layer condition is necessary for considerations of localization (as opposed to classical localization).

EXERCISE 12N. Let R be a prime noetherian ring which is not artinian but such that for every essential right ideal I, the module R/I is artinian. (We say that R has *right Krull dimension one*, according to the notion of Krull dimension to be discussed in the following chapter.)

(a) Show that if P is a prime of R which fails to satisfy the right second layer condition, then there is a finitely generated right R–module M which has an essential submodule U isomorphic to $(R/P)_R$, such that M/U is a faithful simple right R–module.

(b) Let N be a semiprime ideal of R such that the set X of primes minimal over N is right link closed. Show that N is right localizable if and only if every prime in X satisfies the right second layer condition. [Hint: If N is right localizable and $E = E((R/N)_R)$, use Theorem 12.20 to show that $E/\text{ann}_E(N)$ is $\mathscr{C}(N)$–torsionfree.] □

EXERCISE 12O. We give here three well–known open questions. The point of the exercise is not to prove them (though feel free to try) but to show that they are equivalent.

Conjecture (a): If R is a noetherian ring and N is a localizable semiprime ideal in R, then N is classically localizable.

Conjecture (b): If R is a noetherian ring and N is a localizable semiprime ideal in R then each of the primes minimal over N must satisfy the second layer condition.

Conjecture (c): If R is a noetherian ring such that R/J(R) is artinian, then J(R) satisfies the AR–property. [Hint: Lenagan's Theorem.] □

The next two exercises explore some of the distinctions between the AR–property and localizability.

EXERCISE 12P. Give an example of a noetherian ring containing a semiprime ideal which satisfies the right AR–property but is not right localizable. [Hint: The prime radical always satisfies the AR–property.] □

EXERCISE 12Q. The point of this exercise is to construct a localizable semiprime ideal P in a noetherian ring R, such that P does not satisfy the AR–property. In the example, P will be a minimal prime and R will be semiprime, and so P will be clearly localizable (Exercise 9U). An obvious way to have the AR–property fail is to have linked maximal ideals, $N \rightsquigarrow M$, such that $P \leq M$ while $P \nleq N$ (Proposition 11.16). We construct such a ring as follows. Let T be the ring $\begin{pmatrix} \mathbb{Z} & \mathbb{Z} \\ 2\mathbb{Z} & \mathbb{Z} \end{pmatrix}$ and I the ideal $\begin{pmatrix} 2\mathbb{Z} & \mathbb{Z} \\ 2\mathbb{Z} & \mathbb{Z} \end{pmatrix}$. There is a ring homomorphism $\phi : T \to \mathbb{Z}/2\mathbb{Z}$ with kernel I. Let

$$R = \{(t,m) \in T \times \mathbb{Z} \mid \phi(t) = m + 2\mathbb{Z}\},$$

and verify that R is a ring of the desired type, with $P = I \times 0$ and $M = I \times 2\mathbb{Z}$. □

♦ EMBEDDINGS IN ARTINIAN RINGS

As an application of the localization results obtained in the previous section, we show that any noetherian ring satisfying the second layer condition can be embedded in an artinian ring.

LEMMA 12.23. *Let R be a noetherian ring satisfying the second layer condition and let U be a finitely generated uniform right R–module. Let P be the assassinator of U, and assume that $\mathrm{ann}_U(P)$ is torsionfree as a right (R/P)–module. Then the ring $R/\mathrm{ann}_R(U)$ has an artinian classical quotient ring.*

Proof. Without loss of generality, we may assume that U is a faithful R–module. By Lemma 10.8 and Corollary 10.10, it suffices to prove that $\mathscr{C}(N) \subseteq \mathscr{C}(0)$, where N is the prime radical of R.

Choose an affiliated series $0 = U_0 < U_1 < ... < U_n = U$ with corresponding affiliated primes $Q_1,...,Q_n$. Since P is the only associated prime of U, it follows from Theorem 11.6 that each of the primes Q_i is in the right link closure of $\{P\}$ and that each of the factors U_i/U_{i-1} is torsionfree as a right (R/Q_i)–module. Since U is faithful, the set $\{Q_1,...,Q_n\}$ contains all of the minimal primes of R (Proposition 2.14).

By Corollary 12.6, the factors R/Q_i all have the same classical Krull dimension, say α. Now all factors of R modulo minimal primes have classical Krull dimension α, and so we conclude that all of the primes Q_i are minimal. Hence, $\mathscr{C}(N) \subseteq \mathscr{C}(Q_i)$ for all i (Proposition 6.5). As a result, each of the factors U_i/U_{i-1} is $\mathscr{C}(N)$–torsionfree, and thus U is $\mathscr{C}(N)$–torsionfree. Since U is faithful, R_R embeds in a direct product of copies of

U, and hence R_R is $\mathscr{C}(N)$–torsionfree. (This does not yet show that $\mathscr{C}(N) \subseteq \mathscr{C}(0)$.)

Now by Corollary 12.22, $\mathscr{C}(N)$ is an Ore set, and so we may use right reversibility (Proposition 9.9) to conclude from the $\mathscr{C}(N)$–torsionfreeness of R_R that $\mathscr{C}(N) \subseteq \mathscr{C}(0)$, as desired. (The use of Small's Theorem in this proof may be avoided by a more careful analysis of the ring R_N. Since $\mathscr{C}(N) = \mathscr{C}(0)$, it follows that R is an order in R_N, and it just remains to show that R_N is artinian, as in the following exercise.) □

EXERCISE 12R. Let R be a right noetherian ring, N a right localizable semiprime ideal of R, and $P_1,\ldots,P_n$ the primes minimal over N. Show that R_N is right artinian if and only if all of the primes P_i are minimal primes of R. [Hint: Theorem 9.22 and Lemma 12.18.] □

THEOREM 12.24. [Jategaonkar] *Any noetherian ring R satisfying the second layer condition can be embedded in an artinian ring. More precisely, there exist ideals $A_1,\ldots,A_n$ in R such that $A_1 \cap \ldots \cap A_n = 0$ and each of the factor rings R/A_i has an artinian classical quotient ring.*

Proof. We first claim that if U is any uniform right ideal of R, there exists an ideal A such that $U \cap A = 0$ and the ring R/A has an artinian classical quotient ring.

Let P be the assassinator of U, and recall from Proposition 6.11 that $\mathrm{ann}_U(P)$ is either torsion or torsionfree as a right (R/P)–module. It cannot be torsion since, according to Lemma 7.1, R/P can be embedded (as a right module) in a finite direct sum of copies of $\mathrm{ann}_U(P)$. Hence, $\mathrm{ann}_U(P)$ is a torsionfree right (R/P)–module. Now choose a right ideal V of R maximal with respect to the property that $U \cap V = 0$; then U is isomorphic to an essential submodule of R/V. It follows that R/V is uniform, that P is the assassinator of R/V, and that $\mathrm{ann}_{R/V}(P)$ is a torsionfree (R/P)–module. By Lemma 12.23, R/A has an artinian classical quotient ring, where $A = \mathrm{ann}_R(R/V)$. Since $A \leq V$, we have $U \cap A = 0$, and the claim is proved.

If R is nonzero, choose a uniform right ideal U_1. By the claim, there is an ideal A_1 such that $U_1 \cap A_1 = 0$ and R/A_1 has an artinian classical quotient ring. If $A_1 \neq 0$, choose a uniform right ideal $U_2 \leq A_1$. Then there is an ideal A_2 such that $U_2 \cap A_2 = 0$ and R/A_2 has an artinian classical quotient ring. Observe that $(U_1 \oplus U_2) \cap (A_1 \cap A_2) = 0$. Continue this process as long as possible (e.g., next with a uniform right ideal $U_3 \leq A_1 \cap A_2$). Since R_R has finite rank, the process must terminate at some point, and then $A_1 \cap \ldots \cap A_n = 0$ for some n. Since each of the rings R/A_i can be embedded in an artinian ring, so can R. □

EXERCISE 12S. Show that if R is a noetherian ring satisfying the strong second layer condition and U is a finitely generated uniform right R–module, then $R/\text{ann}_R(U)$ has an artinian classical quotient ring. [Hint: Assume that U is faithful. Show that $\mathscr{C}(N)$ is an Ore set, and then consider $UT \cap \text{ann}_U(P)$, where T is the $\mathscr{C}(N)$–torsion ideal of R and P is the assassinator of U.] □

♦ LOCALIZATION AT INFINITE SETS OF PRIME IDEALS

The above results clearly do not exhaust the question of localizability in noetherian rings. What one would like is to be able to start with a prime ideal P in a noetherian ring R and to find a right Ore set (or a two–sided Ore set) in some canonical way such that in the corresponding ring of fractions R' it is true that R'/PR' can be identified with the Goldie quotient ring of R/P. It is clear from Lemma 12.17 that the best chance for such an Ore set is the intersection of the $\mathscr{C}(Q)$ for all Q in the clique of P. This leads us naturally to the notion of a localization at an infinite set of primes. We will discuss this possibility in a series of exercises, since the results in this direction are not yet definitive.

DEFINITION. Let X be a nonempty set of prime ideals in a right noetherian ring R. We define $\mathscr{C}(X) = \bigcap_{Q \in X} \mathscr{C}(Q)$, and in case $\mathscr{C}(X)$ is a right Ore set we denote the corresponding localization by R_X. If X is infinite, we do not have any analog of Lemma 12.18 available, and so we build the desired properties of R_X into our definition of localizability. Thus, we say that X is *right localizable* provided $\mathscr{C}(X)$ is a right Ore set, the rings R_X/QR_X (for $Q \in X$) are all simple artinian, and the ideals QR_X (for $Q \in X$) are the only right primitive ideals of R_X. Further, X is *classically right localizable* if, in addition, for each $Q \in X$ the injective hull of the right R_X–module R_X/QR_X is the union of its socle series. A nonempty set of primes in a noetherian ring is *(classically) localizable* provided it is both right and left (classically) localizable.

To see the necessity of the extra conditions in this definition of localizability, consider first a polynomial ring $R = \mathbb{C}[x,y]$ and let X be the set of all maximal ideals of R except $xR + yR$. Then X satisfies the first two conditions of the localizability definition but not the third. (Here $\mathscr{C}(X)$ is just the set of units of R.) For a second example, consider a polynomial ring $R = \mathbb{Q}[x,y]$ and let X be the set of all prime ideals of the form fR where f is irreducible. Here again $\mathscr{C}(X)$ is the set of units of R, whence $\mathscr{C}(X)$ is an Ore set and $R_X = R$, yet none of the rings R/Q (for $Q \in X$) is artinian.

Exercise 12L shows that if R is a noetherian ring satisfying the second layer condition, X the right link closure of some prime of R, and X is finite, then X is classically right localizable.

EXERCISE 12T. Let $R = \mathbb{C}[x][\theta; x\frac{d}{dx}]$ as in Exercise 11G. As shown there, the set $X = \{M_\alpha \mid \alpha \in \mathbb{Z}\}$ is a clique in Spec(R).

(a) For each $\alpha \in \mathbb{Z}$, show that the module $E_\alpha = E((R/M_\alpha)_R)$ is the union of its socle series, and that E_α is $\mathscr{C}(X)$–torsionfree.

(b) If I is a right ideal of R such that $\mathrm{Hom}_R(R/I, E_\alpha) = 0$ for all $\alpha \in \mathbb{Z}$, show that I contains an element of $\mathscr{C}(X)$. [Hint: Choose generators $t_1, \ldots, t_n$ for I, write each $t_i = f_i + xr_i$ for some $f_i \in \mathbb{C}[\theta]$ and $r_i \in R$, and consider the ideal in $\mathbb{C}[\theta]$ generated by the f_i.]

(c) Show that X is classically localizable. [Hint: To see that R_X has no primitive ideals other than the $M_\alpha R_X$ for $\alpha \in \mathbb{Z}$, show that $J(R_X) = xR_X$.] □

EXERCISE 12U. Let X be a right localizable set of primes in a right noetherian ring R. For $Q \in X$, show that the natural map $R/Q \to R_X/QR_X$ extends to a ring isomorphism from the right Goldie quotient ring of R/Q onto R_X/QR_X. □

DEFINITION. A nonempty set X of prime ideals in a right noetherian ring R satisfies the *right intersection condition* if whenever I is a right ideal of R which contains an element of $\mathscr{C}(Q)$ for each $Q \in X$, then I contains an element of $\mathscr{C}(X)$.

EXERCISE 12V. Show that a finite nonempty set X of incomparable primes in a right noetherian ring R satisfies the right intersection condition. [Hint: Corollary 6.7.] □

EXERCISE 12W. Let X be a nonempty set of incomparable primes in a right noetherian ring R, and assume that $\mathscr{C}(X)$ is a right Ore set.

(a) If X satisfies the right intersection condition, show that for each $Q \in X$ the ring R_X/QR_X is simple artinian. [Hint: If $c \in \mathscr{C}(Q)$, show that $cR + Q$ contains an element of $\mathscr{C}(X)$, and then use Corollary 5.6.]

(b) If X satisfies the right intersection condition, show that the only right primitive ideals in R_X are the ideals QR_X for $Q \in X$; hence, X is right localizable. [Hint: If I is a maximal right ideal of R_X, find $Q \in X$ such that $R/(I^c + Q)$ is not torsion as a right (R/Q)–module, and proceed as in Lemma 12.18.]

(c) If X is right localizable, show that X satisfies the right intersection condition. [Hint: If not, consider a right ideal I such that $I \cap \mathscr{C}(Q)$ is nonempty for each $Q \in X$ while I is maximal with respect to the property that $I \cap \mathscr{C}(X)$ is empty.] □

EXERCISE 12X. Let R be a noetherian ring satisfying the right second layer condition and let X be a nonempty right link closed set of incomparable primes of R satisfying the right intersection condition.

(a) Modify the proof of Theorem 12.21 to show that if $Q \in X$ then $E((R/Q)_R)$ is $\mathscr{C}(X)$–torsionfree and each element of $E((R/Q)_R)$ is annihilated by a product of primes from

X.

(b) Show that X is classically right localizable. [Hint: Modify the proof of Theorem 12.20 to show that $\mathscr{C}(X)$ is a right Ore set.] □

The general theorem in Exercise 12X is due to Jategaonkar [1986, Theorem 7.1.5]. Obviously, it leaves open the question of what sets of primes satisfy the intersection condition. A more concrete recent result is that if R is a noetherian ring satisfying the second layer condition and which is an algebra over an uncountable field, and if X is a countable link–closed set of incomparable prime ideals such that there is a bound on the ranks of the factors R/Q for $Q \in X$, then X is classically localizable. (This was proved by Stafford [1987, Proposition 4.5] and Warfield [1986, Theorem 8].) As we will see in Chapter 14 below, every clique in a noetherian ring is countable, so this theorem is frequently applicable. (Also, by Corollary 12.6, in a noetherian ring satisfying the second layer condition, the primes in a clique are incomparable.) More concretely still, if R is an FBN ring which is an algebra over an uncountable field, every clique in R is classically localizable (see Stafford [1987, Proposition 4.5]).

♦ NOTES

Classical Krull Dimension. The relationship between chains of prime ideals and dimensions of algebraic varieties was explored by Noether. She showed that, first, if X is an irreducible algebraic variety over a field k, and if P is the prime ideal in the appropriate polynomial ring R over k consisting of those polynomials that vanish on X, then the dimension of X equals the transcendence degree over k of the quotient field of R/P [1923, Satz V], and second, (assuming that k is infinite) this dimension equals the maximum length of a chain of prime ideals ascending from P [1923, Satz VII]. Krull then developed this idea into a powerful tool for arbitrary commutative noetherian rings [1928a], and later writers gave the name *(classical) Krull dimension* to the supremum of the lengths of finite chains of prime ideals in a ring. The ordinal–valued definition of classical Krull dimension was introduced by Krause [1970, Definition 11].

Transfer of Classical Krull Dimension Across Noetherian Bimodules. Jategaonkar proved that if R and S are noetherian rings with the second layer condition, and if there exists a faithful noetherian (R,S)–bimodule, then R and S have the same classical Krull dimension [1979, Theorem H; 1982, Corollary 1.6 and Theorem 1.7]. He also gave the consequence that distinct primes in the same clique of a noetherian ring with the second layer condition are incomparable [1979, Theorem H; 1982, Theorem 1.8].

Second Layer Condition Implies Jacobson's Conjecture. This was proved by Jategaonkar [1979, Theorem H; 1982, Theorem 1.8]. The stronger intersection theorem given in Theorem 12.8 is due to the authors.

Contraction of Prime Ideals in Finite Ring Extensions. Theorem 12.12 was first proved by Joseph and Small (with no second layer condition) for certain factor rings of enveloping algebras [1978, Corollary 3.7]. Borho then proved it for noetherian rings with suitable symmetric dimension functions, assuming just that ${}_RRbR$ is finitely generated for all $b \in S$ [1982, Theorem 7.2]. The version we have given was proved by Warfield, assuming the second layer condition and that the bimodules ${}_RRbR_R$ are finitely generated on both sides [1983, Corollary 2].

Transfer of Second Layer Condition in Finite Ring Extensions. Theorem 12.14 is due to Letzter [Pb, Theorem 4.2].

Lying Over in Finite Ring Extensions. Theorem 12.16 was proved by Letzter [Pb, Theorem 4.6].

Localizability of Semiprime Ideals. The localizability criterion given in Theorem 12.20 is due to Jategaonkar [1974a, Theorem 3.2], who also gave an analogous criterion for classical localizability [1974a, Theorem 4.5]. The classical localizability criterion in terms of links (Theorem 12.21) was given for semiprime ideals in FBN rings by Müller [1976b, Theorem 5]. Jategaonkar proved that a semiprime ideal N in a noetherian ring is classically right localizable if and only if the set of primes minimal over N is right link closed and satisfies the right second layer condition [1986, Theorem 7.3.1].

Embeddability in Artinian Rings. Theorem 12.24 is due to Jategaonkar [1986, Theorem 8.3.9 and Proposition 8.3.5].

13 KRULL DIMENSION

Krull dimension is a measurement of size of a ring which has an intrinsic importance of its own, and is also a useful technical tool in the theory of noetherian rings. We have already discussed in the previous chapter the "classical" Krull dimension, which originated in commutative ring theory and is defined using the prime ideals of the ring. In the noncommutative theory we need a notion of Krull dimension which does not depend on prime ideals, but which shares many of the important properties of the classical Krull dimension for commutative rings. For instance, we would like a notion of Krull dimension that gives some useful information even for simple rings. This is done by defining a dimension on modules rather than just on rings, which has the advantage that it in some sense replaces considerations involving two–sided ideals with considerations involving only one–sided ideals. The definition now used is due to Rentschler and Gabriel, and will be defined in detail in the next section. While at first sight it appears completely unrelated to the classical definition, we shall see that it coincides with the classical Krull dimension on commutative noetherian rings, and in fact on FBN rings.

♦ DEFINITIONS AND BASIC PROPERTIES

Many of the results in this and subsequent sections follow fairly directly from the definitions. We have therefore left more of the details than usual to the reader. The Krull dimension to be defined and studied in this chapter is an ordinal–valued invariant defined for some modules but not all. It will in particular be defined for all noetherian modules, and in most of the "standard" examples it will be finite. As with classical Krull dimension, it is convenient to begin our list of ordinals with -1.

DEFINITION. The *Krull dimension* of a module M, when defined, is denoted $\mathrm{K.dim}(M)$. The actual definition is given by a transfinite induction. First, $\mathrm{K.dim}(M) = -1$ if and only if $M = 0$. Second, consider an ordinal $\alpha \geq 0$; assuming that we have already defined which modules have Krull dimension β for ordinals $\beta < \alpha$, we now define what it means for M to have Krull dimension α. Namely, $\mathrm{K.dim}(M) = \alpha$ if and only if (a) we have not already defined $\mathrm{K.dim}(M) = \beta$ for some ordinal $\beta < \alpha$, and (b) for every (countable) descending chain

$$M_0 \geq M_1 \geq M_2 \geq \ldots$$

of submodules of M, we have $K.dim(M_i/M_{i+1}) < \alpha$ for all but finitely many indices i (that is, for all but finitely many i, the Krull dimension of M_i/M_{i+1} has previously been defined, and so is an ordinal less than α). In case the statement "$K.dim(M) = \alpha$" does not hold for any ordinal α, we say that "K.dim(M) is not defined", or that "M does not have Krull dimension".

Observe that a module M has Krull dimension 0 if and only if $M \neq 0$ and for every descending chain $M_0 \geq M_1 \geq \ldots$ of submodules of M, we have $M_i/M_{i+1} = 0$ for all but finitely many indices i, that is, M satisfies the descending chain condition on submodules. In other words, the modules of Krull dimension 0 are precisely the nonzero artinian modules.

To avoid excessive statements about the existence of Krull dimensions, we make the convention that a statement such as "$K.dim(M) = \alpha$" is an abbreviation for "the Krull dimension of M exists and equals α".

EXERCISE 13A. Show that $\mathbb{Z}$ and k[x] (where k is a field and x is an indeterminate), considered as modules over themselves, have Krull dimension 1. □

EXERCISE 13B. If M is a module with $K.dim(M) \leq \alpha$ for some ordinal α, and N is a submodule of M, show that $K.dim(N) \leq \alpha$ and $K.dim(M/N) \leq \alpha$. □

EXERCISE 13C. If a module M contains an infinite direct sum of copies of a nonzero module N, show that M does not have Krull dimension. [Hint: Find submodules $M_1 > M_2 > \ldots$ such that $M_i/M_{i+1} \cong M_1$ for all i.] □

Much of the awkwardness in the definition of Krull dimension stems from the care needed in handling modules that may or may not have Krull dimension. Once a module M is known to have a Krull dimension, however, the definition of K.dim(M) becomes much simpler. Namely, for an ordinal $\alpha \geq 0$ we have $K.dim(M) = \alpha$ if and only if $K.dim(M) \geq \alpha$ but M contains no descending chain $M_0 \geq M_1 \geq \ldots$ of submodules such that $K.dim(M_i/M_{i+1}) \geq \alpha$ for infinitely many i. (The subfactors M_i/M_{i+1} all have Krull dimension, by Exercise 13B.)

EXERCISE 13D. Let M be a module with Krull dimension, and let $\alpha \geq 0$ be an ordinal. Show that $K.dim(M) > \alpha$ if and only if M contains a chain $M_0 \geq M_1 \geq \ldots$ of submodules such that $K.dim(M_i/M_{i+1}) \geq \alpha$ for infinitely many i. □

LEMMA 13.1. *If M is a module and N a submodule, and K.dim(M) is defined, then*

$$K.dim(M) = \max\{K.dim(N), K.dim(M/N)\}.$$

Proof. By Exercise 13B, $K.dim(N)$ and $K.dim(M/N)$ both exist, and if

$$\alpha = \max\{K.dim(N),\ K.dim(M/N)\},$$

then $\alpha \leq K.dim(M)$. We proceed by induction on α to show that $K.dim(M) \leq \alpha$, the case $\alpha = -1$ being clear.

Let $M_0 \geq M_1 \geq \ldots$ be a descending chain of submodules of M. Then the submodules $(M_i + N)/N$ form a descending chain of submodules of M/N, and thus for all but finitely many indices i, we must have

$$K.dim((M_i + N)/(M_{i+1} + N)) < K.dim(M/N) \leq \alpha.$$

Similarly, the submodules $M_i \cap N$ form a descending chain of submodules of N, and so for all but finitely many indices i we must have

$$K.dim((M_i \cap N)/(M_{i+1} \cap N)) < K.dim(N) \leq \alpha.$$

For any i, the kernel of the natural epimorphism $f_i : M_i/M_{i+1} \to (M_i + N)/(M_{i+1} + N)$ is isomorphic to $(M_i \cap N)/(M_{i+1} \cap N)$, and hence $K.dim(\ker(f_i)) < \alpha$ for all but finitely many i. By induction, for all but finitely many indices i we conclude that

$$K.dim(M_i/M_{i+1}) \leq \max\{K.dim(\ker(f_i)),\ K.dim((M_i + N)/(M_{i+1} + N))\} < \alpha,$$

which proves that $K.dim(M) \leq \alpha$. □

COROLLARY 13.2. *If $M_1,\ldots,M_k$ are modules with Krull dimension, then*

$$K.dim(M_1 \oplus \ldots \oplus M_k) = \max\{K.dim(M_1),\ldots,K.dim(M_k)\}. \ \square$$

So far, we have had almost no examples of modules which do have Krull dimension. We generously rectify this omission in the next lemma.

LEMMA 13.3. *If M is a noetherian module, then $K.dim(M)$ is defined.*

Proof. Assume not. By noetherian induction, we may assume that all proper homomorphic images of M have Krull dimension. Let

$$\alpha = \sup\{K.dim(M/N) \mid N \text{ is a nonzero submodule of } M\}.$$

We will show that $K.dim(M) \leq \alpha + 1$ and thereby obtain a contradiction.

Let $M_0 \geq M_1 \geq \ldots$ be submodules of M. We show that $K.dim(M_i/M_{i+1}) \leq \alpha$ for almost all i. If some $M_n = 0$, then $K.dim(M_i/M_{i+1}) = -1$ for almost all i, and so our condition is verified in this case. Otherwise, we may assume that all M_i are nonzero, in which case $K.dim(M_i/M_{i+1}) \leq K.dim(M/M_{i+1}) \leq \alpha$ for all i. Therefore $K.dim(M) \leq \alpha + 1$, as desired. □

DEFINITION. For any ring R, the Krull dimension of the right module R_R (if it exists) is called the *right Krull dimension of* R and is denoted $r.K.dim(R)$. Similarly, the *left Krull dimension of* R is the value $l.K.dim(R) = K.dim({}_RR)$. It remains an unsolved problem whether the left and right Krull dimensions of a noetherian ring are the same, although it is easy to find examples of rings for which only one of the two Krull dimensions is defined. For instance, if $R = \begin{pmatrix} \mathbb{Q} & \mathbb{R} \\ 0 & \mathbb{R} \end{pmatrix}$ then R is right artinian and so r.K.dim(R) = 0, while ${}_RR$ contains an infinite direct sum of copies of a simple left module, and hence it follows from Exercise 13C that l.K.dim(R) is not defined.

PROPOSITION 13.4. *If R is a right noetherian ring and M is a finitely generated right R–module, then*

$$K.dim(M) \leq r.K.dim(R).$$

Proof. Write $M \cong R^k/N$ for some $k \in \mathbb{N}$ and some submodule $N \leq R^k$. Apply, in turn, Lemma 13.3, Corollary 13.2, and Exercise 13B. □

♦ PRIME NOETHERIAN RINGS

PROPOSITION 13.5. *If R is a nonzero right noetherian ring and N is its prime radical, then*

$$r.K.dim(R) = r.K.dim(R/N) = \max\{r.K.dim(R/P) \mid P \text{ is a minimal prime ideal of } R\}.$$

Proof. Clearly, $r.K.dim(R/N) = K.dim((R/N)_R)$, and so

$$r.K.dim(R/N) \leq r.K.dim(R).$$

Similarly, $r.K.dim(R/P) \leq r.K.dim(R/N)$ for any (minimal) prime P, because $P \geq N$.

There are minimal prime ideals $P_1,\dots,P_n$ in R such that $P_1P_2\cdots P_n = 0$. For $i = 1,\dots,n$ we have

$$K.dim((P_1P_2\cdots P_{i-1}/P_1P_2\cdots P_i)_R) = K.dim((P_1P_2\cdots P_{i-1}/P_1P_2\cdots P_i)_{R/P_i}) \leq r.K.dim(R/P_i)$$

by Proposition 13.4, and so from Lemma 13.1 we conclude that

$$r.K.dim(R) \leq \max\{r.K.dim(R/P_1),\dots,r.K.dim(R/P_n)\}.$$

Therefore $r.K.dim(R) \leq r.K.dim(R/P)$ for some minimal prime P. □

LEMMA 13.6. *Let M be a nonzero module with Krull dimension and $f : M \to M$ an injective endomorphism. Then*

$$K.dim(M) \geq K.dim(M/f(M)) + 1.$$

Proof. Let $\alpha = K.dim(M/f(M))$. Since M is nonzero, $K.dim(M) \geq 0$, and so we are done if $\alpha = -1$. Now assume that $\alpha \geq 0$. In the chain of submodules $M \geq f(M) \geq f^2(M) \geq \dots$, all the successive factors $f^i(M)/f^{i+1}(M)$ are isomorphic to $M/f(M)$ and hence have Krull dimension α. Therefore $K.dim(M) > \alpha$. □

EXERCISE 13E. Let R be a prime right noetherian ring. If I is any nonzero right ideal of R, show that K.dim(I) = r.K.dim(R). [Hint: Corollary 6.26.] □

PROPOSITION 13.7. *Let R be a prime right noetherian ring and M a finitely generated right R–module. Then K.dim(M) < r.K.dim(R) if and only if M is a torsion module.*

Proof. Set α = r.K.dim(R). If x is any regular element in R, left multiplication by x defines a monomorphism $R_R \to R_R$, and consequently Lemma 13.6 implies that K.dim(R/xR) < α. It follows that if M is torsion, all cyclic submodules of M have Krull dimension less than α, and thus K.dim(M) < α. If M is not torsion, then by Lemma 6.17, M has a uniform submodule U isomorphic to a right ideal of R, and K.dim(U) = α by Exercise 13E. Therefore K.dim(M) = α in this case. □

EXERCISE 13F. Let R be a right noetherian ring. If P is a prime ideal of R, and I is an ideal with I > P, show that

r.K.dim(R/I) < r.K.dim(R/P).

Consequently, show that Cl.K.dim(R) ≤ r.K.dim(R). □

It is easily possible to have Cl.K.dim(R) < r.K.dim(R). For instance, if R = $A_1(k)$ where k is a field of characteristic zero, then because R is a simple ring, Cl.K.dim(R) = 0. On the other hand, R is not artinian, and so r.K.dim(R) > 0. (We shall see later that r.K.dim(R) = 1.)

♦ CRITICAL MODULES

DEFINITION. Let α be an ordinal, $\alpha \geq 0$. A module M is α–*critical* if K.dim(M) = α and K.dim(M/N) < α for all nonzero submodules N of M. A module is called *critical* if it is α–critical for some ordinal $\alpha \geq 0$. (We do not define "(–1)–critical".)

For example, the 0–critical modules are precisely the simple modules. Note that a module M (with Krull dimension) is 1–critical if and only if M is not artinian whereas all proper factors of M are artinian. For example, $\mathbb{Z}$ is a 1–critical module over itself, as is a polynomial ring in one variable over a field.

EXERCISE 13G. If M is a noetherian module of Krull dimension $\alpha \geq 0$, show that M has a proper submodule N such that M/N is α–critical. □

EXERCISE 13H. If M is an α–critical module and N is a nonzero submodule of M, show that N is α–critical. □

EXERCISE 13I. If M is a critical module, show that M is uniform. □

EXERCISE 13J. Let R be a semiprime right noetherian ring and I a right ideal of R. Show that I is critical if and only if I is uniform. □

EXERCISE 13K. If M is a critical module and $f : M \to M$ is a nonzero endomorphism, show that f is injective. □

EXERCISE 13L. If R is a right noetherian ring and P a completely prime ideal of R, show that R/P is an α–critical right R–module (where $\alpha = \text{r.K.dim}(R/P)$). More generally, if P is any prime ideal of R, and U is a uniform right ideal of R/P, show that U is α–critical, where $\alpha = \text{r.K.dim}(R/P)$. □

LEMMA 13.8. *If M is a nonzero module with Krull dimension, then M has a critical submodule. However, M need not have a critical submodule with the same Krull dimension as M.*

Proof. Choose a nonzero submodule N_0 of M of minimal Krull dimension (which we may do, because ordinals satisfy the descending chain condition.) Say $\text{K.dim}(N_0) = \alpha$. If N_0 is not critical, then N_0 has a nonzero submodule N_1 with $\text{K.dim}(N_0/N_1) = \alpha$. Necessarily, $\text{K.dim}(N_1) = \alpha$, since α is the smallest Krull dimension for nonzero submodules of M (and $\text{K.dim}(N_1) \leq \text{K.dim}(N_0)$ in any case). Now assume that we have constructed nonzero submodules $N_0 \geq N_1 \geq \ldots \geq N_k$ with

$$\text{K.dim}(N_i) = \text{K.dim}(N_{i-1}/N_i) = \alpha$$

for $i = 1,\ldots,k$. If N_k is not critical, there is a nonzero submodule $N_{k+1} \leq N_k$ such that

$$\text{K.dim}(N_{k+1}) = \text{K.dim}(N_k/N_{k+1}) = \alpha.$$

Since $\text{K.dim}(N_0) = \alpha$, this process cannot continue indefinitely, and therefore some N_k is critical.

The final statement of the lemma follows from Exercise 8I, in which there is an example of a noetherian ring R with a non–artinian cyclic module E possessing an essential simple submodule A. Since E is noetherian, it has Krull dimension, but since it is not artinian, $\text{K.dim}(E) > 0$. (In fact, it can be shown that $\text{K.dim}(E) = 1$.) If C is a critical submodule of E, then since A is simple and essential, $A \leq C$. By Exercise 13H, it follows that $\text{K.dim}(C) = \text{K.dim}(A) = 0$. Now C is 0–critical and thus simple, whence $C = A$. Therefore A is the only critical submodule of E, and $\text{K.dim}(A) < \text{K.dim}(E)$. □

EXERCISE 13M. Let M be a nonzero noetherian module and α the smallest ordinal which occurs as the Krull dimension of a nonzero submodule of M. Show that M has an α–critical submodule N such that M/N has no nonzero submodules of Krull dimension less than α. □

DEFINITION. A module M is α*–homogeneous* (for an ordinal α) provided M is nonzero and all its nonzero submodules have Krull dimension α. If M is α-homogeneous for some α, we just say that M is *Krull homogeneous* in case we don't wish to specify α.

For instance, it is immediate from Exercise 13H that any α-critical module is α-homogeneous, and from Exercise 13E that if R is a prime right noetherian ring then R_R is Krull homogeneous. For another example, every nonzero artinian module is 0-homogeneous.

EXERCISE 13N. Show that the $\mathbb{Z}$-module $M = \mathbb{Z} \oplus (\mathbb{Z}/2\mathbb{Z})$ is a sum of two 1-homogeneous (in fact, 1-critical) submodules, yet M is not 1-homogeneous. □

♦ CRITICAL COMPOSITION SERIES

DEFINITION. A *critical composition series* for a module M is a chain

$$0 = M_0 < M_1 < \ldots < M_n = M$$

of submodules of M such that each of the factors M_i/M_{i-1} is critical and such that $\text{K.dim}(M_{i+1}/M_i) \geq \text{K.dim}(M_i/M_{i-1})$ for all $i = 1,\ldots,n-1$.

For instance, any ordinary composition series for a nonzero module of finite length is also a critical composition series.

EXERCISE 13O. Let M be a module which has a critical composition series

$$0 = M_0 < M_1 < \ldots < M_n = M.$$

For $i = 1,\ldots,n$, show that M/M_{i-1} has no nonzero submodules with smaller Krull dimension than M_i/M_{i-1}. In particular, if the factors M_i/M_{i-1} are all α-critical (for the same ordinal α), show that M is α-homogeneous. □

THEOREM 13.9. [Jategaonkar, Gordon] *If M is a nonzero noetherian module, then M has a critical composition series. Furthermore any two critical composition series for M have the same length, and given two such critical composition series, say*

$$0 = M_0 < M_1 < \ldots < M_n = M \qquad \text{and} \qquad 0 = N_0 < N_1 < \ldots < N_n = M,$$

there exists a permutation π of $\{1,2,\ldots,n\}$ such that for each $i = 1,\ldots,n$ the factors M_i/M_{i-1} and $N_{\pi(i)}/N_{\pi(i)-1}$ contain nonzero isomorphic submodules.

Proof. Let α_1 be the smallest ordinal which occurs as the Krull dimension of a nonzero submodule of M. By Exercise 13M, M has an α_1-critical submodule M_1 such that M/M_1 has no nonzero submodules of Krull dimension less than α_1. If $M_1 \neq M$ and α_2 is the smallest ordinal which occurs as the Krull dimension of a nonzero submodule of M/M_1, then $\alpha_2 \geq \alpha_1$, and M/M_1 has an α_2-critical submodule M_2/M_1 such that M/M_2 has no nonzero submodules of Krull dimension less than α_2. If $M_2 \neq M$, continue in the same fashion. Since M is noetherian, this process eventually terminates in a chain of submodules

$$0 = M_0 < M_1 < \dots < M_n = M$$

where each M_i/M_{i-1} is α_i–critical and $\alpha_1 \le \alpha_2 \le \dots \le \alpha_n$. This chain is a critical composition series for M.

Now suppose that we are given two critical composition series

$$0 = M_0 < M_1 < \dots < M_n = M \qquad \text{and} \qquad 0 = N_0 < N_1 < \dots < N_k = M.$$

Consider first the case that all the factors M_i/M_{i-1} and N_j/N_{j-1} are α–critical (for the same α). By the Schreier Refinement Theorem (Theorem 3.10), these two chains of submodules have isomorphic refinements, say

$$0 = A_0 = A_{p(0)} < A_1 < \dots < A_{p(1)} = M_1 < \dots < A_{p(2)} = M_2 < \dots < A_t = M$$

$$0 = B_0 = B_{q(0)} < B_1 < \dots < B_{q(1)} = N_1 < \dots < B_{q(2)} = N_2 < \dots < B_t = M.$$

Since each M_i/M_{i-1} is α–critical, each of the factors $A_{p(i)+1}/A_{p(i)}$ has Krull dimension α, while all other factors A_m/A_{m-1} have Krull dimension less than α. Similarly, each of the factors $B_{q(j)+1}/B_{q(j)}$ has Krull dimension α, while all other factors B_m/B_{m-1} have Krull dimension less than α.

Consequently, the pairing between the successive factors of these two refinements must match each $A_{p(i)+1}/A_{p(i)}$ with some $B_{q(j)+1}/B_{q(j)}$ and vice versa. Thus $n = k$, and there is a permutation π of $\{1,\dots,n\}$ such that

$$A_{p(i-1)+1}/A_{p(i-1)} \cong B_{q(\pi(i)-1)+1}/B_{q(\pi(i)-1)}$$

for $i = 1,\dots,n$. Therefore for $i = 1,\dots,n$ the factors M_i/M_{i-1} and $N_{\pi(i)}/N_{\pi(i)-1}$ have nonzero isomorphic submodules. This concludes the proof in the case that the M_i/M_{i-1} and the N_j/N_{j-1} all have the same Krull dimension.

In the general case, set $\alpha_i = \text{K.dim}(M_i/M_{i-1})$ for $i = 1,\dots,n$ and $\beta_j = \text{K.dim}(N_j/N_{j-1})$ for $j = 1,\dots,k$. Then $\alpha_n = \text{K.dim}(M) = \beta_k$. Let p be the least index for which $\alpha_p = \alpha_n$, and let q be the least index for which $\beta_q = \alpha_n$. Now M/M_{p-1} has a critical composition series

$$0 = M_{p-1}/M_{p-1} < M_p/M_{p-1} < \dots < M_n/M_{p-1} = M/M_{p-1}$$

in which all the successive factors are α_n–critical. By Exercise 13O, M/M_{p-1} is α_n–homogeneous, and similarly M/N_{q-1} is α_n–homogeneous. The minimality of p implies that $\text{K.dim}(M_{p-1}) < \alpha_n$, whence the image of M_{p-1} in M/N_{q-1} has Krull dimension less than α_n. Since M/N_{q-1} is α_n–homogeneous, the image of M_{p-1} must be zero, that is, $M_{p-1} \le N_{q-1}$. The reverse inclusion follows by symmetry, and thus $M_{p-1} = N_{q-1}$.

Applying the case proved above, $n - p + 1 = k - q + 1$ and the factors M_i/M_{i-1} (for $i = p,\dots,n$) and N_j/N_{j-1} (for $j = q,\dots,n$) can be paired in such a way that each pair contain nonzero isomorphic submodules. Now M_{p-1} has a critical composition series of length less than n. By induction on the length of a critical composition series (or by induction on Krull dimension), we may conclude that $p - 1 = q - 1$ and that the factors M_i/M_{i-1} (for

$i = 1,\ldots,p-1$) and N_j/N_{j-1} (for $j = 1,\ldots,q-1$) can be paired in such a way that each pair contain nonzero isomorphic submodules. Therefore $n = k$ and all the factors M_i/M_{i-1} and N_j/N_{j-1} can be paired in the desired fashion. □

Theorem 13.9 depends on the fact that in the definition of a critical composition series, the sequence of Krull dimensions of successive factors is nondecreasing. If we consider arbitrary chains of submodules with successive factors critical, we lose uniqueness of lengths. For instance, given any nonnegative integer n there is a chain of ideals

$$0 < 2^n\mathbb{Z} < 2^{n-1}\mathbb{Z} < \ldots < 2\mathbb{Z} < \mathbb{Z}$$

in $\mathbb{Z}$, of length $n + 1$, such that all successive factors are critical $\mathbb{Z}$–modules.

EXERCISE 13P. If $0 = M_0 < M_1 < \ldots < M_n = M$ and $0 = N_0 < N_1 < \ldots < N_n = M$ are critical composition series for a module M, show that

$$\text{K.dim}(M_i/M_{i-1}) = \text{K.dim}(N_i/N_{i-1})$$

for $i = 1,\ldots,n$. □

COROLLARY 13.10. *Let M be a nonzero submodule of a direct sum $A_1 \oplus \ldots \oplus A_n$ where each A_i is a critical module. Then M has a critical composition series of length at most n, and $M \leq_e A_1 \oplus \ldots \oplus A_n$ if and only if M has a critical composition series of length n.*

Proof. Set $\alpha_i = \text{K.dim}(A_i)$ for $i = 1,\ldots,n$, and re–order the A_i so that

$$\alpha_1 \leq \alpha_2 \leq \ldots \leq \alpha_n.$$

Then set $M_0 = 0$ and $M_i = M \cap (A_1 \oplus \ldots \oplus A_i)$ for $i = 1,\ldots,n$. Since M_i/M_{i-1} embeds in A_i, it is either zero or α_i–critical. Hence, after discarding any possible repetitions in the series

$$0 = M_0 \leq M_1 \leq \ldots \leq M_n = M,$$

we obtain a critical composition series for M, of length at most n. Moreover, if M is essential in $A_1 \oplus \ldots \oplus A_n$, then $M \cap A_i \neq 0$ for each i, whence

$$0 = M_0 < M_1 < \ldots < M_n = M.$$

In this case M has a critical composition series of length exactly n.

Conversely, assume that some (and hence all) critical composition series for M has length n. By the argument of the previous paragraph, M cannot be embedded in a direct sum of $n - 1$ critical modules. Consequently, $M \cap A_i \neq 0$ for all i. Since the A_i are uniform (Exercise 13I), it follows that $M \cap A_i \leq_e A_i$ for all $i = 1,\ldots,n$. Thus

$$(M \cap A_1) \oplus \ldots \oplus (M \cap A_n) \leq_e A_1 \oplus \ldots \oplus A_n,$$

and therefore $M \leq_e A_1 \oplus \ldots \oplus A_n$. □

DEFINITION. Let M be a nonzero noetherian module, of Krull dimension α. Then the *Krull radical* of M, written $J_\alpha(M)$, is the intersection of the kernels of all homomorphisms from M to α-critical modules.

PROPOSITION 13.11. *Let M be a nonzero noetherian module of Krull dimension α and $J_\alpha(M)$ its Krull radical. Then $M/J_\alpha(M)$ is α-homogeneous, and $M/J_\alpha(M)$ is isomorphic to an essential submodule of a finite direct sum of α-critical modules.*

Proof. By Exercise 13G, $J_\alpha(M) < M$. Now without loss of generality, we may assume that $J_\alpha(M) = 0$. Choose a critical composition series

$$0 = M_0 < M_1 < \ldots < M_n = M.$$

There exists a homomorphism f from M to an α-critical module C such that $f(M_1) \neq 0$, and so $\text{K.dim}(M_1) \geq \text{K.dim}(f(M_1)) = \alpha$. On the other hand,

$$\text{K.dim}(M_1) \leq \text{K.dim}(M_2/M_1) \leq \ldots \leq \text{K.dim}(M_n/M_{n-1}) \leq \text{K.dim}(M) = \alpha.$$

Thus $\text{K.dim}(M_i/M_{i-1}) = \alpha$ for all $i = 1,\ldots,n$, and hence, by Exercise 13O, M is α-homogeneous.

Choose a nonzero homomorphism $f_1 : M \to A_1$ where A_1 is α-critical, set $K_1 = \ker(f_1)$, and observe that M/K_1 is α-critical. If $K_1 \neq 0$, choose a homomorphism $f_2 : M \to A_2$ where A_2 is α-critical and $f_2(K_1) \neq 0$, set $K_2 = K_1 \cap \ker(f_2)$, and observe that K_1/K_2 is α-critical. Continue this process as long as possible.

If this process continues beyond n steps, we obtain a chain of submodules

$$M = K_0 > K_1 > \ldots > K_n > 0$$

such that K_{i-1}/K_i is α-critical for $i = 1,\ldots,n$. Since M is α-homogeneous, the successive factors in any critical composition series for K_n must all be α-critical. Hence, combining such a series with the chain above, we obtain a critical composition series for M, of length greater than n, which contradicts Theorem 13.9. Thus our process must terminate, after at most n steps.

Therefore we obtain a critical composition series $0 = K_m < K_{m-1} < \ldots < K_1 < K_0 = M$, for some $m \leq n$. By Theorem 13.9, $m = n$. If f denotes the embedding

$$(f_1,\ldots,f_n) : M \to A_1 \oplus \ldots \oplus A_n,$$

we conclude from Corollary 13.10 that $f(M) \leq_e A_1 \oplus \ldots \oplus A_n$. □

COROLLARY 13.12. *Let M be a nonzero noetherian module having Krull dimension α, let $J_\alpha(M)$ be its Krull radical, and let N be a submodule of M. Then $\text{K.dim}(M/N) < \alpha$ if and only if $(N + J_\alpha(M))/J_\alpha(M)$ is essential in $M/J_\alpha(M)$.*

Proof. If $\text{K.dim}(M/N) < \alpha$ then clearly $\text{K.dim}(M/(N + J_\alpha(M))) < \alpha$. On the other hand, if $\text{K.dim}(M/N) = \alpha$, then by Exercise 13G, M has a submodule N' such that

$N' \geq N$ and M/N' is α–critical. Since $J_\alpha(M) \leq N'$, it follows that in this case $K.dim(M/(N + J_\alpha(M))) = \alpha$. Thus, after replacing M and N by $M/J_\alpha(M)$ and $(N + J_\alpha(M))/J_\alpha(M)$, we may assume that $J_\alpha(M) = 0$. By Proposition 13.11, we may also assume that M is an essential submodule of a finite direct sum $A_1 \oplus ... \oplus A_n$ of α–critical modules.

Suppose first that $K.dim(M/N) < \alpha$. Then none of the α–critical modules A_i can embed in M/N, whence $N \cap A_i \neq 0$ for $i = 1,...,n$. As in the proof of Corollary 13.10, it follows that $N \leq_e A_1 \oplus ... \oplus A_n$, and thus $N \leq_e M$.

Conversely, if $N \leq_e M$ then $N \leq_e A_1 \oplus ... \oplus A_n$, and hence $N \cap A_i \neq 0$ for $i = 1,...,n$. Since each A_i is α–critical, each factor $A_i/(N \cap A_i)$ has Krull dimension less than α. Therefore

$$K.dim((A_1 \oplus ... \oplus A_n)/((N \cap A_1) \oplus ... \oplus (N \cap A_n))) < \alpha,$$

whence $K.dim((A_1 \oplus ... \oplus A_n)/N) < \alpha$, and consequently $K.dim(M/N) < \alpha$. □

♦ FBN RINGS

In this and the following section, we obtain some specific results on Krull dimension which enable us to calculate the actual Krull dimension of several rings, including fully bounded rings, polynomial rings, and Weyl algebras. Our first computation will show that for many important rings, the new notion of Krull dimension agrees with the old (i.e., "classical") one, as defined in the previous chapter.

THEOREM 13.13. [Gabriel, Gordon–Robson, Krause] *If R is a right FBN ring, then $r.K.dim(R) = Cl.K.dim(R)$.*

Proof. We already have $r.K.dim(R) \geq Cl.K.dim(R)$ by Exercise 13F. To prove the reverse inequality, we proceed by induction on the ordinal $\alpha = r.K.dim(R)$, the case in which $\alpha = 0$ being trivial. Now assume that $\alpha > 0$, and that the result holds for right FBN rings with right Krull dimension less than α.

By Proposition 13.5, R has a minimal prime P such that $r.K.dim(R/P) = \alpha$, and it suffices to show that $Cl.K.dim(R/P) \geq \alpha$. Thus, without loss of generality, we may assume that R is prime.

Choose a uniform right ideal U of R, and recall from Exercise 13E that $K.dim(U) = \alpha$. Hence, it suffices to show that $K.dim(U) \leq Cl.dim(R)$. Thus consider a descending chain $U_0 \geq U_1 \geq ...$ of submodules of U; we must show that $K.dim(U_i/U_{i+1}) < Cl.K.dim(R)$ for all but finitely many indices i. As this is clear if some $U_i = 0$, we may assume that $U_i \neq 0$ for all i.

Now each factor U_i/U_{i+1} is torsion (since U is uniform) and hence unfaithful, by Lemma 8.2. If $A_i = ann_R(U_i/U_{i+1})$, then by Exercise 13F, $r.K.dim(R/A_i) < \alpha$. It follows using our induction hypothesis and Exercise 12A that

$$\text{K.dim}(U_i/U_{i+1}) \leq \text{r.K.dim}(R/A_i) = \text{Cl.K.dim}(R/A_i) < \text{Cl.K.dim}(R),$$

as desired. Therefore $\text{r.K.dim}(R) = \text{K.dim}(U) \leq \text{Cl.K.dim}(R)$, and the induction step is established. □

COROLLARY 13.14. *If $R = k[x_1,...,x_n]$ is a polynomial ring over a field k in n independent indeterminates, then $\text{r.K.dim}(R) = n$.*

Proof. Exercise 12B and Theorem 13.13. □

We may also use Theorem 13.13 to see that if R is the commutative noetherian domain constructed in Exercise 12D, then $\text{r.K.dim}(R) = \omega$.

It is immediate from Theorem 13.13 that any FBN ring has the same right and left Krull dimensions. More generally, Krull symmetry holds for noetherian bimodules over FBN rings, as follows.

THEOREM 13.15. [Jategaonkar] *Let R and S be FBN rings and ${}_RB_S$ a bimodule which is finitely generated on both sides. Then $\text{K.dim}({}_RB) = \text{K.dim}(B_S)$.*

Proof. Without loss of generality, we may assume that ${}_RB$ and B_S are faithful. Since R and S satisfy the second layer condition, $\text{Cl.K.dim}(R) = \text{Cl.K.dim}(S)$ by Corollary 12.5. Applying Theorem 13.13, we find that $\text{l.K.dim}(R) = \text{r.K.dim}(S)$. By Lemma 7.1, there exists $n \in \mathbb{N}$ such that S_S can be embedded in B^n (as a right submodule). Consequently,

$$\text{r.K.dim}(S) = \text{K.dim}(B_S).$$

Similarly, $\text{l.K.dim}(R) = \text{K.dim}({}_RB)$, and the theorem is proved. □

♦ POLYNOMIAL RINGS AND WEYL ALGEBRAS

For precise calculations of Krull dimensions of specific rings, it is helpful to produce inequalities relating the Krull dimensions of different rings and modules, for instance the Krull dimensions of a ring S and a subring R. In general, just because a ring R is a subring of a ring S, we cannot conclude that $\text{r.K.dim.}(R) \leq \text{r.K.dim.}(S)$. (For example, let $R = \mathbb{Z}$ and $S = \mathbb{Q}$.) However, if S were a free left R–module, we could draw this conclusion, since the correspondence taking a right ideal I of R to the right ideal IS of S would be injective. (Note that if we write S as a free left R–module, then IS is the set of all elements for which each coordinate is in I.)

The key idea here is to bound the right Krull dimension of R by that of S using a suitable map taking right ideals of R to right ideals of S. In the above example we had an embedding (of ordered sets) from the lattice of right ideals of R into the lattice of right ideals of S, but we can get by with slightly weaker hypotheses, as follows.

EXERCISE 13Q. Let M and N be modules with Krull dimension (possibly over different rings), and let $\mathscr{L}(M)$ and $\mathscr{L}(N)$ be the corresponding lattices of submodules. If there exists a map $g : \mathscr{L}(M) \to \mathscr{L}(N)$ which preserves strict inclusions (that is, whenever $A < B$ in $\mathscr{L}(M)$, then $g(A) < g(B)$ in $\mathscr{L}(N)$), show that $\text{K.dim}(M) \le \text{K.dim}(N)$. [Hint: Prove for all ordinals α that whenever $A < B$ in $\mathscr{L}(M)$ with $\text{K.dim}(B/A) > \alpha$, then $\text{K.dim}(g(B)/g(A)) > \alpha$.] □

EXERCISE 13R. If R is a right noetherian subring of a ring S such that S_R is finitely generated, and M is a finitely generated right S–module, show that $\text{K.dim}(M_S) \le \text{K.dim}(M_R)$. □

EXERCISE 13S. If X is a right denominator set in a right noetherian ring R, show that $\text{r.K.dim}(RX^{-1}) \le \text{r.K.dim}(R)$. □

Exercise 13Q applies in particular to the situation indicated at the beginning of the section, where $R \subseteq S$ are rings with right Krull dimension and S is a free left R–module. We will need a slightly more general form of this situation.

DEFINITION. Let R be a subring of a ring S. We say that S is *left faithfully flat over* R if S is a flat left R–module and if $M \otimes_R S \neq 0$ for every nonzero right R–module M.

EXERCISE 13T. Let R be a subring of a ring S. If S is free as a left R–module, show that S is left faithfully flat over R. If S is flat as a left R–module and $M \otimes_R S \neq 0$ for every simple right R–module M, show that S is left faithfully flat over R. □

EXERCISE 13U. Let R and S be right noetherian rings such that R is a subring of S and S is left faithfully flat over R. If M is any finitely generated right R–module, show that $\text{K.dim}(M) \le \text{K.dim}(M \otimes_R S)$. In particular, $\text{r.K.dim}(R) \le \text{r.K.dim}(S)$. □

EXERCISE 13V. Show that if k is a field and $n \in \mathbb{N}$, then $\text{r.K.dim}(A_n(k)) \ge n$. □

In view of the preceding exercise, we now have examples of simple rings of large Krull dimension, though we have not yet computed the actual Krull dimension.

We now turn to some specific computations of Krull dimension. The rings we will look at are all Ore extensions or iterated Ore extensions. Among these, the polynomial ring $R[x]$ is particularly important, and it also turns out that knowing the Krull dimension of $R[x]$ gives us some control on the Krull dimensions of more general Ore extensions $R[\theta;\alpha,\delta]$. This becomes clear in the next lemma, which also shows that the Krull dimensions of certain modules over $R[x]$ are all that is needed to control the Krull dimension of $R[x]$ and of $R[\theta;\alpha,\delta]$.

We let $T = R[x]$, the polynomial ring over a right noetherian ring R. If M is a right R–module, then there is a corresponding right T–module, namely $M \otimes_R T$. It is convenient to write this module as $M[x]$, since every element of $M \otimes_R T$ can be written as a "polynomial"

$$f = (m_0 \otimes 1) + (m_1 \otimes x) + ... + (m_n \otimes x^n) \equiv m_0 + m_1x + ... + m_nx^n$$

for some $m_i \in M$. If n is the index of the largest nonzero term in an expresion for f, then n is the *degree* of f and m_n is the *leading coefficient*. Similarly, if $S = R[\theta;\alpha,\delta]$, an Ore extension of R, we write the induced module $M \otimes_R S$ as $M[\theta]$, and we define degrees and leading coefficients for elements of $M[\theta]$ as in $M[x]$.

LEMMA 13.16. *Let R be a right noetherian ring, $S = R[\theta;\alpha,\delta]$ an Ore extension of R, and $T = R[x]$ a polynomial ring. Assume that α is an automorphism of R. If M is any finitely generated right R–module, then*

$$K.dim(M[\theta]) \leq K.dim(M[x]).$$

Moreover, if V is a nonzero S–submodule of $M[\theta]$, there exist a nonzero element $m \in M$ and a nonnegative integer n such that

$$K.dim(M[\theta]/V) \leq K.dim(M[x]/mx^nT).$$

Proof. The technique we use is essentially the method of associated graded modules, which we do here only in a special case. We set up a map from the lattice of submodules of $M[\theta]$ to the lattice of submodules of $M[x]$ and then apply Exercise 13Q.

If A is a submodule of $M[\theta]$, then for $i = 0,1,...$ we let $g_i(A)$ be the subset of M consisting of 0 together with the leading coefficients of the nonzero elements of A of degree i. Since α is an automorphism, each $g_i(A)$ is a submodule of M. Note that $g_0(A) \leq g_1(A) \leq ...$. We then let

$$g(A) = g_0(A) + g_1(A)x + g_2(A)x^2 + ...$$

Clearly $g(A)$ is a submodule of $M[x]$, and if $A \leq B \leq M[\theta]$, then $g(A) \leq g(B)$.

We next observe that if A and B are submodules of $M[\theta]$ with $A < B$, then $g(A) < g(B)$. Suppose not; then $g(A) = g(B)$ and so $g_i(A) = g_i(B)$ for all i. Let b be an element of B not in A and choose b to be of least possible degree, say j. Since $g_j(A) = g_j(B)$, there is an element $a \in A$ of the same degree and with the same leading coefficient as b. But then $b - a$ is an element of B of lower degree and not in A, a contradiction. Thus $g(A) < g(B)$, as claimed.

The inequality $K.dim(M[\theta]) \leq K.dim(M[x])$ is now immediate from Exercise 13Q. If V is a nonzero submodule of $M[\theta]$, choose a nonzero element of V with, say, degree n and leading coefficient m, and observe that $mx^nT \leq g(V)$. The final conclusion of the lemma now follows from a second application of Exercise 13Q, using the map $A/V \mapsto g(A)/mx^nT$ from submodules of $M[\theta]/V$ to submodules of $M[x]/mx^nT$. □

THEOREM 13.17. [Rentschler–Gabriel] *Let R be a right noetherian ring, M a nonzero finitely generated right R–module, and x an indeterminate. Then*

$$K.dim(M[x]) = K.dim(M) + 1.$$

In particular, if R is nonzero then $r.K.dim(R[x]) = r.K.dim(R) + 1$.

Proof. [Gordon–Robson] We let $T = R[x]$ and $U = M[x]$, and let $\beta = K.dim(M)$. Now M can be made into a right T–module in a natural way, by letting x act trivially, and if we do this, then $K.dim(M_R) = K.dim(M_T)$. Next note that $Ux^n/Ux^{n+1} \cong M$ and

$$K.dim(Ux^n/Ux^{n+1}) = \beta$$

for all n, whence $K.dim(U) > \beta$. (Here we use the fact that x is a central element of T.)

We now use a critical composition series for M to reduce to the case in which M is a β–critical module, and we may assume by induction that the theorem has been proved for ordinals smaller than β. We will show, in fact, that U is $(\beta+1)$–critical.

To show that U is $(\beta+1)$–critical, it will suffice to show, for every nonzero submodule V of U, that $K.dim(U/V) \leq \beta$. Let us first assume that V has the special form $V = mx^nT$, (i.e., V is generated by a monomial.) Now we already know that $K.dim(U/Ux^n) \leq \beta$, and so we only need to consider the factor Ux^n/mx^nT, which is isomorphic to U/mT. Since $U/mT \cong (M/mR)[x]$, and M is β–critical, it follows by induction that

$$K.dim(Ux^n/mx^nT) = K.dim(U/mT) = K.dim(M/mR) + 1 \leq \beta$$

and hence that $K.dim(U/mx^nT) \leq \beta$.

To prove the theorem, we reduce the general case to this specific case. If we set $\alpha = 1$ and $\delta = 0$ in Lemma 13.16, then $M[\theta] = M[x] = U$, and the lemma provides a nonzero element $m \in M$ and a nonnegative integer n such that $K.dim(U/V) \leq K.dim(U/mx^nT)$. Then $K.dim(U/V) \leq \beta$ by the previous paragraph, which completes the proof of the theorem. □

For example, we may use Theorem 13.17 to extend Corollary 13.14 as follows: If $T = R[x_1,...,x_n]$ is a polynomial ring in n independent indeterminates over a right artinian ring R, then $r.K.dim(T) = n$.

EXERCISE 13W. Let R be a right noetherian ring and let $S = R[\theta;\alpha,\delta]$ be an Ore extension of R. Assume that α is an automorphism of R. Show that if M is a finitely generated right R–module, then

$$K.dim(M) \leq K.dim(M \otimes_R S) \leq K.dim(M) + 1.$$

In particular, $\text{r.K.dim}(R) \le \text{r.K.dim}(S) \le \text{r.K.dim}(R) + 1$. □

EXERCISE 13X. If $S = k[\theta;\alpha,\delta]$ is an Ore extension of a field k, and α is an automorphism of k, show that $\text{r.K.dim}(S) = 1$. □

EXERCISE 13Y. Let $R = A_n(k) = k[x_1,\dots,x_n][\theta_1,\dots,\theta_n; \partial/\partial x_1,\dots,\partial/\partial x_n]$ be a Weyl algebra over a field k.

(a) Show that there is a k–algebra automorphism α of R such that $\alpha(x_i) = \theta_i$ and $\alpha(\theta_i) = -x_i$ for all $i = 1,\dots,n$. [Hint: Exercise 1H.]

(b) Set $X_i = k[x_i] - \{0\}$ and $Y_i = k[\theta_i] - \{0\}$ for $i = 1,\dots,n$. Show that X_i and Y_i are right and left denominator sets in R, and that $RX_i^{-1} \cong RY_i^{-1} \cong A_{n-1}(k(t))[\theta; \partial/\partial t]$ for some indeterminate t (where $A_0(k(t)) = k(t)$). [Hint: Exercise 9S.] □

THEOREM 13.18. [Nouazé–Gabriel, Rentschler–Gabriel] *Let k be a field of characteristic zero. Then $\text{r.K.dim}(A_n(k)) = n$, for each positive integer n.*

Proof. Let $R = A_n(k)$. We proceed by induction on n, the case $n = 0$ (namely, $R = k$) being trivial. Now let $n > 0$ and assume the theorem holds for Weyl algebras of lower degree. We now embed R into $2n$ larger rings, as follows. Adopt the notation of Exercise 13Y, and set $B_i = RX_i^{-1}$ and $C_i = RY_i^{-1}$ for $i = 1,\dots,n$. By induction together with Exercise 13W, we conclude that $\text{r.K.dim}(B_i) \le n$ and $\text{r.K.dim}(C_i) \le n$.

Consider the diagonal embedding

$$R \to S = \left(\prod_{i=1}^{n} B_i\right) \times \left(\prod_{i=1}^{n} C_i\right).$$

It is clear that $\text{r.K.dim}(S) \le n$ (e.g., by Proposition 13.5), and so the induction step will follow via Exercise 13U if we show that S is left faithfully flat over R.

The ring S is a flat left R–module because each B_i and each C_i is (Corollary 9.15). Because of Exercise 13T, it now suffices to show that $(R/I) \otimes_R S \neq 0$ for each maximal right ideal I of R, that is, $S/IS \neq 0$. If not, $IS = S$ and hence $IB_i = B_i$ and $IC_i = C_i$ for all $i = 1,\dots,n$. It follows that there exist monic polynomials $f_i \in k[x_i] \cap I$ and $g_i \in k[\theta_i] \cap I$, for each i. From this we conclude that R/I is finite–dimensional over k. However, R embeds in $\text{End}_k(R/I)$ (because R is a simple algebra), and since R is infinite–dimensional over k we have a contradiction. Therefore $S/IS \neq 0$, as desired, and the induction step is complete. □

EXERCISE 13Z. Show that $\text{r.K.dim}(A_n(k)) = 2n$ for any field k of positive characteristic. [Hint: Show that $A_n(k)$ is faithfully flat over a polynomial ring in $2n$ indeterminates.] □

In particular, we conclude from Theorem 13.18 and Exercise 13Z that for any field k, the Weyl algebras $A_1(k)$, $A_2(k)$, ... are pairwise non–isomorphic.

EXERCISE 13ZA. If k is a field of characteristic zero and D_1 is the Ore quotient ring of $A_1(k)$, show that $\text{r.K.dim}(A_1(D_1)) = 2$. [Hint: Write $A_1(k) = k[x_1][\theta_1; d/dx_1]$ and $A_1(D_1) = D_1[x][\theta; \partial/\partial x]$, and show that left multiplication by $\theta - \theta_1$ induces an injective endomorphism of $A_1(D_1)/(x + x_1)A_1(D_1)$.] □

♦ NOTES

Krull Dimension. Classical Krull dimension is discussed in the notes for Chapter 12. Gabriel introduced an ordinal–valued dimension, which he named "Krull dimension", for objects in an abelian category [1962, p. 382], using a transfinite sequence of localizing subcategories. Rentschler and Gabriel presented the definition we have given (for finite ordinals only) in [1967, p. 712], commenting (without proof) that for a noetherian ring this agrees with Gabriel's definition [1967, Introduction]. The ordinal–valued version of the Rentschler–Gabriel definition was introduced by Krause [1970, Definition 9]. Gordon and Robson gave the name "Gabriel dimension" to Gabriel's original dimension after shifting the finite values by one [1973, p. 16; 1974, p. 461], and gave a proof that the Gabriel dimension of any noetherian module equals its Krull dimension plus one [1974, Proposition 2.3].

Critical Modules. These were introduced by Hart (under the name "restricted modules") [1971, p. 342] and Goldie (using the name "critical") [1972b, p. 162].

Critical Composition Series. Jategaonkar introduced critical composition series for a finitely generated module over an FBN ring (under the name "basic series") [1974b, p. 110], and proved that any two such series are equivalent in the sense of Theorem 13.9 [1974b, Theorem 3.1]. The general case was then developed by Gordon [1974, Corollary 2.8].

Krull Dimension of FBN Rings. For a right T–ring R (see the notes for Chapter 8), Gabriel showed that his version of Krull dimension can be computed in the following manner. If $E_\alpha = \{P \in \text{Spec}(R) \mid \text{K.dim}(R/P) \le \alpha\}$ for each ordinal α, then E_{-1} is empty,

$$E_{\alpha+1} = \{P \in \text{Spec}(R) \mid \text{all primes properly containing } P \text{ lie in } E_\alpha\}$$

for all α, and E_α is the union of the E_β for $\beta < \alpha$ whenever α is a limit ordinal [1962, Corollaire 2, p. 425]. That the later version of Krull dimension agrees with the classical Krull dimension for right FBN rings was obtained independently by Krause [1972, Theorem 2.4] and Gordon and Robson [1973, Theorem 8.12].

Krull Symmetry for Noetherian Bimodules over FBN Rings. Jategaonkar proved that if R and S are FBN rings and there exists a faithful noetherian (R,S)–bimodule, then R and S have the same (right) Krull dimension [1974b, Theorem 2.3].

Krull Dimension of Polynomial Rings. Rentschler and Gabriel showed that if R is a left noetherian ring with finite left Krull dimension, then l.K.dim(R[x]) = l.K.dim(R) + 1 [1967, Application 1°]. The general case of Theorem 13.17 was given by Gordon and Robson [1973, Theorem 9.2].

Krull Dimension of Weyl Algebras. Using Gabriel's definition of Krull dimension, Nouazé and Gabriel showed that the Weyl algebra $A_n(k)$ over a field k of characteristic zero has Krull dimension at least n and at most $2n - 1$ [1967, Proposition, p. 83]. Using the new definition, Rentschler and Gabriel then showed that the Krull dimension of $A_n(k)$ is exactly n [1967, Application 3°].

14 NUMBERS OF GENERATORS OF MODULES

We turn in this chapter to a very "classical" problem – estimating the minimum number of generators needed for a finitely generated module A over a noetherian ring R. In case R is commutative, there is a theorem of Forster from 1964 giving an estimate for the number of generators of A in terms of "local data", namely the values g(A,P) + K.dim(R/P), where P is any prime ideal of R and g(A,P) is the minimum number of generators of the localized module A_P over the local ring R_P. In the noncommutative case, we shall see that an appropriate analog of g(A,P) is the minimum number of generators needed for the tensor product of A with the Goldie quotient ring of R/P. With this adjustment, we shall derive an analog of Forster's theorem for finitely generated modules over any FBN ring.

It order to handle data from all prime ideals at once, it is most convenient to work topologically. Thus we first develop an appropriate topology for the prime spectrum of R, and then we develop a continuity theorem for a normalized version of g(A,P) (considered as a function of P). In case R is FBN, this normalized function turns out to be locally constant, and the estimate for the number of generators of A can be obtained without too much further work. At the end of the chapter, we use the topological methods just developed to prove that all cliques of prime ideals in any noetherian ring are countable.

Some of our motivating discussion relies on the localization theory of commutative noetherian rings, which we have only briefly touched on in this book. The reader who is unfamiliar with this theory may simply ignore these arguments.

♦ TOPOLOGIES ON THE PRIME SPECTRUM

DEFINITION. For each ideal I in a ring R, define

$$V(I) = \{P \in \mathrm{Spec}(R) \mid P \geq I\}$$
$$W(I) = \{P \in \mathrm{Spec}(R) \mid P \ngeq I\}.$$

The family $\mathscr{W} = \{W(I) \mid I$ is an ideal of $R\}$ is closed under finite intersections and arbitrary unions, since

$$W(I_1) \cap \ldots \cap W(I_n) = W(I_1 I_2 \cdots I_n)$$

$$\bigcup_{j \in J} W(I_j) = W(\sum_{j \in J} I_j)$$

for all ideals I_j in R. Moreover, Spec(R) = W(R) and the empty set equals W(0). Therefore $\mathscr{W}$ is the family of open sets for a topology on Spec(R), known as the *Zariski topology* (or the *Stone topology*, or the *hull–kernel topology*). The closed subsets of Spec(R) in this topology are exactly the sets V(I).

EXERCISE 14A. Find all the open sets for the Zariski topology on Spec($\mathbb{Z}$), and show that Spec($\mathbb{Z}$) with this topology is not Hausdorff. □

EXERCISE 14B. Given a ring R, show that in the Zariski topology on Spec(R), all the closed subsets are compact. Moreover, if R has the ACC on ideals, show that all subsets of Spec(R) are compact. □

As we shall see in the following section, the Zariski topology is not suitable for some of our purposes, and so we introduce a second topology, with more open sets than the Zariski topology.

DEFINITION. Let R be a ring, and let $\mathscr{V}$ be the family of all subsets of Spec(R) of the form V(I) ∪ W(J) where V(I) is any Zariski–closed subset of Spec(R) and W(J) is any Zariski–compact Zariski–open subset of Spec(R). Clearly $\mathscr{V}$ is closed under finite unions, and $\mathscr{V}$ contains Spec(R) and the empty set, since Spec(R) equals V(0) ∪ W(0) and the empty set equals V(R) ∪ W(0). Therefore $\mathscr{V}$ is a basis for the family of closed sets of a topology on Spec(R), known as the *patch topology* or the *constructible topology*. (Thus the closed sets for the patch topology are precisely the intersections of sets from $\mathscr{V}$.)

We immediately restrict attention to the case that R has ACC on ideals. Now all subsets of Spec(R) are Zariski–compact (Exercise 14B), and so

$$\mathscr{V} = \{V(I) \cup W(J) \mid I,J \text{ are any ideals of } R\}.$$

Taking complements, we obtain the family

$$\mathscr{U} = \{V(J) \cap W(I) \mid I,J \text{ are any ideals of } R\},$$

which is a basis for the *open* sets of the patch topology, i.e., the patch–open subsets of Spec(R) are precisely the unions of sets from $\mathscr{U}$.

Any patch–neighborhood of a point P ∈ Spec(R) must contain a neighborhood from $\mathscr{U}$, that is, a neighborhood of the form V(J) ∩ W(I) where $P \geq J$ and $P \ngeq I$. Since P ∈ V(P) and V(P) ⊆ V(J), we may replace V(J) by V(P). Then, since V(P) ∩ W(I) = V(P) ∩ W(P + I), we may replace I by P + I. What we have shown is that every patch–neighborhood of P contains a neighborhood of the form V(P) ∩ W(I) where I is an ideal properly containing P, that is, the patch–neighborhoods V(P) ∩ W(I) form a base for the patch–neighborhoods of P.

EXERCISE 14C. Find all the open sets for the patch topology on Spec($\mathbb{Z}$), and show that Spec($\mathbb{Z}$) with this topology is homeomorphic to the one–point compactification of $\mathbb{N}$. □

PROPOSITION 14.1. *If R is a ring with ACC on ideals, then Spec(R) with the patch topology is a compact, Hausdorff, totally disconnected space.*

Proof. Given distinct points $P,Q \in$ Spec(R), either $P \nsubseteq Q$ or $Q \nsubseteq P$, say $P \nsubseteq Q$. Then $V(P) \cap W(R)$ is a patch–neighborhood of P and $V(Q) \cap W(P)$ is a patch–neighborhood of Q, and these neighborhoods are disjoint because V(P) and W(P) are disjoint. Therefore Spec(R) is Hausdorff in the patch topology.

For any ideal I of R, observe that the sets V(I) and W(I) are both patch–open, since $V(I) = V(I) \cap W(R)$ and $W(I) = V(0) \cap W(I)$. Since V(I) and W(I) are complements of each other, they are both patch–closed as well. Thus the basic open sets $V(I) \cap W(J)$, for ideals I,J of R, are all patch–closed. Therefore the patch topology on Spec(R) has a basis of open sets which are also closed, and hence Spec(R) is totally disconnected in this topology.

It remains to show that Spec(R) is patch–compact. Let X be a family of patch–open sets covering Spec(R), and suppose that no finite subfamily of X covers Spec(R). Since Spec(R) = V(0), we may use the ACC on ideals to choose an ideal Q maximal with respect to the property that no finite subfamily of X covers V(Q). If A and B are ideals properly containing Q, there must be a finite subfamily Y of X that covers both V(A) and V(B). Then Y covers V(AB), whence $V(AB) \nsupseteq V(Q)$ and so $AB \nsubseteq Q$. Thus Q must be a prime ideal.

Choose $U \in X$ such that $Q \in U$. Then Q must have a patch–neighborhood $V(Q) \cap W(I)$, for some ideal $I > Q$, such that $V(Q) \cap W(I) \subseteq U$. Now V(I) can be covered by some finite subfamily Y' of X. But

$$V(Q) - V(I) = V(Q) \cap W(I) \subseteq U,$$

and so V(Q) can be covered by $Y' \cup \{U\}$, contrary to our choice of Q. Thus there must exist a finite subfamily of X which covers Spec(R).

Therefore Spec(R) is compact in the patch topology. □

EXERCISE 14D. If R is a ring with ACC on ideals and $X \subseteq$ Spec(R), show that the patch–closure of X is the set of those primes Q which can be obtained as intersections of subsets of X. □

♦ LOCAL NUMBERS OF GENERATORS

Given a finitely generated module A over a commutative noetherian ring R, we may localize A with respect to the various prime ideals P of R and look at the number of generators required for each of these localized modules A_P. These "local numbers of generators" are of course bounded by the number of generators needed for A, and, conversely, we may try to use these local numbers of generators to estimate the number of generators for A. In order to find a suitable analog for the noncommutative case, we adjust the point of view a bit. First note that by Nakayama's Lemma, the minimum number of generators needed for A_P as an R_P–module is the same as the minimum number of generators needed for A_P/PA_P as an (R_P/PR_P)–module, since $PR_P = J(R_P)$. (See Theorem 2.17 ff.) Now R_P/PR_P is isomorphic to the quotient field Q_P of R/P, and A_P/PA_P is correspondingly isomorphic to $Q_P \otimes_R A$. Thus the number of generators needed for the localized module A_P is the same as the dimension of the vector space $Q_P \otimes_R A$. Since we have an appropriate noncommutative analog of the quotient field of a domain – namely the Goldie quotient ring of a prime noetherian ring – we may define local numbers of generators in the following fashion.

DEFINITION. Let R be a right noetherian ring, P a prime ideal of R, and A a finitely generated right R–module. Let Q_P be the right Goldie quotient ring of R/P. The *local number of generators of* A *at* P, denoted g(A,P), is the minimum number of generators needed for $A \otimes_R Q_P$ as a right Q_P–module. (By convention, the minimum number of generators needed for the zero module is defined to be 0.)

EXERCISE 14E. Let R be a right noetherian ring, P a prime ideal of R, and A a finitely generated right R–module. Show that g(A,P) = 0 if and only if A/AP is torsion as a right (R/P)–module. □

In the commutative case, fixing a module and computing the local number of generators at varying prime ideals results in a function with partial continuity properties, as follows.

PROPOSITION 14.2. *If* A *is a finitely generated module over a commutative noetherian ring* R, *the function* $g(A,-) : Spec(R) \to \mathbb{Z}$ *is upper semicontinuous with respect to the Zariski topology. Moreover, if* A *is projective this function is Zariski–continuous.*

Proof. Upper semicontinuity means that for each $m \in \mathbb{Z}$, the set

$$\{P \in Spec(R) \mid g(A,P) < m\}$$

must be open. Thus consider any $P \in Spec(R)$ for which $g(A,P) < m$; we shall find a Zariski–neighborhood U of P such that $g(A,Q) < m$ for all $Q \in U$.

Since $g(A,P) < m$, the R_P–module A_P can be generated by $m-1$ elements, say $a_1/r_1,\ldots,a_{m-1}/r_{m-1}$ where the $a_i \in A$ and $r_i \in R-P$. If B denotes the submodule of A generated by $a_1,\ldots,a_{m-1}$, then $B_P = A_P$, whence $(A/B)_P = 0$. Since A is finitely generated, there exists $s \in R-P$ such that $s(A/B) = 0$. Now let $U = W(sR)$, which is a Zariski–neighborhood of P in $\mathrm{Spec}(R)$. For $Q \in U$, we have $s \notin Q$ and $s(A/B) = 0$, whence $(A/B)_Q = 0$ and $A_Q = B_Q$, yielding $g(A,Q) \le m-1$. This proves the upper semicontinuity of $g(A,-)$.

If A is projective, $A \oplus C \cong R^n$ for some finitely generated module C and some positive integer n. Then $g(A,-) + g(C,-)$ equals the constant function n, and since $g(C,-)$ is upper semicontinuous by our previous result, $g(A,-)$ is lower semicontinuous. Therefore $g(A,-)$ is continuous with respect to the Zariski topology in this case. □

The function $g(A,-)$ in Proposition 14.2 need not be Zariski–continuous. For instance, if $R = \mathbb{Z}$ and $A = \mathbb{Z}/2\mathbb{Z}$, then $g(A,2\mathbb{Z}) = 1$ and $g(A,0) = 0$, although $\{0\}$ is a Zariski–dense subset of $\mathrm{Spec}(\mathbb{Z})$. (Alternatively, $g(A,p\mathbb{Z}) = 0$ for all odd primes p, and the set $\{p\mathbb{Z} \mid p \text{ odd}\}$ is Zariski–dense in $\mathrm{Spec}(\mathbb{Z})$.) Note that in this example, $g(A,-)$ is continuous with respect to the patch topology, since the singleton $\{2\mathbb{Z}\}$ is a patch–open subset of $\mathrm{Spec}(\mathbb{Z})$.

In the noncommutative case, even the semicontinuity of Proposition 14.2 is lost. For instance, let $R = \begin{pmatrix} \mathbb{Z} & \mathbb{Z} \\ 2\mathbb{Z} & \mathbb{Z} \end{pmatrix}$ and consider the right R–module $A = \begin{pmatrix} \mathbb{Z} & \mathbb{Z} \\ \mathbb{Z} & \mathbb{Z} \end{pmatrix}$, which is finitely generated and projective (because it is isomorphic to the direct sum of two copies of $\begin{pmatrix} 1 & 0 \\ 0 & 0 \end{pmatrix}R$). For each odd prime integer p, the ideal pR is a maximal ideal of R, and it is easily checked that A/Ap is a cyclic right (R/pR)–module, so that $g(A,pR) = 1$. There are two other maximal ideals of R, namely the ideals $M = \begin{pmatrix} \mathbb{Z} & \mathbb{Z} \\ 2\mathbb{Z} & 2\mathbb{Z} \end{pmatrix}$ and $N = \begin{pmatrix} 2\mathbb{Z} & \mathbb{Z} \\ 2\mathbb{Z} & \mathbb{Z} \end{pmatrix}$, and $R/M \cong R/N \cong \mathbb{Z}/2\mathbb{Z}$. On one hand, $A = AM$ and so $g(A,M) = 0$, while on the other hand $A/AN \cong (R/N)^2$ and so $g(A,N) = 2$. Since the set $\{pR \mid p \text{ odd}\}$ is Zariski–dense in $\mathrm{Spec}(R)$, we conclude that the function $g(A,-)$ is neither upper nor lower semicontinuous with respect to the Zariski topology.

As in the previous case, continuity could be saved by using the patch topology in this example. However, switching over to the patch topology is still not quite enough, as the following example shows.

EXERCISE 14F. Let $R = A_1(\mathbb{Z}) = \mathbb{Z}[x][\theta; d/dx]$ and $A = R/\theta R$. For all prime integers q, let $P_q = qR + x^qR + \theta^qR$, which is a maximal ideal of R (Exercise 2ZG). Show that $g(A,P_q) = 1$ for all q, while $g(A,0) = 0$. Show also that 0 is in the patch–closure of the set $\{P_q \mid q \text{ prime}\}$. Thus $g(A,-)$ is not patch–continuous. □

EXERCISE 14G. Continuing the notation of Exercise 14F, show that each A/AP_q is a simple right (R/P_q)–module, whereas R/P_q is a simple artinian ring of length q (Exercise 2ZG). □

In view of Exercises 14F/G, we might say that the lack of continuity of the function $g(A,-)$ in this example is due to the fact that $g(A,P_q)$ is too crude a measure of the size of A/AP_q. Instead, we might say that relative to the ring R/P_q, the "size" of A/AP_q is just $1/q$. This "normalization" is all that remains to do to obtain a continuity theorem in general.

DEFINITION. Let R be a right noetherian ring, $P \in \mathrm{Spec}(R)$, and Q_P the right Goldie quotient ring of R/P. For any finitely generated right R–module A, the *normalized rank of* A *at* P is the rational number

$$r_P(A) = \mathrm{length}(A \otimes_R Q_P)/\mathrm{length}(Q_P),$$

where "length" refers to composition series length of right Q_P–modules.

EXERCISE 14H. Let R be a right noetherian ring, $P \in \mathrm{Spec}(R)$, and A a finitely generated right R–module.

(a) Show that $\mathrm{length}(A \otimes_R Q_P) = \rho_{R/P}(A/AP)$.

(b) Show that $g(A,P)$ is the smallest integer greater than or equal to $r_P(A)$.

(c) Show that $r_P(A) = 0$ if and only if A/AP is torsion as an (R/P)–module. □

PROPOSITION 14.3. *Let R be a right noetherian ring, $P \in Spec(R)$, and $B \leq A$ finitely generated right R–modules. Then*

$$r_P(A) \leq r_P(B) + r_P(A/B).$$

If either $AP = 0$ or B is a direct summand of A, then

$$r_P(A) = r_P(B) + r_P(A/B).$$

Proof. Let $f : B \to A$ and $g : A \to A/B$ be the inclusion and quotient maps. A basic property of tensor products is that the sequence of maps

$$B \otimes_R Q_P \xrightarrow{f \otimes 1} A \otimes_R Q_P \xrightarrow{g \otimes 1} (A/B) \otimes_R Q_P \to 0$$

is *exact,* meaning that the image of $f \otimes 1$ equals the kernel of $g \otimes 1$ and $g \otimes 1$ is surjective. Hence, if $K = (f \otimes 1)(B \otimes_R Q_P)$, then $(A \otimes_R Q_P)/K \cong (A/B) \otimes_R Q_P$, and so

$$\begin{aligned}\mathrm{length}(A \otimes_R Q_P) &= \mathrm{length}(K) + \mathrm{length}((A/B) \otimes_R Q_P)\\ &\leq \mathrm{length}(B \otimes_R Q_P) + \mathrm{length}((A/B) \otimes_R Q_P).\end{aligned}$$

Dividing by $\mathrm{length}(Q_P)$, we obtain $r_P(A) \leq r_P(B) + r_P(A/B)$.

If $AP = 0$, then since Q_P is flat as a left (R/P)–module (Exercise 6I or Corollary 9.15), $f \otimes 1$ is injective and so $K \cong B \otimes_R Q_P$. On the other hand, if B is a direct summand of A there exists a homomorphism $h : A \to B$ such that hf is the identity map on B. Then $(h \otimes 1)(f \otimes 1)$ is the identity map on $B \otimes_R Q_P$, whence again $f \otimes 1$ is injective and $K \cong B \otimes_R Q_P$. Therefore in these cases

$$\text{length}(A \otimes_R Q_P) = \text{length}(B \otimes_R Q_P) + \text{length}((A/B) \otimes_R Q_P)$$

and hence $r_P(A) = r_P(B) + r_P(A/B)$. □

♦ PATCH–CONTINUITY OF NORMALIZED RANKS

Our aim in this section is a noncommutative continuity theorem: given a right noetherian ring R and a finitely generated right R–module A, we shall prove that the normalized rank $r_P(A)$ is a patch–continuous function of P.

LEMMA 14.4. *Let R be a right noetherian ring and A a finitely generated, fully faithful right R–module. If $K.dim(A) < r.K.dim(R)$, then A has an essential cyclic submodule.*

Proof. Every nonzero submodule of A contains a critical submodule (Lemma 13.8), and so A has an essential submodule which is a direct sum of critical modules. Hence, after replacing A by this submodule, we may assume that $A = A_1 \oplus ... \oplus A_n$ where each A_i is α_i–critical for some α_i. We may also assume that $\alpha_1 \geq \alpha_2 \geq ... \geq \alpha_n$.

We next construct elements $x_i \in A_i$ such that if $I_0 = R$ and $I_j = \text{ann}_R(\{x_1,...,x_j\})$, then I_{j-1}/I_j is α_j–critical for each $j = 1,...,n$. First choose any nonzero element $x_1 \in A_1$ and observe that $I_0/I_1 \cong x_1R$, which is α_1–critical because A_1 is α_1–critical. Now suppose that $x_1,...,x_j$ have been chosen, for some $j < n$. As

$$K.dim(R/I_j) \leq K.dim(A) < r.K.dim(R),$$

we find that $I_j \neq 0$. Since A is fully faithful, A_{j+1} is faithful, and so there must be an element $x_{j+1} \in A_{j+1}$ with $x_{j+1}I_j \neq 0$. It follows that $I_j/I_{j+1} \cong x_{j+1}I_j$, which is α_{j+1}–critical because A_{j+1} is. This completes our inductive construction of the elements x_i.

Now let $x = x_1 + ... + x_n$ and observe that $\text{ann}_R(x) = I_n$ (since the A_i are independent). Further, $xI_{j-1}/xI_j \cong I_{j-1}/I_j$ for each $j = 1,...,n$, whence xI_{j-1}/xI_j is α_j–critical. Thus

$$0 = xI_n < xI_{n-1} < ... < xI_1 < xI_0 = xR$$

is a critical composition series for xR, of length n. Therefore, by Corollary 13.10, $xR \leq_e A$. □

EXERCISE 14I. Let R be a right noetherian ring and A a simple right R–module. If $R/\text{ann}_R(A)$ is not artinian, show that for every positive integer n, the direct sum of n copies of A is cyclic. More generally, if B is a right R–module of finite length, and $R/\text{ann}_R(A)$ is not artinian for any composition factor A of B, show that B is cyclic. [Hint: Replace B by B modulo the intersection of its maximal submodules.] □

PROPOSITION 14.5. *Let R be a prime right noetherian ring, A a finitely generated torsion right R–module, and ε a positive real number. Then there exists a nonzero ideal I in R such that $r_P(A) < \varepsilon$ for all primes $P \in W(I)$.*

Proof. We proceed by induction on the ordinal $\alpha = \text{K.dim}(A)$, the case $\alpha = -1$ being trivial. Now let $\alpha \geq 0$, and assume the result holds for finitely generated torsion modules of smaller Krull dimension. Choose a critical composition series

$$0 = A_0 < A_1 < \ldots < A_m = A.$$

If, for each $j = 1,\ldots,m$, there exists a nonzero ideal I_j in R such that $r_P(A_j/A_{j-1}) < \varepsilon/m$ for all $P \in W(I_j)$, it follows from Proposition 14.3 that $r_P(A) < \varepsilon$ for all $P \in W(I_1I_2\cdots I_n)$. Hence, there is no loss of generality in assuming that A is α–critical.

Suppose first that A contains a nonzero unfaithful submodule B, and set $I_1 = \text{ann}_R(B)$. If $P \in W(I_1)$, then B/BP is an unfaithful (R/P)–module (because $(I_1 + P)/P \neq 0)$ and so is torsion over R/P, whence $r_P(B) = 0$ (Exercise 14H). As $\text{K.dim}(A/B) < \alpha$ (because A is α–critical), we obtain from the induction hypothesis a nonzero ideal I_2 in R such that $r_P(A/B) < \varepsilon$ for all $P \in W(I_2)$. Thus $r_P(A) < \varepsilon$ for all $P \in W(I_1I_2)$ (Proposition 14.3 again), completing the induction step in this case.

Finally, suppose that A is fully faithful, and note that $\text{r.K.dim}(R) > \alpha$, because A is a torsion module (Proposition 13.7). Choose an integer $n > 2/\varepsilon$. Then A^n is a fully faithful right R–module, and so by Lemma 14.4, it has an essential cyclic submodule C. Observe that A^n/C is a homomorphic image of a direct sum of proper factors of A; since A is α–critical, it follows that $\text{K.dim}(A^n/C) < \alpha$. Hence, by induction, R contains a nonzero ideal I such that $r_P(A^n/C) < 1$ for all $P \in W(I)$. For any such P, we have $r_P(C) \leq 1$ because C is cyclic, whence $r_P(A^n) < 2$ (Proposition 14.3). Therefore $r_P(A) < 2/n < \varepsilon$ for all $P \in W(I)$, and the induction step is complete. □

THEOREM 14.6. [Stafford, Goodearl] *Let A be a finitely generated right module over a right noetherian ring R. Then the rule assigning to each prime ideal P the normalized rank $r_P(A)$ is a patch–continuous function from Spec(R) to $\mathbb{Q}$.*

Proof. It suffices to show that given $Q \in \text{Spec}(R)$ and a positive real number ε, there exists an ideal I properly containing Q such that

$$r_Q(A) - \varepsilon < r_P(A) < r_Q(A) + \varepsilon$$

for all primes $P \in V(Q) \cap W(I)$. As $r_P(A) = r_{P/Q}(A/AQ)$ for all $P \in V(Q)$, we may reduce to the case that $Q = 0$, with no loss of generality.

By Corollary 6.26, there is a positive integer n such that A^n has a free submodule F with A^n/F torsion. It follows that $r_0(A^n/F) = 0$. If $F \cong R^k$, then $F \otimes_R Q_0 \cong Q_0^k$, and so $r_0(F) = k$. Thus $r_0(A^n) = r_0(F) = k$ and $r_0(A) = k/n$, by Proposition 14.3. In view of

Proposition 14.5, there is a nonzero ideal I_1 in R such that $r_P(A^n/F) < n\varepsilon$ for all $P \in W(I_1)$. Using Proposition 14.3 once again, we obtain

$$r_P(A^n) \leq r_P(F) + r_P(A^n/F) < k + n\varepsilon$$

for all $P \in W(I_1)$, and hence $r_P(A) < (k/n) + \varepsilon = r_0(A) + \varepsilon$ for all $P \in W(I_1)$.

To prove the remaining inequality, note first that $A \cong R^m/K$ for some positive integer m and some submodule $K \leq R^m$. Then

$$m = r_0(R^m) = r_0(K) + r_0(A).$$

Applying the argument of the previous paragraph to K, there is a nonzero ideal I_2 in R such that $r_P(K) < r_0(K) + \varepsilon$ for all $P \in W(I_2)$. Finally,

$$m = r_P(R^m) \leq r_P(K) + r_P(A)$$

for any prime P, and thus

$$r_P(A) \geq m - r_P(K) > m - r_0(K) - \varepsilon = r_0(A) - \varepsilon$$

for all $P \in W(I_2)$.

Therefore $|r_P(A) - r_0(A)| < \varepsilon$ for all $P \in W(I_1I_2)$. □

EXERCISE 14J. Let R be a prime right noetherian ring, $n \in \mathbb{N}$. Show that there is a nonzero ideal I in R such that for all primes $P \in W(I)$, either rank(R) divides rank(R/P) or rank(R/P) > n. [Hint: If $r = \text{rank}(R)$ and U is a uniform right ideal of R, then $r_0(U) = 1/r$. Apply Theorem 14.6.] □

♦ GENERATING MODULES OVER SIMPLE NOETHERIAN RINGS

Our first application of the results of the previous section is to show that the number of generators for certain modules over simple noetherian rings can be estimated just from the Krull dimension of the module or the ring.

THEOREM 14.7. [Stafford] *Let R be a right noetherian ring and let A be a finitely generated right R–module such that all nonzero factor modules A/B are fully faithful. If $K.dim(A) = m < \infty$ and $m < r.K.dim(R)$, then A can be generated by $m + 1$ elements.*

Proof. The theorem is automatic in case $m = -1$. Now let $m \geq 0$ and assume the theorem holds for modules with Krull dimension less than m. If $J_m(A)$ is the Krull radical of A, then by hypothesis the factor module $A/J_m(A)$ is fully faithful, and

$$\text{K.dim}(A/J_m(A)) \leq \text{K.dim}(A) < \text{r.K.dim}(R).$$

Thus the hypotheses of Lemma 14.4 are satisfied for $A/J_m(A)$, and hence there is an element $x \in A$ such that $(xR + J_m(A))/J_m(A)$ is essential in $A/J_m(A)$. It follows from Corollary 13.12 that $\text{K.dim}(A/xR) < m$, and so by the induction hypothesis, A/xR can be generated by m elements. Therefore A can be generated by $m + 1$ elements. □

COROLLARY 14.8. [Stafford] *Let R be a simple right noetherian ring.*

(a) If A is a finitely generated torsion right R–module with $K.dim(A) = m < \infty$, then A can be generated by $m + 1$ elements.

(b) If $r.K.dim(R) = n < \infty$, then every right ideal of R can be generated by $n + 1$ elements.

Proof. (a) All nonzero right R–modules are fully faithful because R is simple, and $m < r.K.dim(R)$ because A is torsion (Proposition 13.7). Hence, Theorem 14.7 applies.

(b) Any right ideal I of R is a direct summand of an essential right ideal J, and it suffices to show that J can be generated by $n + 1$ elements. Hence, we may as well assume that $I \leq_e R_R$. Now there exists a regular element $x \in I$ (Proposition 5.9), and I/xR is a torsion module. If $m = K.dim(I/xR)$, then $m < n$ by Proposition 13.7. By part (a), I/xR can be generated by $m + 1$ elements, and hence by n elements. Therefore I can be generated by $n + 1$ elements. □

The situation for a Weyl algebra $A_n(k)$ over a field k of characteristic zero is even better than that in Corollary 14.8(b). Although $A_n(k)$ has Krull dimension n (Theorem 13.18), Stafford has proved that every right ideal of $A_n(k)$ can be generated by two elements [1978, Corollary 3.2].

EXERCISE 14K. If R is a simple right noetherian ring with $r.K.dim(R) = n < \infty$ and A is a finitely generated right R–module, show that A can be generated by $g(A,0) + n$ elements. □

EXERCISE 14L. If $R = k[x,y]$ is a polynomial ring over a field k, show that there is no finite bound on the number of generators needed for ideals of R. [Hint: If $M = xR + yR$, look at M^i/M^{i+1} for $i \in \mathbb{N}$.] □

♦ GENERIC REGULARITY

For certain rings, e.g. FBN rings, the patch–continuity theorem (Theorem 14.6) can be improved to say that the functions $P \mapsto r_P(A)$ are actually locally constant (i.e., constant on open sets) with respect to the patch topology. If this is to happen, regularity modulo prime ideals must define patch–open sets, in the following manner. Consider a right noetherian ring R, a prime ideal Q of R, and an element $x \in \mathscr{C}(Q)$. Then $R/(xR + Q)$ is torsion as a right (R/Q)–module, whence $r_Q(R/xR) = 0$. If the map $P \mapsto r_P(R/xR)$ is locally constant in the patch topology, the set

$$U = \{P \in Spec(R) \mid r_P(R/xR) = 0\}$$

must be patch–open. For all $P \in U$, observe that $R/(xR + P)$ is torsion as a right (R/P)–module, whence $(xR + P)/P$ is an essential right ideal of R/P, and so $x \in \mathscr{C}(P)$.

Since x is regular modulo all the prime ideals in the patch–open set U, we may say that x is "generically" regular (with respect to the patch topology).

For technical reasons, we define "generic regularity" for sets of prime ideals as well as for the full prime spectrum.

DEFINITION. Let R be a ring. A subset X of Spec(R) satisfies the *generic regularity condition* (alternative terminology: X is *sparse*) provided that for any prime ideal $Q \in \text{Spec}(R)$ (not necessarily in X) and any element $x \in \mathscr{C}(Q)$, there exists a patch–neighborhood U of Q such that $x \in \mathscr{C}(P)$ for all $P \in X \cap U$.

EXERCISE 14M. If $R = A_1(\mathbb{Z})$, show that Spec(R) does not satisfy the generic regularity condition. [Hint: Exercise 14F.] □

EXERCISE 14N. If R is a right noetherian ring and X is the set of completely prime ideals of R (that is, $X = \{P \in \text{Spec}(R) \mid R/P \text{ is a domain}\}$), show that X satisfies the generic regularity condition. □

PROPOSITION 14.9. *If R is a right noetherian ring such that either R is right fully bounded or $r.K.dim(R) \leq 1$, then Spec(R) satisfies the generic regularity condition.*

Proof. Consider $Q \in \text{Spec}(R)$ and $x \in \mathscr{C}(Q)$, and note that $(xR + Q)/Q$ is an essential right ideal of R/Q.

If R is right fully bounded, $(xR + Q)/Q$ contains a nonzero ideal I/Q. Given any P in $V(Q) \cap W(I)$, we have

$$xR + P \geq xR + Q \geq I$$

and so $(xR + P)/P$ contains the ideal $(I + P)/P$, which is nonzero because $P \in W(I)$. Thus $(xR + P)/P$ is an essential right ideal of R/P, whence $x \in \mathscr{C}(P)$. This verifies generic regularity in the fully bounded case.

Now assume that $r.K.dim(R) \leq 1$. If Q is a maximal ideal of R, then $\{Q\} = V(Q)$ and so $\{Q\}$ is a patch–neighborhood of Q, and we are done. Thus we may assume that Q is not maximal. Because of Krull dimension 1, all proper prime factor rings of R/Q and all finitely generated torsion (R/Q)–modules have Krull dimension 0 (Proposition 13.7) and so are artinian. In particular, all prime ideals properly containing Q are maximal ideals, and $R/(xR + Q)$ must have finite length. Hence, there are at most finitely many maximal ideals $M_i > Q$ such that M_i annihilates a composition factor of $R/(xR + Q)$. Multiplying these M_i together and adding Q, we obtain an ideal $I > Q$ such that each $M_i \geq I$.

Let $P \in V(Q) \cap W(I)$. Since $x \in \mathscr{C}(Q)$ already, suppose that $P \neq Q$. Then P is a maximal ideal of R. If A is any composition factor of $R/(xR + Q)$, the annihilator of A is either Q or one of the M_i. In either case, $P \neq \text{ann}_R(A)$ (because $P \neq Q$ and $P \ngeq I$), whence $P \nsubseteq \mathrm{ann}_R(A)$. Since P does not annihilate any composition factor of $R/(xR+Q)$, it follows that

$$[R/(xR+Q)]P = R/(xR+Q),$$

and hence $xR + P = R$. Therefore $x + P$ is a unit in R/P and so $x \in \mathscr{C}(P)$, verifying generic regularity in this case. □

PROPOSITION 14.10. *Let R be a right noetherian ring and $n \in \mathbb{N}$. Then the set*

$$X = \{P \in \mathrm{Spec}(R) \mid \mathrm{rank}(R/P) \le n\}$$

satisfies the generic regularity condition.

Proof. For any $P \in X$, we have $\mathrm{length}(Q_P) = \mathrm{rank}(R/P) \le n$ by Exercise 5G. Consequently, if A is any finitely generated right R–module, either $r_P(A) = 0$ or $r_P(A) \ge 1/n$.

Given $Q \in \mathrm{Spec}(R)$ and $x \in \mathscr{C}(Q)$, we have $r_Q(R/xR) = 0$. By Theorem 14.6, Q has a patch–neighborhood U such that $r_P(R/xR) < 1/n$ for all $P \in U$. Thus for $P \in X \cap U$, we have $r_P(R/xR) = 0$ and therefore $x \in \mathscr{C}(P)$. □

THEOREM 14.11. *Let R be a right noetherian ring, X a subset of $\mathrm{Spec}(R)$ satisfying the generic regularity condition, and A a finitely generated right R–module. Given any $Q \in \mathrm{Spec}(R)$, there is a patch–neighborhood U of Q such that $r_P(A) = r_Q(A)$ for all $P \in X \cap U$.*

Proof. Since $V(Q)$ is a patch–neighborhood of Q and $r_P(A) = r_{P/Q}(A/AQ)$ for all $P \in V(Q)$, we may replace R, A, and Q by R/Q, A/AQ, and 0. Thus there is no loss of generality in assuming that $Q = 0$.

We first consider the case that $r_0(A) = 0$, whence A is torsion as a right R–module. Choose generators $a_1,\dots,a_n$ for A. Then there exists $x \in \mathscr{C}(0)$ such that $a_i x = 0$ for $i = 1,\dots,n$. By generic regularity, 0 has a patch–neighborhood U such that $x \in \mathscr{C}(P)$ for all $P \in X \cap U$. For any such P, it follows from the equations $a_i x = 0$ that A/AP is torsion as a right (R/P)–module, and thus $r_P(A) = 0$. This verifies the theorem in case $r_0(A) = 0$.

In the general case we parallel the proof of Theorem 14.6. By Corollary 6.26, there exist positive integers n,k such that A^n has a submodule F with $F \cong R^k$ and A^n/F torsion. Then $r_0(F) = k$ and $r_0(A^n/F) = 0$, while $r_0(A) = k/n$. By the result of the previous paragraph, there is a patch–neighborhood U' of 0 such that $r_P(A^n/F) = 0$ for all $P \in X \cap U'$. Since $r_P(F) = k$ for all primes P, it follows using Proposition 14.3 that $r_P(A^n) \le r_P(F) = k$ for all $P \in X \cap U'$, and therefore $r_P(A) \le k/n = r_0(A)$ for all

$P \in X \cap U'$.

Now $A \cong R^m/K$ for some positive integer m and some submodule $K \leq R^m$. Then $m = r_0(R^m) = r_0(K) + r_0(A)$. Applying the result of the last paragraph to the finitely generated module K, there exists a patch–neighborhood U'' of 0 such that $r_P(K) \leq r_0(K)$ for all $P \in X \cap U''$. For any prime P, we have $m = r_P(R^m) \leq r_P(K) + r_P(A)$. Consequently,

$$r_P(A) \geq m - r_P(K) \geq m - r_0(K) = r_0(A)$$

for all $P \in X \cap U''$.

Therefore $r_P(A) = r_0(A)$ for all $P \in X \cap U' \cap U''$. □

COROLLARY 14.12. [Goodearl–Warfield] *If R is a right FBN ring and A is a finitely generated right R–module, the rule $P \mapsto r_P(A)$ defines a function from Spec(R) to $\mathbb{Q}$ that is locally constant with respect to the patch topology.*

Proof. Proposition 14.9 and Theorem 14.11. □

EXERCISE 14O. Let R be a prime right noetherian ring and X a subset of Spec(R) satisfying the generic regularity condition. Show that there is a nonzero ideal I in R such that rank(R) divides rank(R/P) for all $P \in X \cap W(I)$. [Cf. Exercise 14J.] □

♦ GENERATING MODULES OVER FBN RINGS

We have now developed most of the tools needed to derive estimates for the number of generators of a finitely generated module A over an FBN ring R, in terms of the "local data" $g(A,P)$ (for $P \in \text{Spec}(R)$). We do not expect the maximum of the numbers $g(A,P)$ to bound the number of generators needed for A, but rather the maximum of numbers such as $g(A,P) + \text{r.K.dim}(R/P)$. Since R is FBN, $\text{r.K.dim}(R/P) = \text{Cl.K.dim}(R/P)$ (Theorem 13.13). In fact, we may replace $\text{Cl.K.dim}(R/P)$ with a (smaller) number obtained from considering just chains of semiprimitive prime ideals, for which the following notation is used.

DEFINITION. A *J–prime ideal* in a ring R is any prime J–ideal P, that is, any prime ideal P for which $J(R/P) = 0$. The *J–spectrum* of R, denoted J–Spec(R), is the collection of all J–prime ideals of R. The *J–dimension* of R, denoted J–dim(R), is the supremum of the lengths of all finite chains of J–prime ideals in R.

Obviously $\text{J–dim}(R) \leq \text{Cl.K.dim}(R)$ if $\text{Cl.K.dim}(R)$ is finite. If R is local in any sense – for instance, assume only that $J(R)$ is a maximal ideal of R – then $J(R)$ is the only J–prime of R and so $\text{J–dim}(R) = 0$, regardless of the Krull dimension of R. Thus J–dim(R) may be considerably less than Cl.K.dim(R).

LEMMA 14.13. *If R is a ring with ACC on ideals, J–Spec(R) is a patch–compact subset of Spec(R). Moreover, J–Spec(R) is the patch–closure of the set of right primitive ideals in Spec(R).*

Proof. The second statement is immediate from Exercise 14D. Then since Spec(R) is patch–compact (Proposition 14.1), the first statement follows. □

DEFINITION. Let R be a right noetherian ring and A a finitely generated right R–module. For $P \in \text{Spec}(R)$, recall that the local number of generators $g(A,P)$ is the minimum number of generators for the Q_P–module $A \otimes_R Q_P$, and that $g(A,P)$ is the smallest integer greater than or equal to $r_P(A)$ (Exercise 14H). If $r_P(A) = 0$, set $b(A,P) = 0$, while if $r_P(A) > 0$, set

$$b(A,P) = g(A,P) + \text{J–dim}(R/P).$$

We shall use the numbers $b(A,P)$ to estimate the number of generators of A, in case R is fully bounded.

LEMMA 14.14. *Let R be a right noetherian ring and A a finitely generated right R–module, and set*

$$b = \sup\{b(A,P) \mid P \in \text{J–Spec}(R)\}.$$

If $0 < b < \infty$ and J–Spec(R) satisfies the generic regularity condition, there are only finitely many J–primes P for which $b(A,P) = b$.

Proof. Set $X = \text{J–Spec}(R)$. We claim that each $Q \in X$ has a patch–neighborhood $U(Q)$ such that $b(A,P) < b$ for all $P \neq Q$ in $X \cap U(Q)$.

If $r_Q(A) = 0$, then by Theorem 14.11, Q has a patch–neighborhood $U(Q)$ such that $r_P(A) = 0$ for all $P \in X \cap U(Q)$. In this case, $b(A,P) = 0 < b$ for all $P \in X \cap U(Q)$.

Now assume that $r_Q(A) > 0$, and set $k = \text{J–dim}(R/Q)$. Note that

$$g(A,Q) + k = b(A,Q) \leq b < \infty,$$

whence $k < \infty$. By Theorem 14.11, Q has a patch–neighborhood $U(Q)$ such that $r_P(A) = r_Q(A)$ for all $P \in X \cap U(Q)$, and there is no loss of generality in assuming that $U(Q) \subseteq V(Q)$. Given any $P \in X \cap U(Q)$, note from $r_P(A) = r_Q(A)$ that $g(A,P) = g(A,Q)$. If $P \neq Q$, then $P > Q$ and so $\text{J–dim}(R/P) < k$ (since P and Q are both J–primes), whence

$$b(A,P) \leq g(A,P) + k - 1 = g(A,Q) + k - 1 = b(A,Q) - 1 < b.$$

This completes the proof of the claim.

Since X is patch–compact (Lemma 14.13), there exist $Q_1,\ldots,Q_n \in X$ such that $U(Q_1),\ldots,U(Q_n)$ cover X. Therefore any $P \in X$ for which $b(A,P) = b$ must be one of $Q_1,\ldots,Q_n$. □

Lemma 14.14 is the key to an induction step, in which we must be able to reduce the values $b(A,P)$ at finitely many P simultaneously. This is accomplished by means of the next two lemmas.

LEMMA 14.15. *Let R be a right noetherian ring, $P \in Spec(R)$, and A a finitely generated right R–module. Let $x \in A$ and $B \le A$ such that $r_P(A/(xR + B)) = 0$. Then there exists $y \in B$ such that either*

$$g(A/(x + y)R, P) = 0 \qquad \text{or} \qquad g(A/(x + y)R, P) < g(A,P).$$

Proof. We may replace R by R/P and A by A/AP modulo its torsion submodule. Hence, there is no loss of generality in assuming that $P = 0$ and that A is torsionfree. Since $r_0(A/(xR + B)) = 0$, we know that $A/(xR + B)$ is a torsion module (Exercise 14H), and because A is torsionfree it follows that $xR + B \le_e A$ (Proposition 3.27). Choose $y \in B$ such that $\text{rank}((x + y)R)$ is as large as possible, and replace x by $x + y$. Thus we may assume that $\text{rank}(xR) \ge \text{rank}((x + z)R)$ for all $z \in B$, and we shall prove that the lemma holds with $y = 0$.

Choose a submodule $B' \le B$ maximal with respect to the property $B' \cap xR = 0$. Then $B' \oplus (xR \cap B)$ is essential in B, whence $(xR + B)/(xR + B')$ is torsion, and so $A/(xR + B')$ is torsion. Thus we may replace B by B', that is, there is no loss of generality in assuming that $B \cap xR = 0$.

We next show that $B \cdot \text{ann}_R(x) = 0$. Consider any $z \in B$, and note that the projection $xR \oplus B \to xR$ maps $(x + z)R$ onto xR, with kernel $K = (x + z)R \cap B$. Since A is torsionfree, $(x + z)R/K$ is torsionfree, and hence it follows from Exercise 4N that

$$\text{rank}((x + z)R) = \text{rank}(K) + \text{rank}(xR).$$

But $\text{rank}(xR) \ge \text{rank}((x + z)R)$ by assumption, whence $\text{rank}(K) = 0$ and so $K = 0$. It follows that

$$z \cdot \text{ann}_R(x) = (x + z) \cdot \text{ann}_R(x) \le K = 0.$$

Therefore $B \cdot \text{ann}_R(x) = 0$.

If $\text{ann}_R(x) \ne 0$, then $\text{ann}_R(B) \ne 0$, and so $\text{ann}_R(B)$ is a nonzero ideal in the prime ring R. In this case, $\text{ann}_R(B)$ contains a regular element and B is torsion. Since A is torsionfree, we must have $B = 0$. In this case $xR \le_e A$, whence A/xR is torsion, and therefore $g(A/xR, 0) = 0$.

On the other hand, if $\text{ann}_R(x) = 0$, then $xR \cong R$ and $r_0(xR) = 1$. In this case,

$$r_0(A/xR) = r_0(A) - r_0(xR) = r_0(A) - 1,$$

and consequently $g(A/xR, 0) = g(A,0) - 1$. □

LEMMA 14.16. *Let R be a right noetherian ring, X a finite subset of $Spec(R)$, and A a finitely generated right R–module. Then there exists an element $z \in A$ such that for all $P \in X$, either*

$$g(A/zR, P) = 0 \qquad or \qquad g(A/zR, P) < g(A,P).$$

Proof. If X contains only one prime, this follows from the case of Lemma 14.15 in which $x = 0$ and $B = A$. Now assume that X contains more than one prime, and that the result holds for smaller sets of primes. Choose P minimal among the primes in X, and set $Y = X - \{P\}$. By the induction hypothesis, there exists $x \in A$ such that for all $Q \in Y$, either $g(A/xR, Q) = 0$ or $g(A/xR, Q) < g(A,Q)$.

Let I be the product of the ideals in Y. Since P is minimal in X, it contains none of the ideals in Y, and so $P \not\supseteq I$. Then $(I + P)/P$ is a nonzero ideal of R/P, and so I contains an element of $\mathcal{C}(P)$. Since

$$(A/AI)/(A/AI)P \cong A/(AI + AP) \cong (A/AP)/(A/AP)I,$$

it follows that $(A/AI)/(A/AI)P$ is torsion as an (R/P)–module, and so $r_P(A/AI) = 0$ (Exercise 14H). Consequently, $r_P(A/(xR + AI)) = 0$.

By Lemma 14.15, there exists $y \in AI$ such that either $g(A/(x + y)R, P) = 0$ or $g(A/(x + y)R, P) < g(A,P)$. For $Q \in Y$, we have $y \in AQ$ and so

$$g(A/(x + y)R, Q) = g(A/xR, Q).$$

Therefore either $g(A/(x + y)R, Q) = 0$ or $g(A/(x + y)R, Q) < g(A,Q)$, and the induction step is established. □

THEOREM 14.17. [Forster, Swan, Warfield] *Let R be a right FBN ring and A a finitely generated right R–module, and set*

$$b = sup\{b(A,P) \mid P \in J\text{–}Spec(R)\}.$$

Then A can be generated by b elements.

Proof. We may obviously assume that $b < \infty$. Suppose first that $b = 0$, so that $g(A,P) = 0$ for all $P \in J\text{–Spec}(R)$. If $A \neq 0$, there exists an epimorphism of A onto a simple right R–module B. If $P = \text{ann}_R(B)$, then P is a maximal ideal and R/P is a simple artinian ring (Corollary 8.5); in particular, $P \in J\text{–Spec}(R)$. Now R/P is its own Goldie quotient ring, and so $g(B,P) = 1$. But then $g(A,P) \geq 1$, contradicting the assumption that $g(A,P) = 0$. Thus $A = 0$, and A can be generated by b elements in this case.

Now let $b > 0$ and assume the theorem holds for modules with smaller values of b. By Proposition 14.9, $J\text{–Spec}(R)$ satisfies the generic regularity condition, and hence Lemma 14.14 shows that the set

$$X = \{P \in \text{J-Spec}(R) \mid b(A,P) = b\}$$

is finite. Then by Lemma 14.16 there exists $z \in A$ such that for all $P \in X$, either $g(A/zR, P) = 0$ or $g(A/zR, P) < g(A,P)$. Since the first case implies $r_P(A/zR) = 0$ and so $b(A/zR, P) = 0$, we obtain $b(A/zR, P) < b$ for all $P \in X$. On the other hand, for all other $P \in \text{J-Spec}(R)$ we have

$$b(A/zR, P) \leq b(A,P) \leq b - 1.$$

Therefore $\sup\{b(A/zR, P) \mid P \in \text{J-Spec}(R)\} \leq b - 1$.

By the induction hypothesis, A/zR can be generated by $b - 1$ elements. Therefore A can be generated by b elements. □

In the commutative case, the estimate given in Theorem 14.17 immediately yields an estimate coming from the local data at maximal ideals, since if P is a prime ideal and M is a maximal ideal containing P, then A_P is a localization of A_M, whence $g(A,P) \leq g(A,M)$. However, such inequalities do not always hold in the noncommutative case, as the following example shows, and so some further work is needed to obtain an estimate from the numbers $g(A,M)$.

EXERCISE 14P. If R is the ring constructed in Exercise 2ZC, and M is the maximal ideal θS, show that $g(M,0) = 1$ whereas $g(M,M) = 0$. □

LEMMA 14.18. *Let R be a right noetherian ring, $P \in$ J-Spec(R), and A a finitely generated right R-module. Then there exists a right primitive ideal M containing P such that $g(A,P) \leq g(A,M)$.*

Proof. Since $g(A,P)$ is the smallest integer greater than or equal to $r_P(A)$, we have $r_P(A) > g(A,P) - 1$. By Theorem 14.6, there is an ideal I, properly containing P, such that $r_Q(A) > g(A,P) - 1$ for all $Q \in V(P) \cap W(I)$. As R/P is semiprimitive, P is an intersection of right primitive ideals, and so there exists a right primitive ideal $M \geq P$ such that $M \not\geq I$. Thus $M \in V(P) \cap W(I)$ and so

$$g(A,M) \geq r_M(A) > g(A,P) - 1.$$

Since $g(A,M)$ and $g(A,P)$ are integers, $g(A,M) \geq g(A,P)$. □

THEOREM 14.19. [Forster, Swan, Warfield] *Let R be a right FBN ring and A a finitely generated right R-module, and set*

$$m = \max\{g(A,M) \mid M \text{ is a maximal ideal of } R\}.$$

Then A can be generated by m + J-dim(R) elements.

Proof. Because of Theorem 14.17, it suffices to show that $b(A,P) \leq m + \text{J-dim}(R)$ for any $P \in \text{J-Spec}(R)$. By Lemma 14.18, there exists a right primitive ideal $M \geq P$ such that $g(A,P) \leq g(A,M)$, and M is a maximal ideal by Proposition 8.4. Therefore

$$b(A,P) \leq g(A,P) + \text{J-dim}(R/P) \leq g(A,M) + \text{J-dim}(R) \leq m + \text{J-dim}(R). \square$$

In the case of a finitely generated right module A over a right noetherian ring R which is not necessarily fully bounded, it is also possible to estimate the number of generators for A in terms of the local data, but J–dimension must be replaced by Krull dimension. The proofs require a mixture of the methods used in the simple case (Theorem 14.7) and the FBN case (Theorem 14.17), in order to deal with the appearance of fully faithful torsion modules over prime factor rings of R. Stafford has proved that the number of generators needed for A is bounded by each of the following:

$$\sup\{g(A,P) + \text{r.K.dim}(R/P) \mid P \in \text{J-Spec}(R)\};$$

$$\sup\{g(A,P) \mid P \text{ is a right primitive ideal of } R\} + \text{r.K.dim}(R/J(R))$$

[1981, Corollaries 3.7 and 4.6]. (See also McConnell–Robson [1987, Corollary 11.7.10].)

♦ COUNTABILITY OF CLIQUES

The patch–continuity and generic regularity results developed above have applications to parts of noetherian ring theory other than estimating numbers of generators of modules. Perhaps the most important such application to date has been Stafford's theorem that cliques in noetherian rings are always countable. We conclude the chapter with a proof of this theorem. Note that while the proof may appear to rely entirely on generic regularity and local constantness (Theorem 14.11), the full patch–continuity theorem (Theorem 14.6) is needed to show that in any noetherian ring R the sets $\{P \in \text{Spec}(R) \mid \text{rank}(R/P) \leq n\}$ satisfy generic regularity (Proposition 14.10).

LEMMA 14.20. *Let P and Q be prime ideals in a noetherian ring R, such that $Q \nsubseteq P$. Then $r_P(Q) \geq 1$, and $r_P(Q) = 1$ if and only if $(Q \cap P)/QP$ is torsion as a right (R/P)–module. In particular, if $Q \rightsquigarrow P$ then $r_P(Q) > 1$. (In computing $r_P(Q)$ here, we view Q as a right R–module.)*

Proof. Since Q/QP is a right (R/P)–module, we have

$$r_P(Q) = r_P(Q/QP) = r_P(Q/(Q \cap P)) + r_P((Q \cap P)/QP).$$

Note that $Q/(Q \cap P) \cong (Q + P)/P$, which is a nonzero ideal of R/P (because $Q \nsubseteq P$). Hence, $(Q + P)/P$ is an essential right ideal of R/P, and so $(Q + P)/P$ has the same rank as R/P. Thus

$$r_P(Q/(Q \cap P)) = r_P((Q + P)/P) = r_P(R/P) = 1.$$

Consequently, $r_P(Q) \geq 1$, and $r_P(Q) = 1$ if and only if $r_P((Q \cap P)/QP) = 0$, which occurs if and only if $(Q \cap P)/QP$ is torsion as a right (R/P)–module. If $Q \rightsquigarrow P$, the right (R/P)–module $(Q \cap P)/QP$ has a nonzero torsionfree factor, and therefore $r_P(Q) > 1$. □

PROPOSITION 14.21. *Let Q be a prime ideal in a noetherian ring R, and let X be a subset of Spec(R) satisfying the generic regularity condition. There are at most finitely many primes $P \in X$ such that either $Q \rightsquigarrow P$ or $P \rightsquigarrow Q$.*

Proof. By symmetry, it is enough to show that Q is linked to at most finitely many primes in X. Assume that this conclusion fails. By noetherian induction, we may assume that the conclusion holds in all proper factor rings of R. Note that if I is any ideal of R, the set

$$\{P/I \mid P \in X \text{ and } P \geq I\}$$

of primes of R/I satisfies the generic regularity condition.

By assumption, there is an infinite set $Y \subseteq X$ such that $Q \rightsquigarrow P$ for all $P \in Y$, and we may replace X by Y. Following the argument of Theorem 11.19, we find that R must be a prime ring, that $\cap X = 0$, and that $Q \neq 0$.

Lemma 14.20 now shows that $r_0(Q) = 1$. By Theorem 14.11, there exists a patch–neighborhood U of 0 such that $r_P(Q) = 1$ for all $P \in X \cap U$. There is no loss of generality in assuming that $U = W(I)$ for some nonzero ideal I.

Now $IQ \neq 0$. Since $\cap X = 0$, there must exist a prime $P \in X$ such that $P \not\geq IQ$. Then $P \in X \cap U$, whence $r_P(Q) = 1$. However, as $Q \not\leq P$ and $Q \rightsquigarrow P$, this contradicts Lemma 14.20.

Therefore Q cannot be linked to infinitely many primes from X. □

THEOREM 14.22. [Stafford] *Let Q be a prime ideal in a noetherian ring R. Assume that R is right fully bounded, or that r.K.dim(R) ≤ 1, or that there exists a positive integer n such that rank$(R/P) \leq n$ for all prime ideals P in R. Then there are at most finitely many primes P in R for which either $Q \rightsquigarrow P$ or $P \rightsquigarrow Q$.*

Proof. By either Proposition 14.9 or 14.10, Spec(R) satisfies the generic regularity condition. Hence, the theorem follows from Proposition 14.21. □

THEOREM 14.23. [Stafford] *In any noetherian ring R, all cliques of prime ideals are countable.*

Proof. It suffices to show that for any prime Q in R, there are at most countably many primes linked to or from Q. For $n \in \mathbb{N}$, let

$$X_n = \{P \in \text{Spec}(R) \mid \text{rank}(R/P) \leq n\},$$

and recall from Proposition 14.10 that X_n satisfies the generic regularity condition. Proposition 14.21 then shows that there are at most finitely many primes from each X_n linked to or from Q. Therefore there are at most countably many primes linked to or from Q, as desired. □

♦ NOTES

Zariski Topology. Stone topologized the prime spectrum of a Boolean ring R by declaring the open sets to be those of the form W(I) (in our notation) [1937, Theorem 1]. His idea was then used to topologize the set of maximal ideals of an arbitrary ring by Gelfand and Kolmogoroff [1939, p. 11], and to topologize the set of right primitive ideals of an arbitrary ring by Jacobson [1945b, p. 334]. In commutative algebra, this topology seems to have germinated from work of Zariski, who topologized a projective algebraic variety by declaring the closed subsets to be the algebraic subvarieties [1944, p. 684]; the affine analog of this is equivalent to topologizing a polynomial ring over a field by declaring the closed sets to be those of the form V(I).

Patch Topology. This was introduced by Hochster for a "spectral space", meaning any topological space that is homeomorphic to the prime spectrum of a commutative ring with the Zariski topology [1969, p. 45]. He showed that the patch topology on any spectral space is compact, Hausdorff, and totally disconnected [1969, Theorem 1 and Proposition 4].

Patch–Continuity of Normalized Ranks. This was first proved for finitely generated projective right modules over right FBN rings and over right noetherian rings with right Krull dimension at most one, by Goodearl and Warfield [1981, Propositions 4.4 and 4.10]. Theorem 14.6 was proved by Stafford over a two–sided noetherian ring [1981, Theorem 4.5, Lemma 6.1, and Corrigendum]. Goodearl then showed that the result also holds over a one–sided noetherian ring [1986, Theorem 1.4].

Generating Modules over Simple Noetherian Rings. Theorem 14.7 and Corollary 14.8 are due to Stafford [1976, Theorem 1.3 and Corollary 1.5].

Generic Regularity. This was introduced by Goodearl [1986, p. 89] (see also Warfield [1986, p. 180]).

Local Patch–Constantness of Normalized Ranks. A special case of Corollary 14.12, for a finitely generated torsionless right module over a right bounded prime right Goldie ring, is contained in a result of Warfield [1980, Theorem 4]. The corollary was then proved for finitely generated projective modules by Goodearl and Warfield, along with the corresponding result over a right noetherian ring with right Krull dimension at most one [1981, Propositions 4.4 and 4.10].

Generating Modules over FBN Rings. Theorems 14.17 and 14.19 were first proved by

Forster for finitely generated modules over a commutative noetherian ring, using Krull dimension in place of J–dimension [1964, Sätze 1,2]. Swan then proved analogs for finitely generated modules over a module–finite algebra over a commutative J–noetherian ring R, using J–dimension and data from localizations at J–primes of R [1967, Theorems 1,2]. The theorems as we have stated them (with an extra left noetherian hypothesis in the case of Theorem 14.19) were proved by Warfield [1979b, Theorems A, B; 1980, Theorems A, C, 5].

Countability of Cliques. Theorems 14.22 and 14.23 are due to Stafford [1987, Corollaries 3.10, 3.13].

15 TRANSCENDENTAL DIVISION ALGEBRAS

In this chapter we study another very "classical" topic, namely transcendental division algebras (that is, division rings which are not algebraic over their centers). While at first glance it may not appear that the general theory of noetherian rings has anything to say about division rings, we shall see that much concrete information can be gained by applying noetherian methods to polynomial rings over division rings, in particular by applying what we have learned in previous chapters about injective modules, Ore localizations, and Krull dimension. We shall, for instance, derive analogs of the Hilbert Nullstellensatz for polynomial rings over division rings and over fully bounded rings. Information about a division ring D with center k will then be obtained by developing connections among the transcendence degree of D over k, the question of primitivity of a polynomial ring $D[x_1,...,x_n]$, and the Krull dimension of $D \otimes_k k(x_1,...,x_n)$, as well as connections between the noetherian condition on $D \otimes_k D$ and the question of finite generation of subfields of D. For technical reasons, and in order to be able to apply some of these results to Goldie quotient rings, we actually derive most of the results in the chapter for simple artinian rings rather than for division rings.

The transcendental division algebras most accessible to us are the so-called *Weyl division algebras,* which we label as follows. For any field k and any positive integer n, the Weyl algebra $A_n(k)$ is a noetherian domain (Corollary 1.13), and so it has an Ore quotient division ring, which we shall denote by $D_n(k)$. In case $\mathrm{char}(k) = 0$, we have seen that the center of $D_1(k)$ is just k and that $D_1(k)$ is transcendental over k (Exercise 5K), and similar methods can be used to derive the same conclusions for any $D_n(k)$ (Exercise 15M).

♦ POLYNOMIALS OVER DIVISION RINGS

While polynomial rings over noncommutative rings are in some ways very similar to polynomial rings over commutative rings, there are important differences that should be noted. The most important difference is that one cannot simply evaluate a polynomial at a ring element. If $R[x]$ is a polynomial ring over a commutative ring R, then we can define a ring homomorphism from $R[x]$ to a commutative ring S by specifying a homomorphism from R to S and choosing an element $s \in S$ to be the image of x. The homomorphism $R[x] \to S$ is then defined by "evaluation", taking a polynomial in x to the corresponding polynomial in s. This is used particularly often when $S = R$. In the noncommutative case, the indeterminate x still commutes with the elements of R (and so x is in the center of R[x]). Hence, for any ring homomorphism $R[x] \to S$, the image of x must commute with the image of R. In particular, we can only "evaluate" polynomials from R[x] at an element $r \in R$ if r is in the center of R. (An alternate viewpoint: the right (or left) ideal of R[x] generated by $x - r$ is an ideal only if r is central.)

If k is a field, the chief importance of modules over a polynomial ring k[x] is in the study of linear transformations, since if V is a vector space over k and f is a linear transformation of V into itself, then V can be made into a k[x]–module by letting x act as the linear transformation f. (Conversely, every k[x]–module arises in this fashion.) In the noncommutative case we can proceed similarly, but care is necessary. Suppose then that we have a division ring D and a right vector space V over D. If f is a linear transformation on V, we let f act (as usual) on the left of V; this emphasizes the requirement that f commute with scalar multiplications by elements of D (which act on the right of V). Because f commutes with scalar multiplications, we can make V into a right module over D[x] by letting x act (as before) as f. The resulting D[x]–module is sometimes denoted (V,f). In particular, if V is finite–dimensional we may identify it with D^n for some n, in which case f will be given by an $n \times n$ matrix α. However, since we are viewing D^n as a *right* vector space, we should regard its elements as columns, and f should act as *left* multiplication by the matrix α. Thus we obtain a right D[x]–module (D^n, α) in which D acts on the right via scalar multiplications while x acts on the left via α.

We do not want to develop the theory of linear transformations over a division ring here, although that point of view is occasionally useful. However, this point of view suggests a question which can be used as a motivation for our subsequent discussion: does every linear transformation on a finite–dimensional vector space satisfy a (minimal) polynomial? In module–theoretic terms, are the right D[x]–modules (D^n, α) discussed above necessarily unfaithful? It turns out that this is not the case if D is transcendental over its center, as we shall see shortly (Exercise 15A).

In order to use torsion terminology, we note that if S is a prime right noetherian ring, then any polynomial ring $S[x_1, \ldots, x_n]$ is also prime and right noetherian.

PROPOSITION 15.1. *Let S be a simple artinian ring with center k, and S[x] a polynomial ring.*

(a) Every finitely generated torsion S[x]–module has finite length as an S–module, and hence also finite length as an S[x]–module.

(b) Every simple S[x]–module has finite length as an S–module.

(c) Every ideal of $S[x]$ is generated by a central element, that is, an element of $k[x]$.

Proof. (a) Since it suffices to show this for all cyclic torsion $S[x]$–modules, it is enough to look at the case of a module $S[x]/J$ where J is an essential right ideal of $S[x]$.

Now $S_S = A_1 \oplus ... \oplus A_m$ for some simple right ideals A_i, and

$$S[x] = A_1S[x] \oplus ... \oplus A_mS[x].$$

Each of the intersections $J \cap A_iS[x]$ is nonzero, and since $S[x]/J$ is an epimorphic image of $\oplus A_iS[x]/(J \cap A_iS[x])$, it suffices to show that each of the factors $A_iS[x]/(J \cap A_iS[x])$ has finite length as an S–module. Relabelling, let A be a simple right ideal of S and let K be a nonzero right ideal of $S[x]$ contained in $AS[x]$; we shall show that $AS[x]/K$ has finite length as a right S–module. Since S is semisimple, $A = eS$ for some idempotent e.

Choose a nonzero polynomial $f \in K$, say with degree n and leading coefficient b. Since b is a nonzero element of A, there exists $s \in S$ such that $bs = e$, and we may replace f by fs. Hence, we may assume that the leading coefficient of f is e. Via the usual division algorithm, it follows that

$$AS[x] = K + A + Ax + ... + Ax^{n-1},$$

and therefore $AS[x]/K$ has finite length, as desired.

(b) We first observe that no right ideal H of $S[x]$ can be a simple $S[x]$–module, for if $H \neq 0$, then Hx is a proper nonzero submodule of H. Hence, all maximal right ideals of $S[x]$ must be essential, and thus all simple right $S[x]$–modules are torsion.

(c) Let I be a nonzero ideal of $S[x]$, and let f be a nonzero polynomial of least degree in I, say with degree n and leading coefficient c. Since S is simple, there are elements $s_i, t_i \in S$, for $i = 1,...,m$, such that

$$\sum_{i=1}^{m} s_i c t_i = 1.$$

Hence, replacing f by $\sum s_i f t_i$, we may assume that $c = 1$. It now follows from the division algorithm that $I = fS[x] = S[x]f$. For any $s \in S$, we have $sf - fs \in I$, and since f is monic, it is clear that $sf - fs$ has lower degree than f. Hence, $sf - fs = 0$, and we conclude that $f \in k[x]$. □

EXERCISE 15A. Let S be a simple artinian ring, x an indeterminate, $n \in \mathbb{N}$, and $\alpha \in M_n(S)$. Show that the right $S[x]$–module (S^n, α) is unfaithful if and only if α is algebraic over the center of S. □

We now want to discuss the question of when a polynomial ring over a division ring D is primitive. As the previous exercise suggests, this requires a discussion of algebraic and transcendental elements, not only in D but in matrix rings over D.

DEFINITION. An algebra R over a field k is called *matrix–algebraic over* k if all the matrix algebras $M_n(R)$ are algebraic over k.

The reader should verify that a matrix ring $M_m(D)$ over a k–algebra D is matrix–algebraic over k if and only if D is matrix–algebraic over k. It is an unsolved problem (raised by Jacobson in 1945) whether a division ring algebraic over its center is necessarily matrix–algebraic. (For division rings with uncountable center, this was proved by Amitsur in [1956, Theorem 9].)

THEOREM 15.2. [Jacobson] *Let S be a simple artinian ring with center k, and let x be an indeterminate. Then the following conditions are equivalent:*

(a) S is matrix–algebraic over k.

(b) $S \otimes_k k(x)$ is a simple artinian ring.

(c) S[x] is right bounded.

(d) S[x] is not right primitive.

Proof. (a) ⇒ (b): We may assume that $S = M_m(D)$ for some $m \in \mathbb{N}$ and some division ring D, and that k is the center of D. Then we may identify $S \otimes_k k(x)$ with $M_m(D \otimes_k k(x))$. Since S is matrix–algebraic over k, so is D. We shall show that $D \otimes_k k(x)$ is a division ring, from which it is immediate that $S \otimes_k k(x)$ is simple artinian. Note that $D \otimes_k k(x)$ is (isomorphic to) an Ore localization $D[x]X^{-1}$ where $X = k[x] - \{0\}$. Hence, to prove that $D \otimes_k k(x)$ is a division ring, it suffices to show that every nonzero element of D[x] has a right inverse in $D[x]X^{-1}$.

Now let f be a nonzero polynomial in D[x], and let n = deg(f). The cyclic right module $M = D[x]/fD[x]$ is isomorphic, as a right vector space over D, to D^n, and under this isomorphism right multiplication by x on M corresponds to left multiplication by some matrix $\alpha \in M_n(D)$. (Actually, α is the usual "companion matrix" of ordinary linear algebra.) In other words, $M \cong (D^n, \alpha)$.

By assumption, α is algebraic over k, and so by Exercise 15A, M is unfaithful. The nonzero ideal $\mathrm{ann}_{D[x]}(M)$ is generated by a nonzero central polynomial $g \in k[x]$, and since $Mg = 0$ we have $g = fh$ for some $h \in D[x]$. Thus hg^{-1} is a right inverse for f in $D[x]X^{-1}$, as desired.

(b) ⇒ (c): We note again that $S \otimes_k k(x) \cong S[x]X^{-1}$, where $X = k[x] - \{0\}$. If I is any essential right ideal of S[x], then we see by Theorem 9.17 that the extension I^e is an essential right ideal of $S[x]X^{-1}$. Since $S[x]X^{-1}$ is simple artinian, $I^e = S[x]X^{-1}$, from which it follows that $I \cap X$ is nonempty. If $f \in I \cap X$, then fS[x] is a nonzero ideal of S[x] contained in I. As S[x] is a prime ring, this proves that S[x] is right bounded.

(c) ⇒ (d): As in Proposition 8.4, any right noetherian ring which is both right bounded and right primitive must be simple artinian. Therefore S[x] cannot be right primitive.

(d) $\Rightarrow$ (a): By assumption, all simple right $S[x]$–modules are unfaithful, and consequently all right $S[x]$–modules of finite length are unfaithful. Given a matrix $\alpha \in M_n(S)$, form the right $S[x]$–module (S^n,α), and observe that since (S^n,α) has finite length as a right S–module, it also has finite length as a right $S[x]$–module. Hence, (S^n,α) is unfaithful, and so by Exercise 15A, α is algebraic over k. Therefore S is matrix–algebraic over k. □

For example, if k is a field of characteristic zero, then $D_1(k)$ is transcendental over its center, and hence Theorem 15.2 shows that the polynomial ring $D_1(k)[x]$ is right (and left) primitive.

As an application of Theorem 15.2, we can give the following criterion for boundedness of more general polynomial rings. The question of primitivity for polynomial rings over bounded prime rings is partially answered in Theorem 15.10.

THEOREM 15.3. *Let R be a right bounded prime right noetherian ring with right Goldie quotient ring Q. Then the polynomial ring $R[x]$ is right bounded if and only if Q is matrix–algebraic over its center.*

Proof. Because of Theorem 15.2, this amounts to showing that $R[x]$ is right bounded if and only if $Q[x]$ is right bounded. Note that $Q[x]$ may be identified with the right Ore localization $R[x]X^{-1}$ where X is the set of regular elements of R.

Assume first that $R[x]$ is right bounded. If I is an essential right ideal of $Q[x]$, then $I \cap R[x]$ is an essential right ideal of $R[x]$, and so it contains a nonzero ideal J. Obviously $JQ[x] \leq I$, and by Theorem 9.20, $JQ[x]$ is an ideal of $Q[x]$. Thus $Q[x]$ is right bounded.

Conversely, suppose that $Q[x]$ is right bounded, and let I be an essential right ideal of $R[x]$. Then $IQ[x]$ is an essential right ideal of $Q[x]$, and so it contains a nonzero ideal J. By Proposition 15.1, J can be generated by some polynomial $f \in k[x]$, where k is the center of Q. Because $f \in J \leq IQ[x]$ and $Q[x] = R[x]X^{-1}$, there is an element $c \in X$ such that $fc \in I$. Now cR is an essential right ideal of R (Proposition 5.9), and since R is right bounded, cR contains a nonzero ideal B. Note that $fB \subseteq fcR \subseteq I$, whence $fB[x] \leq I$. Since f is central in $Q[x]$, we conclude that $fB[x]$ is an ideal of $R[x]$. This proves that $R[x]$ is right bounded, as required. □

EXERCISE 15B. Let R be a right noetherian ring such that for each prime ideal P in R, the right Goldie quotient ring of R/P is matrix–algebraic over its center. If Q is a non–induced prime ideal in the polynomial ring $R[x]$ (that is, $Q > (Q \cap R)[x]$), show that the right Goldie quotient ring of $R[x]/Q$ is matrix–algebraic over its center. □

EXERCISE 15C. Let R be a right noetherian ring with center S.

(a) If R is *integral over* S (that is, every element of R satisfies a monic polynomial with coefficients from S), show that R is right fully bounded.

(b) If R is prime and all matrix rings $M_t(R)$ are integral over S, show that the polynomial ring R[x] is right bounded. [Hint: Consider the localization of R with respect to $S - \{0\}$.] □

♦ MORE VARIABLES

If D is a division ring with center k, the ring $D \otimes_k k(x)$ appearing in Theorem 15.2 is sometimes called the ring of *rational functions* in one variable over D. It is not (usually) the quotient division ring of D[x], since by the theorem $D \otimes_k k(x)$ is not a division ring unless D is matrix–algebraic over k. As noted in that proof, however, it can be thought of as a ring of fractions with numerators from D[x] and denominators from k[x]; thus it is a subring of the quotient division ring of D[x]. Even when $D \otimes_k k(x)$ is not a division ring, it is nevertheless a simple noetherian ring, as is clear from the localization viewpoint just discussed together with Proposition 15.1. This gives, therefore, a new family of simple noetherian rings which we have not previously seen.

We can generalize this to polynomials in several variables, but it is convenient to first do the following lemma, which we state in greater generality than presently needed. In its proof the reader should notice a similarity to the proof of Proposition 15.1(c), which, of course, is a special case.

LEMMA 15.4. *Let k be a field, A a simple k–algebra with center k, and B any k–algebra. Then the ideals of $A \otimes_k B$ are precisely the ideals $A \otimes_k I$ where I is an ideal of B. Moreover, the prime ideals of $A \otimes_k B$ are precisely the ideals $A \otimes_k P$ where P is a prime ideal of B.*

Proof. Let $C = A \otimes_k B$. Since k is a field, we may identify A and B with the subalgebras $A \otimes 1$ and $1 \otimes B$ inside C. Then all elements of A commute with all elements of B, and all elements of C are sums of products ab where $a \in A$ and $b \in B$. Moreover, any elements $b_i \in B$ that are linearly independent over k are also linearly independent over A, since

$$\Sigma Ab_i = A \otimes (\Sigma kb_i) = A \otimes (\oplus kb_i) \cong \oplus (A \otimes kb_i) = \oplus Ab_i.$$

It is clear that if I is an ideal of B then AI is an ideal of C, and that $C/AI \cong A \otimes_k (B/I)$.

If J is an ideal of C, we want to show that $J = A(J \cap B)$. Passing to the ring

$$C/A(J \cap B) \cong A \otimes_k (B/(J \cap B)),$$

we see that it suffices to assume that $J \cap B = 0$ and then show that $J = 0$.

Assuming that $J \neq 0$, choose a nonzero element $c \in J$ which can be written in the form

$$c = \sum_{i=1}^{m} a_i b_i$$

(where $a_i \in A$ and $b_i \in B$) with m as small as possible. It follows that the b_i must be linearly independent over k. (For instance, if $b_1 = \alpha_2 b_2 + \ldots + \alpha_m b_m$ for some $\alpha_i \in k$, then

$$c = \sum_{i=2}^{m} (a_i + \alpha_i a_1) b_i,$$

which contradicts the minimality of m.) For the same reason, $a_1 \neq 0$. Since A is simple, there are elements $r_j, s_j \in A$, for $j = 1,\ldots,n$, such that

$$\sum_{j=1}^{n} r_j a_1 s_j = 1.$$

Now observe that the element

$$c' = \sum_{j=1}^{n} r_j c s_j = \sum_{i=1}^{m} \Big(\sum_{j=1}^{n} r_j a_i s_j \Big) b_i$$

is a nonzero element of J. (That $c' \neq 0$ follows from the linear independence of the b_i.) Hence, replacing c by c', we may assume that $a_1 = 1$.

Now for any $a \in A$, the element

$$ac - ca = \sum_{i=2}^{m} (aa_i - a_i a) b_i$$

lies in J, and so $ac - ca = 0$, by the minimality of m. Since the b_i are linearly independent, $aa_i - a_i a = 0$ for all i, and we conclude that all the $a_i \in k$. But then $c \in J \cap B$, a contradiction.

This proves that all ideals of C have the desired form. The statement about prime ideals is a direct consequence. □

PROPOSITION 15.5. *Let S be a simple artinian ring with center k, and let $x_1,\ldots,x_n$ be independent indeterminates.*

(a) The center of the ring $S[x_1,\ldots,x_n]$ is exactly $k[x_1,\ldots,x_n]$.

(b) Every ideal of $S[x_1,\ldots,x_n]$ is generated by elements in the center.

(c) Extension and contraction provide inverse lattice isomorphisms between the lattice of ideals of $k[x_1,\ldots,x_n]$ and the lattice of ideals of $S[x_1,\ldots,x_n]$. These maps restrict to inverse bijections between $Spec(k[x_1,\ldots,x_n])$ and $Spec(S[x_1,\ldots,x_n])$.

(d) The ring $S \otimes_k k(x_1,\ldots,x_n)$ is a simple noetherian ring.

Proof. (a) is clear, (b) and (c) are immediate from Lemma 15.4, and (d) follows from Corollary 9.18 and Proposition 9.19. □

Proposition 15.5 should be thought of as the n–variable analog of Proposition 15.1. Of course, the question of what kind of simple noetherian ring we obtain in (d) arises, as does the question of an analog for Theorem 15.2. It was conjectured at one time that a polynomial ring in several indeterminates over S would be primitive if and only if S[x] were primitive, and it was even believed that this had been proved. The error was discovered by Resco, who conjectured the more subtle result that we will discuss below (Theorem 15.13). However, we must prove a noncommutative Nullstellensatz first.

♦ THE NULLSTELLENSATZ

The Hilbert Nullstellensatz is traditionally stated in several different ways. One says that in a polynomial ring $R = k[x_1,...,x_n]$ over a field k, every prime ideal is an intersection of maximal ideals. A second statement is that if M is a maximal ideal of R, then R/M is a finite–dimensional extension of k. If we assume the statement for fields as known, then the first of these statements, for polynomials over a division ring, follows trivially from Proposition 15.5. There are two analogs of the second statement. Namely, if $R = D[x_1,...,x_n]$ is a polynomial ring over a division ring D, we could ask whether all factors of R by maximal ideals are finite–dimensional over D, or whether all factors of R by maximal one–sided ideals are finite–dimensional over D, that is, whether all simple R–modules are finite–dimensional. The first analog is again an easy consequence of Proposition 15.5; we give it as Exercise 15D in order to have the statement recorded. The second analog of the second form of the Nullstellensatz, however, is more difficult to prove, but also, as it turns out, far more useful. This result was first conjectured by Resco and then proved by Amitsur and Small.

EXERCISE 15D. If $R = S[x_1,...,x_n]$ is a polynomial ring over a simple artinian ring and M is a maximal ideal of R, show that R/M has finite length as a right or left S–module. □

We first make an observation that will allow us to reduce to the case of division ring coefficients.

EXERCISE 15E. Let $T = M_t(R)$ for a ring R and some $t \in \mathbb{N}$, and identify R with the subring of scalar matrices in T. Show that any simple T–module A is finitely generated and semisimple when viewed as an R–module. [Hint: Multiply A by the standard matrix units.] □

If S is a simple artinian ring, we may assume that $S = M_t(D)$ for some division ring D and some positive integer t, and then we may identify the polynomial ring $S[x_1,...,x_n]$ with $M_t(D[x_1,...,x_n])$. According to Exercise 15E, any simple module over $S[x_1,...,x_n]$ is finitely generated and semisimple over $D[x_1,...,x_n]$. Hence, if we know that all simple $D[x_1,...,x_n]$–modules are finite–dimensional over D, then every simple $S[x_1,...,x_n]$–module is finite–dimensional over D and hence has finite length over S.

Therefore in proving the Nullstellensatz, we may reduce to the case that S is a division ring. This makes it easier to work with the rings $S[x_i]$ which, for instance, are principal right and left ideal domains (Theorem 1.11).

EXERCISE 15F. If $D[x]$ is a polynomial ring over a division ring, show that there are infinitely many pairwise non–isomorphic unfaithful simple right $D[x]$–modules. [Hint: Let k be the center of D, and use either Euclid's method or Exercise 2W to find infinitely many pairwise non–associate irreducible polynomials in $k[x]$.] □

EXERCISE 15G. If D is a division ring and M is a right ideal in the polynomial ring $D[x_1,...,x_n]$ such that $M \cap D[x_i] \neq 0$ for all $i = 1,...,n$, show that $D[x_1,...,x_n]/M$ is finite–dimensional over D. □

THEOREM 15.6. [Amitsur–Small] *If S is a simple artinian ring, then every simple module over the polynomial ring $S[x_1,...,x_n]$ has finite length as an S–module.*

Proof. As observed above, it is enough to deal with simple modules over a polynomial ring $T = D[x_1,...,x_n]$ where D is a division ring. The case $n = 1$ follows directly from the division algorithm (or Proposition 15.1). Now let $n > 1$, and assume that the theorem holds over polynomial rings in $n - 1$ indeterminates.

Let $A = T/M$ where M is a maximal right ideal of T. By Exercise 15G, it suffices to show that $M \cap D[x_i] \neq 0$ for all $i = 1,...,n$, and by symmetry it is enough to show that $M \cap D[x_1] \neq 0$. Set $R = D[x_1]$, and suppose that $M \cap R = 0$.

Now R embeds in A, and so the R–module A_R is not torsion. Since the torsion submodule of A_R is invariant under all endomorphisms of A_R, it must be invariant under multiplication by each x_i, and so it is a T–submodule. By simplicity, the torsion submodule of A_R must be zero, and hence A_R is torsionfree. Similarly, if k is the center of D and p is a nonzero polynomial in $k[x_1]$, then Ap is a nonzero T–submodule of A, whence $Ap = A$. Thus A is divisible as a $k[x_1]$–module.

The set $X = R - \{0\}$ is a right and left denominator set in both R and T, and TX^{-1} is isomorphic to $(RX^{-1})[x_2,...,x_n]$. Since A_R is torsionfree, AX^{-1} is nonzero, and so it is a simple module over TX^{-1}. As RX^{-1} is a division ring, it follows from our induction hypothesis that AX^{-1} is finite–dimensional over RX^{-1}, and from this we conclude that A

has finite rank as an R–module.

Now A has an essential R–submodule which is a finite direct sum of uniform submodules, and each of these, being torsionfree, contains a copy of R_R. Hence, A has an essential R–submodule B which is finitely generated and free.

Observe using Proposition 15.1 that any nonzero prime ideal P of R is a maximal ideal, that R/P is simple artinian, and that $P = pR$ for some nonzero $p \in k[x_1]$. Since A is divisible as a $k[x_1]$–module, it contains an R–submodule C such that $Cp = B$, and then since A is torsionfree over R we see that $C/B \cong B/Bp$, which is in turn isomorphic to a direct sum of copies of R/P. Therefore A/B contains at least one simple R–submodule annihilated by P. In view of Exercise 15F, it follows that A/B has infinitely many pairwise non–isomorphic (unfaithful) simple R–submodules.

Set $B_0 = B$, and then set $B_{i+1} = B_i + B_ix_1 + ... + B_ix_n$ for all $i = 0,1,...$. These B_i are finitely generated R–submodules of A, and their union is a T–submodule, whence $\bigcup_{i=0}^{\infty} B_i = A$. Since B_1/B_0 is a finitely generated torsion R–module, it has finite length, by Proposition 15.1. Let $\mathscr{S}$ be the set of isomorphism types of composition factors of B_1/B_0, and keep in mind that $\mathscr{S}$ is finite. Multiplication by any x_j induces a homomorphism $B_1/B_0 \to B_2/B_1$, and the sum of the images of these homomorphisms is all of B_2/B_1. In other words, B_2/B_1 is an epimorphic image of $(B_1/B_0)^n$, and consequently all composition factors of B_2/B_1 lie in $\mathscr{S}$. Likewise, all composition factors of each of the modules B_{i+1}/B_i lie in $\mathscr{S}$. Since A is the union of the B_i, it follows that the isomorphism types of all simple submodules of A/B are in $\mathscr{S}$. However, this contradicts the fact that A/B contains infinitely many pairwise non–isomorphic simple submodules.

Therefore $M \cap R \neq 0$, as desired. □

One form of the commutative Nullstellensatz is obviously contained in Theorem 15.6: if k is a field and M is a maximal ideal in the polynomial ring $k[x_1,...,x_n]$, then $k[x_1,...,x_n]/M$ is finite–dimensional over k. A second form may be obtained from this as in the following exercise.

EXERCISE 15H. If k is a field and P is a prime ideal in the polynomial ring $R = k[x_1,...,x_n]$, show that P is an intersection of maximal ideals. [Hint: If not, assume that P is maximal with respect to this property, observe that the intersection of the nonzero prime ideals of R/P is nonzero, and show that the quotient field of R/P is a homomorphic image of R[x].] □

EXERCISE 15I. This is an extended exercise which indicates how uncountability assumptions can simplify the proofs of both the commutative Nullstellensatz and the Amitsur–Small Nullstellensatz. Throughout, let k be an uncountable field and D a division ring with center k.

(a) Show that the field $k(x)$ of rational functions over k has uncountable dimension as a vector space over k. [Hint: The elements $1/(x - \alpha)$, for $\alpha \in k$, are linearly independent.]

(b) Show that if M is a maximal ideal in the polynomial ring $k[x_1,\ldots,x_n]$, then $k[x_1,\ldots,x_n]/M$ is finite–dimensional over k. [Hint: Use Exercise 15G.]

(c) Show that the ring $D \otimes_k k(x_1)$ has uncountable dimension over D. [Note that it is useless to show that the dimension over k is uncountable, since this will already be true of D itself if D is transcendental over k.]

(d) Show that $D \otimes_k k[x_1]$ is an essential $k[x_1]$–submodule of $D \otimes_k k(x_1)$.

(e) Show that if M is a maximal right ideal in the polynomial ring $R = D[x_1,\ldots,x_n]$, then R/M is finite–dimensional over D. [Hint: If not, then by Exercise 15G we may assume that $M \cap D[x_1] = 0$. Make R/M into a $(k[x_1],D)$–bimodule, and use the fact that the endomorphism ring of R/M is a division ring to make R/M into a $(k(x_1),D)$–bimodule. Convert this bimodule structure into a right $D[x_1]$–module homomorphism $D \otimes_k k(x_1) \to R/M$ which is injective on $D \otimes_k k[x_1]$. Finally, use part (d) and part (c).] □

♦ FULLY BOUNDED G–RINGS

We now use the Amitsur–Small Nullstellensatz to study primitive ideals in polynomial rings. (We should point out that although Proposition 15.5 seems to say a great deal about the ideals in a polynomial ring over a simple artinian ring, it does not say anything about which ideals are primitive.) The approach we shall take goes by way of the notion of a G–ring, which is also important in some standard commutative treatments of the Nullstellensatz.

DEFINITION. A *G–ring* is a prime ring in which the intersection of the nonzero prime ideals is nonzero. A *G–ideal* in a ring R is any prime ideal P such that R/P is a G–ring.

For instance, any simple ring is a G–ring. The local ring

$$\mathbb{Z}_{(2)} = \{a/b \mid a,b \in \mathbb{Z} \text{ and } b \text{ is odd}\}$$

is a G–ring (since its only nonzero prime ideal is $2\mathbb{Z}_{(2)}$), while $\mathbb{Z}$ itself is not a G–ring. It can be proved that a commutative noetherian domain is a G–ring if and only if it has only finitely many prime ideals. In the noncommutative case, however, this no longer holds: for example, the noetherian domain S considered in Exercise 2Y is a G–ring with infinitely many prime ideals.

A commutative domain R is a G–ring if and only if the quotient field of R can be obtained by inverting a single element b, that is, if every element of the quotient field can be written as a fraction with numerator from R and denominator a power of b. While the noncommutative analog of this equivalence is not generally true (Exercise 15J), it does hold for bounded prime Goldie rings, as Lemma 15.7 shows.

EXERCISE 15J. If $R = A_1(k)$ for a field k of characteristic zero, show that there does not exist a nonzero element $b \in R$ such that $D_1(k) = \{ab^{-n} \mid a \in R,\ n \in \mathbb{Z}^+\}$. [Hint: Observe that R/bR has finite length. If R is expressed in the form $k[x][\theta; d/dx]$, then for $\alpha \in k$ show that $R/(x-\alpha)R$ is a simple module and that it is a composition factor of some R/b^nR.] □

LEMMA 15.7. *Let R be a right bounded prime right Goldie ring and Q its right Goldie quotient ring. Then the following conditions are equivalent:*

(a) R is a G–ring.

(b) There is a regular element $b \in R$ such that every essential right ideal of R contains a power of b.

(c) There is a regular element $b \in R$ such that every element of Q can be put in the form ab^{-n} for $a \in R$ and $n \in \mathbb{Z}^+$.

(d) There is a nonzero ideal I in R such that every element of Q/R can be annihilated on the right by a power of I.

Proof. (a) ⇒ (b): The intersection of the nonzero prime ideals of R is a nonzero ideal I, and it must contain a regular element b.

If J is an essential right ideal of R, then since R is right bounded, J contains a nonzero ideal K. Suppose that K is disjoint from the set $B = \{1, b, b^2, \ldots\}$, and enlarge K to an ideal P maximal with respect to being disjoint from B. According to Lemma 2.5, P is prime. But then $b \in I \leq P$, a contradiction. Thus there must be some power of b contained in K, and hence in J.

(b) ⇒ (c): Any element of Q has the form cd^{-1} for some $c, d \in R$ with d regular. Then dR is an essential right ideal of R (Proposition 5.9), and so dR contains a power of b, say $b^n = de$ for some $n \in \mathbb{Z}^+$, $e \in R$. Then $d^{-1} = eb^{-n}$ and thus $cd^{-1} = ceb^{-n}$.

(c) ⇒ (d): Since bR is an essential right ideal of R, by our right boundedness hypothesis there is a nonzero ideal I contained in bR. For all $n \in \mathbb{N}$, observe that

$$I^n \leq bRI^{n-1} = bI^{n-1} \leq b^2RI^{n-2} = b^2I^{n-2} \leq \ldots \leq b^nR.$$

Statement (d) follows from this and statement (c).

(d) ⇒ (a): Any nonzero prime ideal P of R is essential as a right ideal, and so P contains a regular element c. Then $c^{-1}I^n \leq R$ for some $n \in \mathbb{N}$, whence $I^n \leq cR \leq P$, and consequently $I \leq P$. Therefore the intersection of the nonzero prime ideals of R contains I. □

EXERCISE 15K. If the hypotheses and the equivalent conditions of Lemma 15.7 hold, show that the set $X = \{1, b, b^2, \ldots\}$ is a right denominator set in R, and that RX^{-1} is the right Goldie quotient ring of R. □

EXERCISE 15L. If P is a prime ideal in a polynomial ring $R[x_1, \ldots, x_n]$, show that $P \cap R$ is a prime ideal of R. □

LEMMA 15.8. *If R is a G–ring which is either right or left Goldie, then in the polynomial ring $T = R[x_1, \ldots, x_n]$ (where $n > 0$) there is a right primitive ideal P such that $P \cap R = 0$.*

Proof. The intersection of all the nonzero primes of R is a nonzero ideal, which must contain a regular element b. Observe that $1 - bx_1$ is not right invertible in T. Let M be a maximal right ideal of T containing $(1 - bx_1)T$, and let P be the annihilator of the simple right T–module T/M. Then P is a right primitive ideal of T, and hence prime, and so by Exercise 15L, $P \cap R$ is a prime ideal of R. Since $1 - bx_1 \in M$, we must have $b \notin M$, whence $b \notin P \cap R$. Therefore $P \cap R = 0$. □

The converse of Lemma 15.8 is false in general, as we shall see below (Exercise 15N), but we shall prove it when R is bounded and Goldie on both sides.

LEMMA 15.9. *Let R be a right and left bounded, prime, right and left Goldie ring, and let A be a simple right module over the polynomial ring $T = R[x_1, \ldots, x_n]$. If A_R is faithful, then it is a torsionfree injective module of finite rank.*

Proof. If A_R is not torsionfree, choose a nonzero torsion element $a \in A$. There is an essential right ideal J in R such that $aJ = 0$, and since R is right bounded there is a nonzero ideal I contained in J, whence $aI = 0$. Now $aT = A$, because A is a simple T–module. But then $AI = aTI = aIT = 0$, which contradicts the assumption that A_R is faithful. Thus A_R must be torsionfree.

Since A_R is now torsionfree, to prove that it is injective it is sufficient to prove that it is divisible (Proposition 6.12). Let c be a regular element of R. As Rc is an essential left ideal of R and R is left bounded, there is a nonzero ideal H contained in Rc, and $AH \neq 0$ because A_R is faithful. Observe that AH is a T–submodule of A. Since A is simple over T, we conclude that $A = AH \leq Ac$. Therefore A is divisible, and hence injective, as an R–module.

If Q is the Goldie quotient ring of R, then by Proposition 6.13, A has a unique right Q–module structure compatible with its right R–module structure. For any regular element $b \in R$, since multiplication by b on A commutes with multiplication by each x_i, so does multiplication by b^{-1}. It follows that the Q–module structure on A extends to a $Q[x_1,...,x_n]$–module structure, compatible with the T–module structure. Obviously A is simple as a module over $Q[x_1,...,x_n]$. Applying Theorem 15.6, we conclude that A has finite length as a Q–module. Therefore A_R has finite rank. □

THEOREM 15.10. [Resco–Stafford–Warfield] *Let R be a right and left bounded, prime, right and left Goldie ring, and let $R[x_1,...,x_n]$ be a polynomial ring where $n > 0$. Then R is a G–ring if and only if there is a right primitive ideal P in $R[x_1,...,x_n]$ such that $P \cap R = 0$.*

Proof. One implication is given by Lemma 15.8. Conversely, assume that there is a right primitive ideal P in $R[x_1,...,x_n]$ such that $P \cap R = 0$. There is a simple right $R[x_1,...,x_n]$–module A whose annihilator is P, and then A is faithful as an R–module. By Lemma 15.9, A_R is torsionfree injective of finite rank. Hence, A becomes a finitely generated semisimple module over the Goldie quotient ring Q of R. If A has length s and Q has length t, then $A^t \cong Q^s$, and so A^t has an essential R–submodule B such that $A^t/B \cong (Q/R)^s$. We may choose a finitely generated essential R–submodule $C \leq_e A$ such that $C^t \leq B$, and then $(Q/R)^s$ is an epimorphic image of $(A/C)^t$.

We shall find a nonzero ideal I in R such that every element of A/C is annihilated by a power of I. The same is then true in Q/R (on the right), and by Lemma 15.7 this is enough to show that R is a G–ring.

Since for each index $i = 1,...,n$ the R–module $(C + Cx_i)/C$ is finitely generated and torsion, the right boundedness of R implies that there is a nonzero ideal I in R which annihilates all of these modules (see Lemma 8.2). In other words, $Cx_iI \leq C$ for all $i = 1,...,n$. It follows that if f is any polynomial in $R[x_1,...,x_n]$ of total degree d, then $CfI^d \leq C$. Since A is simple as an $R[x_1,...,x_n]$–module, $CR[x_1,...,x_n] = A$. Therefore every element of A/C is annihilated by a power of I, as desired. □

Recall that a *Jacobson ring* is a ring in which all the prime ideals are semiprimitive. For instance, by the commutative Nullstellensatz every finitely generated commutative algebra over a field is a Jacobson ring.

COROLLARY 15.11. [Resco–Stafford–Warfield] *Let R be a fully bounded noetherian Jacobson ring, and let A be a simple right module over the polynomial ring $R[x_1,...,x_n]$, where $n > 0$. Then there is a maximal ideal M of R such that $AM = 0$, and as an R–module A is finitely generated and semisimple.*

Proof. If $M = \mathrm{ann}_R(A)$, then A is a simple right module over the polynomial ring $(R/M)[x_1,...,x_n]$, and A is faithful as an (R/M)–module. The annihilator of A in $(R/M)[x_1,...,x_n]$ is a right primitive ideal P such that $P \cap (R/M) = 0$. Theorem 15.10 then implies that R/M is a G–ring. Since R is a Jacobson ring, M is an intersection of right primitive ideals, and since R/M is a G–ring this is only possible if M itself is right primitive. As R is fully bounded, M must be a maximal ideal and R/M must be simple artinian (Proposition 8.4). It follows that A is semisimple as an R–module. Lemma 15.9 implies that A_R has finite rank, and therefore it is finitely generated. □

♦ PRIMITIVITY AND TRANSCENDENCE DEGREE

We return to the question of when a polynomial ring over a division ring D (or more generally over a simple artinian ring) can be primitive. The case of a single indeterminate has been treated in Theorem 15.2: $D[x]$ is primitive if and only if at least one of the matrix rings over D is transcendental over the center of D. In order to handle polynomial rings in many indeterminates, we need an appropriate notion of transcendence degree.

DEFINITION. Let R be an algebra over a field k. The *transcendence degree of* R *over* k is the supremum of those nonnegative integers n for which there exists a k–algebra embedding of the polynomial ring $k[x_1,...,x_n]$ into R. (By convention, the polynomial ring in 0 indeterminates is just k. Hence, R has transcendence degree 0 over k precisely when R is algebraic over k.) The *matrix–transcendence degree of* R *over* k is the supremum of the transcendence degrees of all matrix rings $M_t(R)$.

The following lemma is frequently useful, and shows that for simple artinian rings, the notion of matrix–transcendence degree just defined agrees with another reasonable notion.

LEMMA 15.12. *Let S be a simple artinian algebra over a field k, and let $x_1,...,x_n$ be independent indeterminates. Then S has matrix–transcendence degree at least n over k if and only if there is a k–algebra embedding of the rational function field $k(x_1,...,x_n)$ into some matrix ring $M_t(S)$.*

Proof. Obviously if $k(x_1,...,x_n)$ embeds in $M_t(S)$, so does the polynomial ring $R = k[x_1,...,x_n]$, whence the matrix–transcendence degree of S over k is at least n.

Conversely, if this matrix–transcendence degree is at least n, there is a k–algebra embedding of R into some matrix ring $M_m(S)$. We use this to produce a bimodule as follows. We regard S^m as a left S–module and as a right module over the matrix ring $M_m(S)$. Since $M_m(S)$ is a simple ring, S^m is faithful as a right $M_m(S)$–module. Using the embedding of R into $M_m(S)$, we can turn S^m into an (S,R)–bimodule, finitely generated free on the left and faithful on the right.

According to Lemma 7.1, R_R embeds in a finite direct sum of copies of S^m, and so $(S^m)_R$ is not torsion. Hence, if T is the torsion submodule of S^m as a right R–module, $S^m/T \neq 0$. Now T is a left S–submodule of S^m, and S^m/T is a finitely generated semisimple left S–module. If S has length s, then $(S^m/T)^s$ is a free left S–module, say of rank t, and so its endomorphism ring is isomorphic to $M_t(S)$. Thus there is a k–algebra embedding of the endomorphism ring of $_S(S^m/T)$ into $M_t(S)$.

Now the bimodule $_S(S^m/T)_R$ is torsionfree on the right and has finite length on the left. By Lemma 7.8, its right R–module structure extends to a right $k(x_1,...,x_n)$–module structure, and S^m/T becomes an $(S, k(x_1,...,x_n))$–bimodule. Therefore $k(x_1,...,x_n)$ embeds in the endomorphism ring of $_S(S^m/T)$, and hence in $M_t(S)$. □

THEOREM 15.13. [Resco, Amitsur–Small] *Let S be a simple artinian ring with center k. Then the polynomial ring $S[x_1,...,x_n]$ is right primitive if and only if the matrix–transcendence degree of S over k is at least n.*

Proof. If A is a faithful simple right $S[x_1,...,x_n]$–module, then according to Theorem 15.6, A has finite length as an S–module. There are positive integers s,t such that $A^s \cong S^t$, whence $End_S(A)$ embeds in $M_t(S)$ (as a k–algebra). Since $k[x_1,...,x_n]$ is the center of $S[x_1,...,x_n]$, there is an induced k–algebra map

$$k[x_1,...,x_n] \to End_S(A) \to M_t(S),$$

and this map is injective because A is faithful over $S[x_1,...,x_n]$. Thus the matrix–transcendence degree of S over k is at least n.

Conversely, if this matrix–transcendence degree is at least n, then by Lemma 15.12, some matrix ring $M_t(S)$ contains a k–subalgebra isomorphic to $k(x_1,...,x_n)$. This matrix ring is isomorphic to the endomorphism ring of the right S–module S^t, thus making S^t into a $(k(x_1,...,x_n), S)$–bimodule. Since $k(x_1,...,x_n)$ is commutative, this bimodule structure can be used to make S^t into a right module over the tensor product $k(x_1,...,x_n) \otimes_k S$ (as in Exercise 1E). Because $k(x_1,...,x_n) \otimes_k S$ is a simple ring (Proposition 15.5), the module we have just constructed is necessarily faithful. Thus, using the natural embedding of $S[x_1,...,x_n]$ into $k(x_1,...,x_n) \otimes_k S$, we have made S^t into a faithful right $S[x_1,...,x_n]$–module. It clearly has finite length, and so at least one of its composition factors must be faithful. Therefore $S[x_1,...,x_n]$ is right primitive. □

Our next result is a remarkable theorem of Resco which connects the transcendence degree of an algebra with the Krull dimension of rational function rings over the algebra. It not only gives concrete information about the transcendence degree, but also leads to unexpected information about the primitivity of polynomial rings.

THEOREM 15.14. [Resco] *Let k be a field and let R be a right noetherian k–algebra which contains a subfield of transcendence degree n over k. If $x_1,\ldots,x_n$ are independent indeterminates, then*

$$r.K.dim(R \otimes_k k(x_1,\ldots,x_n)) \geq n.$$

Proof. Note that $R \otimes_k k(x_1,\ldots,x_n)$ is a right Ore localization of $R[x_1,\ldots,x_n]$, whence it is right noetherian and so does have right Krull dimension.

Now R contains a rational function field $K = k(t_1,\ldots,t_n)$. Let $L = k(x_1,\ldots,x_n)$, and identify $K \otimes_k L$ with the localization of $K[x_1,\ldots,x_n]$ with respect to the multiplicative set $X = k[x_1,\ldots,x_n] - \{0\}$. For $m = 1,\ldots,n$, let P_m be the ideal of $K[x_1,\ldots,x_n]$ generated by $\{x_i - t_i \mid i = 1,\ldots,m\}$, and note that P_m is a prime ideal of $K[x_1,\ldots,x_n]$. The map $k[x_1,\ldots,x_n] \to K$ given by evaluating each x_i at t_i is injective because the t_i are algebraically independent over k. Since this map can be factored as

$$k[x_1,\ldots,x_n] \to K[x_1,\ldots,x_n] \to K[x_1,\ldots,x_n]/P_n \to K,$$

we conclude that P_n is disjoint from X. Hence, we have a chain of prime ideals

$$0 < P_1 < P_2 < \ldots < P_n$$

in $K[x_1,\ldots,x_n]$, all of which are disjoint from X. By Theorem 9.22, there is a corresponding chain of prime ideals of length n in $K \otimes_k L$, whence

$$Cl.K.dim(K \otimes_k L) \geq n.$$

Consequently, $r.K.dim(K \otimes_k L) \geq n$ (Exercise 13F).

Since R is free as a left K–module, we see that $R \otimes_k L$ is free as a left module over $K \otimes_k L$, and hence left faithfully flat. It follows from Exercise 13U that

$$r.K.dim(R \otimes_k L) \geq r.K.dim(K \otimes_k L) \geq n,$$

which proves the theorem. □

COROLLARY 15.15. [Resco] *Let S be a simple artinian algebra over a field k. If the matrix–transcendence degree of S over k is at least n, then*

$$r.K.dim(S \otimes_k k(x_1,\ldots,x_n)) = n.$$

Proof. We first observe that $S \otimes_k k(x_1,\ldots,x_n)$ is a right Ore localization of the polynomial ring $S[x_1,\ldots,x_n]$, which has right Krull dimension n by Theorem 13.17. Hence, by Exercise 13S, $r.K.dim(S \otimes_k k(x_1,\ldots,x_n)) \leq n$. On the other hand, by Lemma 15.12 some matrix ring $M_t(S)$ contains a subfield of transcendence degree n over k, and hence the previous theorem shows that

$$\text{r.K.dim}(M_t(S) \otimes_k k(x_1,\dots,x_n)) \geq n.$$

Since $M_t(S) \otimes_k k(x_1,\dots,x_n)$ is finitely generated as a right module over $S \otimes_k k(x_1,\dots,x_n)$, we conclude from Exercise 13R that the Krull dimension of $M_t(S) \otimes_k k(x_1,\dots,x_n)$ as a right $(S \otimes_k k(x_1,\dots,x_n))$–module is at least n, and therefore

$$\text{r.K.dim}(S \otimes_k k(x_1,\dots,x_n)) \geq n. \ \square$$

EXERCISE 15M. If k is a field of characteristic zero and $m \in \mathbb{N}$, show that the center of $D_m(k)$ is k. [Hint: Exercises 5J/K.] □

THEOREM 15.16. [Resco, Amitsur–Small] *Let k be a field of characteristic zero, let $m \in \mathbb{N}$, and let $x_1,\dots,x_n$ be independent indeterminates.*

(a) The matrix–transcendence degree of $D_m(k)$ over k is exactly m.

(b) $\text{r.K.dim}(D_m(k) \otimes_k k(x_1,\dots,x_n)) = \min\{m,n\}$.

(c) $D_m(k)[x_1,\dots,x_n]$ is right primitive if and only if $n \leq m$.

Proof. By Exercise 15M, the center of $D_m(k)$ is k.

(b) Since $D_m(k)$ obviously contains a subfield of transcendence degree m over k, Corollary 15.15 shows that $\text{r.K.dim}(D_m(k) \otimes_k k(x_1,\dots,x_n)) = n$ when $n \leq m$.

Now suppose that $n > m$, and put $K = k(x_1,\dots,x_m)$. Then

$$\text{r.K.dim}(D_m(k) \otimes_k K) = m.$$

Since $k(x_1,\dots,x_n)$ is a free K–module, $D_m(k) \otimes_k k(x_1,\dots,x_n)$ is free as a left module over $D_m(k) \otimes_k K$, and hence left faithfully flat. It follows from Exercise 13U that

$$\text{r.K.dim}(D_m(k) \otimes_k k(x_1,\dots,x_n)) \geq \text{r.K.dim}(D_m(k) \otimes_k K) \geq m.$$

On the other hand, since $D_m(k)$ is a right Ore localization of $A_m(k)$, we see that $D_m(k) \otimes_k k(x_1,\dots,x_n)$ is a right Ore localization of $A_m(k) \otimes_k k(x_1,\dots,x_n)$, which is isomorphic to $A_m(k(x_1,\dots,x_n))$. We know from Theorem 13.18 that

$$\text{r.K.dim}(A_m(k(x_1,\dots,x_n))) = m,$$

and thus $\text{r.K.dim}(D_m(k) \otimes_k k(x_1,\dots,x_n)) \leq m$, by Exercise 13S.

(a) and (c) now follow from Corollary 15.15 and Theorem 15.13. □

COROLLARY 15.17. [Gelfand–Kirillov] Let k,k' be fields of characteristic zero, and let m,n be positive integers. Then $D_m(k) \cong D_n(k')$ if and only if $k \cong k'$ and $m = n$. □

EXERCISE 15N. Use Theorem 15.16 to give an example of a noetherian ring R such that the polynomial ring $R[x]$ is right primitive but R is not a G–ring. (Compare Theorem 15.10.) □

EXERCISE 15O. This exercise is designed to show that a division ring can be transcendental over its center and yet have maximal subfields which are algebraic. Let p be a prime integer, K the algebraic closure of the field $k = \mathbb{Z}/p\mathbb{Z}$, and α the Frobenius automorphism of K, given by the rule $\alpha(x) = x^p$. Let D be the Ore quotient ring of $K[\theta;\alpha]$.

(a) Show that the center of D is k. [Hint: Compare Exercise 5J.]

(b) Show that K (which is algebraic over k) is a maximal subfield of D. Observe on the other hand that D is transcendental over k (e.g., θ is transcendental). □

EXERCISE 15P. Let $L \supset k$ be a finitely generated field extension of finite transcendence degree t, let $\delta_1,...,\delta_m$ be commuting k–linear derivations on L, and let D be the Ore quotient ring of the domain $T = L[\theta_1,...,\theta_m; \delta_1,...,\delta_m]$. Show that the matrix–transcendence degree of D over k is at most $t + m$. Derive the same conclusion in case $T = L[\theta_1,...,\theta_m; \alpha_1,...,\alpha_m]$ where $\alpha_1,...,\alpha_m$ are commuting k–linear automorphisms of L. □

♦ FINITE GENERATION OF SUBFIELDS

In this final section we develop another direction in which "external" information about a division ring D can be used to yield "internal" information about subfields of D. We study the question of whether subfields of D containing the center k need to be finitely generated as field extensions of k. (A field L containing k is a finitely generated field extension if and only if there exists a finite subset $E \subseteq L$ such that no proper sub*field* of L contains $k \cup E$. Note that this occurs if and only if L is finite–dimensional over a purely transcendental subfield of finite transcendence degree over k.) We begin with another characterization of finitely generated extension fields.

THEOREM 15.18. [Vámos] *For fields $L \supseteq k$, the following conditions are equivalent:*

(a) L is a finitely generated extension field of k.

(b) L satisfies the ACC on subfields containing k.

(c) $L \otimes_k L$ is a noetherian ring.

Proof. (a) ⇒ (c): As L is finitely generated over k, it is finite–dimensional over a rational function field $k(x_1,...,x_n)$. Since $L \otimes_k k(x_1,...,x_n)$ is a localization of the polynomial ring $L[x_1,...,x_n]$, it must be noetherian. Moreover, $L \otimes_k L$ is finitely generated as a module over $L \otimes_k k(x_1,...,x_n)$, and therefore $L \otimes_k L$ is noetherian.

(c) ⇒ (b): If not, there is a strictly ascending chain $L_1 \subset L_2 \subset ...$ of fields between k and L. For $n = 1,2,...$, let I_n be the kernel of the natural ring homomorphism from

$L \otimes_k L$ to $L \otimes_{L_n} L$; then $I_1 \leq I_2 \leq \ldots$ is an ascending chain of ideals in $L \otimes_k L$. If $\alpha \in L_{n+1} - L_n$, then the element $\alpha \otimes 1 - 1 \otimes \alpha$ in $L \otimes_k L$ lies in I_{n+1} but not in I_n (since 1 and α are linearly independent over L_n). Hence, $I_1 < I_2 < \ldots$ is a strictly ascending chain of ideals in $L \otimes_k L$, contradicting the noetherian assumption.

(b) $\Rightarrow$ (a): This is clear. □

EXERCISE 15Q. Let $R \subseteq S$ be rings such that S is left faithfully flat over R. If S is right noetherian, show that R is right noetherian. □

PROPOSITION 15.19. *Let D be a division algebra over a field k, and let L be a subfield of D containing k. Then $D \otimes_k L$ is right noetherian if and only if L is a finitely generated extension field of k.*

Proof. If L is finitely generated over k, we may proceed as in the proof of the implication (a) $\Rightarrow$ (c) in the previous theorem to see that $D \otimes_k L$ is right (and left) noetherian. Conversely, assume that $D \otimes_k L$ is right noetherian. Since D is a free left L-module, $D \otimes_k L$ is a free left $(L \otimes_k L)$-module, and hence left faithfully flat. Then $L \otimes_k L$ is (right) noetherian by Exercise 15Q, and thus Theorem 15.18 shows that L is finitely generated over k. □

COROLLARY 15.20. *Let D be a division algebra over a field k. If either $D \otimes_k D$ or $D \otimes_k D^{op}$ is a right noetherian ring, then all subfields of D containing k are finitely generated extension fields of k.*

Proof. This follows from Proposition 15.19 and the observation that if L is a subfield of D containing k, then $D \otimes_k D$ and $D \otimes_k D^{op}$ are left faithfully flat modules over $D \otimes_k L$. □

THEOREM 15.21. [Resco–Small–Wadsworth] *Let $L \supseteq k$ be fields, let $\delta_1, \ldots, \delta_n$ be commuting k-linear derivations on L, and let D be the Ore quotient ring of the domain*

$$T = L[\theta_1, \ldots, \theta_n; \delta_1, \ldots, \delta_n].$$

Then the following conditions are equivalent:

(a) $D \otimes_k D$ is right noetherian.

(b) L is a finitely generated extension field of k.

(c) All subfields of D containing k are finitely generated extension fields of k.

Proof. (a) $\Rightarrow$ (c) by Corollary 15.20, and (c) $\Rightarrow$ (b) a priori.

(b) $\Rightarrow$ (a): Observe first that the maps $1 \otimes \delta_i : D \otimes_k L \to D \otimes_k L$ are commuting derivations on $D \otimes_k L$. Next, observe that

$$D \otimes_k T \cong (D \otimes_k L)[\theta_1,...,\theta_n; 1 \otimes \delta_1,...,1 \otimes \delta_n].$$

Since $D \otimes_k L$ is right noetherian (Proposition 15.19), we know from Corollary 1.13 that $D \otimes_k T$ is right noetherian. Thus, since $D \otimes_k D$ is a right Ore localization of $D \otimes_k T$, it too must be right noetherian. □

We have omitted to mention $D \otimes_k D^{op}$ in Theorem 15.21 because it is isomorphic to $D \otimes_k D$, as follows.

EXERCISE 15R. In the situation of Theorem 15.21, show that $D^{op} \cong D$ as k–algebras. □

COROLLARY 15.22. [Resco–Small–Wadsworth] *If k is a field and $n \in \mathbb{N}$, then all subfields of $D_n(k)$ containing k are finitely generated field extensions of k.*

Proof. Write $A_n(k) = k[x_1,...,x_n][\theta_1,...,\theta_n; \partial/\partial x_1,...,\partial/\partial x_n]$. Then $D_n(k)$ is also the Ore quotient ring of the domain $k(x_1,...,x_n)[\theta_1,...,\theta_n; \partial/\partial x_1,...,\partial/\partial x_n]$, and Theorem 15.21 applies. □

EXERCISE 15S. Let k be a field of characteristic zero and L a subfield of $D_1(k)$ that properly contains k. Show that any subfield of $D_1(k)$ that contains L must be finite–dimensional over L. □

♦ NOTES

Boundedness of Polynomial Rings over Division Rings. An argument given by Jacobson in [1956, p. 241] shows that if D is a division ring matrix–algebraic over its center, then every polynomial in $D[x]$ is a factor of a central polynomial (which shows that $D[x]$ is right and left bounded).

Hilbert Nullstellensatz. In the original version, Hilbert proved that if $f_1, f_2, ...$ and $F_1, F_2, ...$ are homogeneous polynomials in n variables over an algebraically closed field k, and if the F_j vanish at all points of k^n where the f_i vanish, then there exists $r \in \mathbb{N}$ such that every product of r of the F_j lies in the homogeneous ideal generated by the f_i [1893, § 3, pp. 320, 321]. The inhomogeneous analog of this amounts to saying that if J is the intersection of the maximal ideals containing an ideal I in $k[x_1,...,x_n]$, then some power of J is contained in I; consequently, if I is semiprime it must be an intersection of maximal ideals.

Amitsur–Small Nullstellensatz. This was proved for a polynomial ring over a division ring in [1978, Theorem 1].

Fully Bounded G–Rings. Theorem 15.10 and Corollary 15.11 were proved by Resco, Stafford and Warfield for a ring R in which all prime factors are bounded and Goldie on both sides [1986, Theorem 4 and Corollary 4.1].

Primitivity of Polynomial Rings over Division Rings. Resco proved that if D is a division ring of transcendence degree at least n over its center, then $D[x_1,...,x_n]$ is primitive [1979, Theorem 3.13], and then Amitsur and Small proved that $D[x_1,...,x_n]$ is primitive if and only if the matrix–transcendence degree of D is at least n [1978, Theorem 2].

Krull Dimension of Tensor Products with Rational Function Fields. Theorem 15.14 and Corollary 15.15 are due to Resco [1979, Theorems 3.15, 3.16, Remark 3.17].

Transcendence Degree of Weyl Division Algebras. Parts (a) and (b) of Theorem 15.16 are due to Resco [1979, Theorem 4.2], part (c) to Amitsur and Small [1978, Theorem 3].

Non–Isomorphic Weyl Division Algebras. Gelfand and Kirillov proved that if k is a field of characteristic zero and m,n,p,q are positive integers, then the quotient division rings of $A_m(k[x_1,...,x_p])$ and $A_n(k[x_1,...,x_q])$ are isomorphic as k–algebras if and only if $m = n$ and $p = q$ [1966, Théorème 2].

Finite Generation of Field Extensions. Theorem 15.18 was proved by Vámos [1978, Theorem 11].

Finite Generation of Subfields of Division Algebras. Theorem 15.21 was proved in the case of a single derivation by Resco, Small and Wadsworth, along with Corollary 15.22 [1979, Theorems 5,4].

APPENDIX. SOME TEST PROBLEMS FOR NOETHERIAN RINGS

In this appendix we briefly sketch some open questions in the theory of noetherian rings. These are not necessarily problems whose solutions would significantly advance the theory. Rather, we have concentrated on problems that seem to be good test questions, in the sense that a well–developed structure theory for a class of noetherian rings (e.g., FBN rings, or noetherian rings with the second layer condition) ought to be strong enough to answer some of these questions within that class. The questions to follow all have positive answers in the class of commutative noetherian rings, and all are unsolved (as of this writing) in the class of (two–sided) noetherian rings, but many of them have been answered negatively for one–sided noetherian rings. The reader who has not studied homological algebra may wish to skip over questions 5–8.

1. Jacobson's Conjecture. *Is the intersection of the finite powers of the Jacobson radical in a noetherian ring R equal to zero, i.e., is* $\bigcap_{n=1}^{\infty} J(R)^n = 0$ *?*

That this holds in a commutative noetherian ring R is a well–known consequence of the Krull Intersection Theorem (see e.g. Kaplansky [1970, Theorem 79] or Matsumura [1980, (11.D), Corollary 2]). The noncommutative question was posed for one–sided noetherian rings R by Jacobson in [1956, p. 200]; he had earlier introduced transfinite powers of J(R) (the intersection of the finite powers being $J(R)^{(\omega)}$) and had shown that some transfinite power of J(R) must be zero [1945a, Theorem 11]. Counterexamples to the one–sided question were presented by Herstein [1965] and Jategaonkar [1968, Example 1], and Jategaonkar constructed counterexamples showing that arbitrarily high transfinite powers of J(R) are needed [1969, Theorem 4.6].

The question in the two–sided noetherian case has been answered positively for FBN rings (see Theorem 8.12) by Cauchon [1974, Théorème 5; 1976, Théorème I 2, p. 36] and Jategaonkar [1973, Theorem 8; 1974b, Theorem 3.7] (see also Schelter [1975, Corollary]); for noetherian rings of Krull dimension one by Lenagan [1977, Theorem 4.4]; and for noetherian rings satisfying the second layer condition (see Theorem 12.8) by Jategaonkar

[1979, Theorem H; 1982, Theorem 1.8].

2. The Artin–Rees Property for Jacobson Radicals. *When does the Jacobson radical of a noetherian ring R satisfy the Artin–Rees property? In particular, does this occur if either R/J(R) is artinian or R is prime?*

This of course occurs if R is commutative, since then all ideals of R have the AR–property (Theorem 11.13). In general, J(R) need not satisfy the AR–property. For a one–sided noetherian example in which R/J(R) is artinian, use the ring of Exercise 8L. A two–sided noetherian example is the ring $\begin{pmatrix} S/yS & S/yS \\ 0 & S \end{pmatrix}$ where $S = k[[x]][y]$ for some field k.

This problem is closely linked with the previous one, for if R is a (right) noetherian ring in which J(R) satisfies the (right) AR–property, then Jacobson's Conjecture holds for R. (To see this, combine the AR–property with Nakayama's Lemma.) In the reverse direction, Jategaonkar has shown that if R is a noetherian ring such that R/J(R) is artinian and all factor rings of R satisfy Jacobson's conjecture, then J(R) satisfies the AR–property [1981, Proposition 5.8]. Some other consequences of the AR–property for Jacobson radicals are discussed in Exercise 12O.

3. The Descending Chain Condition for Prime Ideals. *(a) Does a noetherian ring R satisfy the descending chain condition on prime ideals? (b) Does every prime ideal in R have finite height? (c) Does every non–minimal prime ideal in R contain a prime ideal of height one?* (The *height* of a prime ideal P is the supremum of the lengths of all finite chains $P > P_1 > \dots > P_n$ of prime ideals descending from P.)

In case R is commutative, these are standard consequences of Krull's Generalized Principal Ideal Theorem (see e.g. Atiyah–Macdonald [1969, Corollary 11.12], Kaplansky [1970, Theorem 152], or Matsumura [1980, (12.I), Theorem 18]). For one–sided noetherian rings, the general answers are negative, as shown by examples of Jategaonkar [1969, Theorem 4.6].

These questions do have obvious positive answers in case R has finite Krull dimension, or just finite classical Krull dimension. There are also positive answers for noetherian P.I. rings (see Rowen [1980, Theorem 5.2.19] or McConnell–Robson [1987, Proposition 13.7.15]).

4. Countability of Chains of Ideals or Submodules. *(a) In a noetherian ring R, are all chains (i.e., totally ordered families) of ideals countable? (b) In a finitely generated R–module, are all chains of submodules countable?*

Both properties were proved for commutative noetherian rings by Bass [1971, Theorem

1.1]. The examples of Jategaonkar mentioned above [1969, Theorem 4.6] show that they do not generally hold in one–sided noetherian rings.

5. Local Rings of Finite Global Dimension. *(a) If R is a noetherian ring of finite global dimension which is local in the sense that R/J(R) is simple artinian, is R a prime ring? (b) If R is a noetherian ring of finite global dimension which is local in the stricter sense that R/J(R) is a division ring, is R a domain?*

That (b) holds in the commutative case is part of the Auslander–Buchsbaum–Nagata–Serre Theorem that every commutative noetherian regular local ring is a unique factorization domain (see e.g. Kaplansky [1970, Theorem 184], Matsumura [1980, (19.B), Theorem 48], or Rotman [1979, Theorem 9.64]).

Under one–sided assumptions as in (a) or (b), Ramras showed that the answers are positive in case R has an artinian classical quotient ring [1974, Theorem 4 and Corollary 5], and he raised the question of whether R must have an artinian classical quotient ring [1974, p. 586]. He also provided positive answers in case R has left and right global dimension 2 [1974, Proposition 7]. (An example of Stafford in Chatters–Hajarnavis [1980, Example 10.10] shows, however, that a right noetherian local ring with right global dimension 2 need not be semiprime.) Both questions were answered positively by Walker in case J(R) has a regular normalizing set of generators, as well as question (b) in case R is nonsingular [1972, Theorems 2.7, 2.9]. Later, question (a) was answered positively in case R is either nonsingular or integral over its center, by Brown, Hajarnavis, and MacEacharn [1982, Corollary 3.3; 1983, Theorem 6.7]. Snider has verified (b) in case gl.dim(R) = 3 [1988, Theorem].

6. Krull Versus Global Dimension. *If R is a noetherian ring of finite global dimension, is $r.K.dim(R) \leq r.gl.dim(R)$?*

In case R is commutative, more is true: K.dim(R) = gl.dim(R) (see e.g. Matsumura [1980, (18.G), proof of Theorem 45] or Northcott [1962, p. 208, Theorem 24 and Corollary] for the local case; the general result follows via localization). The inequality is not in general valid for one–sided noetherian rings, as seen by looking at Jategaonkar's examples [1969, Theorem 4.6] once again: these can be principal left ideal domains, in which case they have left global dimension 1, yet they have arbitrarily high left Krull dimension.

Many cases are known in which the inequality holds. In case R is semiprime and $r.gl.dim(R) \leq 1$, it follows from a result of Webber [1970, Theorem 4], and with a modification of his argument the semiprime hypothesis may be dropped (see Chatters–Hajarnavis [1980, Theorem 8.21]). Roos has proved the inequality in case R has a nonnegative filtration such that the associated graded ring is a commutative noetherian ring with finite global dimension. (This result was not published for many years, but finally appeared in a paper of Björk [1985, §§ 1.4, 1.8].) In a sequence of results, the inequality was verified for semiprime noetherian P.I. rings by Resco, Small, and Stafford [1982, Theorem 3.2]; for FBN rings which are algebras over uncountable fields by Brown and Warfield [1984, Corollary 12]; and for arbitrary noetherian P.I. rings by Goodearl and Small [1984, Theorem D]. For an FBN ring R which is an algebra over an uncountable field, Brown has shown that r.K.dim(R) is actually bounded by the injective dimension of the module R_R (if it's finite) [P, Theorem B].

7. Global Dimension via Simple Modules. *Is the (right) global dimension of a noetherian ring R equal to the supremum of the projective dimensions of the simple (right) R–modules?*

In the commutative case, this follows from the fact that for any maximal ideal M of R, the global dimension of R_M equals the projective dimension of R/M (see Matsumura [1980, (18.B), Lemma 5 and Theorem 41], Northcott [1962, p. 195, Theorem 19], or Rotman [1979, Theorem 9.52 and Corollary 9.55]). A counterexample in the one–sided noetherian case was constructed by Fields [1970, p. 348].

The equality is known for rings which are finitely generated modules over their noetherian centers (see Bass [1968, Proposition III.6.7(a)]). It was established for noetherian rings of finite global dimension by Bhatwadekar [1976, Proposition 1.1] and Goodearl [1975, Theorem 16]. Thus only the case of infinite global dimension remains open, and the problem becomes: if there is a finite bound on the projective dimensions of the simple right R–modules, is the global dimension of R finite? This has been verified for prime noetherian rings of Krull dimension one by Stafford [1982b, Lemma 2.1]; for FBN rings which are algebras over uncountable fields by Warfield [1986, Corollary 14]; and then for arbitrary FBN rings as well as all noetherian rings of Krull dimension one by Rainwater [1987, Theorem 8 and Corollary 4].

8. Finite Projective Dimension for Finitely Generated Modules. *If all simple (right) modules over a noetherian ring R have finite projective dimension, do all finitely generated (right) R–modules have finite projective dimension?*

In case R is commutative this follows from a result of Bass and Murthy [1967, Lemma 4.5]. It can also be proved in case R is a module–finite algebra over a commutative noetherian ring S, the key step being found in Bass [1968, Corollary III.6.6], which reduces the problem to the case that S is local.

9. Krull Symmetry. *(a) Do the right and left Krull dimensions of a noetherian ring R coincide? (b) Do the right and left Krull dimensions of a noetherian bimodule coincide?*

The zero–dimensional case was proved by Lenagan (Corollary 7.11): a noetherian bimodule has Krull dimension zero on one side if and only if it has Krull dimension zero on the other side [1975, Proposition]. That (a) holds in case R is fully bounded follows immediately from the fact that the right and left Krull dimensions of R then equal the classical Krull dimension (see Theorem 13.13). Jategaonkar verified (b) for noetherian bimodules over FBN rings (Theorem 13.15) [1974b, Theorem 2.3; 1986, Theorem 8.2.17]. Krull symmetry for factor rings of the enveloping algebra U(g) of a finite–dimensional solvable Lie algebra g over a field k of characteristic zero has been proved by Heinicke [1981, Theorem 1], and Krull symmetry for noetherian (U(g),U(g))–bimodules has been established in case either g is algebraic, by Brown and Smith [1985, Theorem 3.3], or k is algebraically closed, by Polo [1987, Théorème 2.7].

10. Transfer Across Noetherian Bimodules. *(a) If R and S are noetherian rings and ${}_RB_S$ is a bimodule which is finitely generated and faithful on each side, what properties transfer from R (or ${}_RB$) to S (or B_S)? (b) What if, in addition, R and S are prime and B is torsionfree on each side?*

A number of transferable properties are known in various cases. That ${}_RB$ is artinian if and only if B_S is artinian (Corollary 7.11) was proved by Lenagan [1975, Proposition]. In case R and S are fully bounded, Jatagaonkar proved that they must have the same Krull dimension (Theorem 13.15) [1974b, Theorem 2.3; 1986, Theorem 8.2.17], while if R and S satisfy the second layer condition, he showed that they must have the same classical Krull dimension (Corollary 12.5) [1982, Theorems 1.5, 1.7; 1986, Theorem 8.2.8]. Under assumption (b), Jategaonkar proved that R is semiprimitive if and only if S is semiprimitive [1981, Theorem 6.1; 1986, Theorem 5.2.15], while Letzter proved that if R is right primitive then so is S [Pa, Lemma 1.3] (see Theorem 7.16 for proofs of both results using common methods). For some other transferable properties, see Jategaonkar [1986, Section 5.2].

11. Incomparability in Cliques. *(a) Are distinct prime ideals in the same clique of a noetherian ring R always incomparable? (b) Are distinct linked prime ideals in R always incomparable?*

Both statements of course hold if R is commutative, since then all cliques are singletons (Exercise 11D). They also hold in case R satisfies the second layer condition (Corollary 12.6), which was proved by Jategaonkar [1982, Theorem 1.8; 1986, Theorem 8.2.4]. A positive answer to question 9(b) would imply positive answers to these questions, since from the bimodule Krull symmetry it would follow that K.dim(R/P) = K.dim(R/Q) for prime ideals P and Q in the same clique.

12. Localizability of Cliques. *Under what conditions is a clique X of prime ideals in a noetherian ring R (classically) localizable?*

It is an open question whether localizable cliques are necessarily classically localizable. The main characterization of classical localizability is due to Jategaonkar: X is classically localizable if and only if X satisfies the second layer condition and the intersection condition [1986, Theorem 7.2.2]. Hence, the key problem is the intersection condition. In case R is an algebra over an uncountable field and X satisfies the generic regularity condition, the intersection condition was proved by Stafford [1987, Lemma 4.4] and Warfield [1986, Theorem 8]. In particular, if R is an algebra over an uncountable field, X satisfies the second layer condition, and either R is fully bounded or there is a finite bound on the ranks of R/P for $P \in X$, then X is classically localizable (see Jategaonkar [1986, Theorem 7.2.15] and Stafford [1987, Proposition 4.5]). Müller has proved that if R is a P.I. ring and a finitely generated algebra over a field (not necessarily uncountable), then all cliques in R are classically localizable [1985, Theorem 10].

13. Prime Middle Annihilators. *(a) In a noetherian ring R, are there only finitely many prime middle annihilator ideals? (b) Are there only finitely many maximal (proper) middle annihilator ideals?* (A *middle annihilator ideal* in a ring R is an ideal of the form $\{r \in R \mid ArB = 0\}$ where A and B are ideals of R.)

Positive answers have been obtained for (b) by Krause in case R satisfies certain Krull symmetry and primary decomposition conditions (which hold for FBN rings) [1980, Theorem 6], and for (a) by Small and Stafford in case R is fully bounded [1982, p. 417]; these cases were then subsumed by Dean's result that any ring embeddable in a left artinian ring has only finitely many prime middle annihilators [1988, Theorem]. In addition, Small and Stafford answered these questions positively in case $R \otimes_k R^{op}$ (where k is the center of R) has ACC on ideals [1982, Theorem 6.5], and Goldie and Krause gave positive answers in case the lattice of left annihilator ideals in R is modular [1987, Corollary 1.8].

14. Nilpotence Modulo One–Sided Ideals. *If I and J are right ideals in a noetherian ring R, and if each element of I has a power which lies in J, is some power of I contained in J?*

In case J is a two–sided ideal, the answer is positive by Levitzki's Theorem (see Theorem 5.18 and Exercise 5L). The one–sided property was conjectured by Herstein, who proved it in case either R is a P.I. ring or there is a fixed positive integer n such that J contains the n–th. power of each element of I [1966, Theorems 1,2]. He also proved it in case R is either right or left artinian (not necessarily noetherian on the other side) [1986, Theorems 1.1, 1.5]. Some further cases were verified by Dean and Stafford [unpublished], namely when $RI = R$, or R is right fully bounded, or $(I + J)/J$ is artinian.

15. Extension of a Base Field. *If R is a finitely generated noetherian algebra over a field k, is $R \otimes_k K$ noetherian for every extension field K of k?*

This is easily checked when K is a finitely generated extension field. In case R is a P.I. algebra, the question has been answered positively by Small [1980, Proposition 53]. Small's result is actually more general: if R is a finitely generated noetherian P.I. algebra and K is any noetherian algebra, then $R \otimes_k K$ is noetherian.

16. Tensor Products of Noetherian Algebras. *If R and S are finitely generated noetherian algebras over a field k, is $R \otimes_k S$ noetherian?*

This has been proved in case either R or S is P.I. by Small [1980, Proposition 53].

17. Classical Krull Dimension of Polynomial Rings. *If R is a nonzero noetherian ring and x an indeterminate, is $\text{Cl.K.dim}(R[x]) = \text{Cl.K.dim}(R) + 1$?*

This is well–known for commutative noetherian rings of finite Krull dimension (see e.g. Matsumura [1980, (14.A), Theorem 22]). In the noncommutative case, certainly

$$\text{Cl.K.dim}(R[x]) \geq \text{Cl.K.dim}(R) + 1$$

(use Lemma 12.2). If R is fully bounded, the reverse inequality also holds, since then $\text{Cl.K.dim}(R) = \text{r.K.dim}(R)$ (Theorem 13.13), while

$$\text{Cl.K.dim}(R[x]) \leq \text{r.K.dim}(R[x]) = \text{r.K.dim}(R) + 1$$

in any case (Theorem 13.17).

18. Symmetry of Primitivity. *Is every right primitive noetherian ring also left primitive?*

This fails for non–noetherian rings, as shown by an example of Bergman [1964].

19. Generating Right Ideals in Simple Noetherian Rings. *Can every right ideal in a simple noetherian ring be generated by two elements?*

Stafford has proved this for the Weyl algebras $A_n(k)$ over any field k of characteristic zero [1978, Corollary 3.2]. Recall also that in a simple noetherian ring of right Krull dimension n, every right ideal can be generated by $n + 1$ elements (Corollary 14.8).

20. Torsionfree Modules over Simple Noetherian Rings. *Over a simple noetherian ring R, is every finitely generated torsionfree right module isomorphic to a direct sum of right ideals of R?*

Stafford has proved this for the Weyl algebras $A_n(k)$ over any field k of characteristic zero [1978, Theorem 3.3].

BIBLIOGRAPHY

[1956] A. S. Amitsur, Algebras over infinite fields. Proc. Amer. Math. Soc. 7 (1956) 35–48.

[1978] S. A. Amitsur and L. W. Small, Polynomials over division rings. Israel J. Math. 31 (1978) 353–358.

[1927] E. Artin, Zur Theorie der hyperkomplexen Zahlen. Abhandlungen Math. Sem. Hamburg. Univ. 5 (1927) 251–260.

[1939] K. Asano, Arithmetische Idealtheorie in nichtkommutativen Ringen. Japanese J. Math. 16 (1939) 1–36.

[1949] K. Asano, Über die Quotientenbildung von Schiefringen. J. Math. Soc. Japan 1 (1949) 73–78.

[1969] M. F. Atiyah and I. G. Macdonald, Introduction to Commutative Algebra. Reading (1969) Addison–Wesley.

[1951] G. Azumaya, On maximally central algebras. Nagoya Math. J. 2 (1951) 119–150.

[1940] R. Baer, Abelian groups that are direct summands of every containing abelian group. Bull. Amer. Math. Soc. 46 (1940) 800–806.

[1962] H. Bass, Injective dimension in noetherian rings. Trans. Amer. Math. Soc. 102 (1962) 18–29.

[1968] H. Bass, Algebraic K–Theory. New York (1968) Benjamin.

[1971] H. Bass, Descending chains and the Krull ordinal of commutative noetherian rings. J. Pure Applied Algebra 1 (1971) 347–360.

[1967] H. Bass and M. P. Murthy, Grothendieck groups and Picard groups of abelian group rings. Annals of Math. 86 (1967) 16–73.

[1987] A. D. Bell, Localization and ideal theory in iterated differential operator rings. J. Algebra 106 (1987) 376–402.

[P] A. D. Bell, Notes on localization in noetherian rings. (to appear).

[1964] G. M. Bergman, A ring primitive on the right but not on the left. Proc. Amer. Math. Soc. 15 (1964) 473–475.

[1976] S. M. Bhatwadekar, On the global dimension of some filtered algebras. J. London Math. Soc. (2) 13 (1976) 239–248.

[1985] J.–E. Björk, Non–commutative noetherian rings and the use of homological algebra. J. Pure Applied Algebra 38 (1985) 111–119.

[1982] W. Borho, On the Joseph–Small additivity principle for Goldie ranks. Compositio Math. 47 (1982) 3–29.

[1981a] K. A. Brown, Module extensions over Noetherian rings. J. Algebra 69 (1981) 247–260.

[1981b] K. A. Brown, The structure of modules over polycyclic groups. Math. Proc. Cambridge Phil. Soc. 89 (1981) 257–283.

[1983] K. A. Brown, Localisation, bimodules and injective modules for enveloping algebras of solvable Lie algebras. Bull. Sci. Math. (2) 107 (1983) 225–251.

[1984] K. A. Brown, Ore sets in enveloping algebras. Compositio Math. 53 (1984) 347–367.

[1985] K. A. Brown, Ore sets in noetherian rings. In Séminaire d'Algèbre P. Dubreil et M.–P. Malliavin (1983–84), pp. 355–366, Lecture Notes in Math. No. 1146, Berlin (1985) Springer–Verlag.

[1986] K. A. Brown, Localisation at cliques in group rings. J. Pure Appl. Algebra 41 (1986) 9–16.

[P] K. A. Brown, Fully bounded noetherian rings of finite injective dimension. Quart. J. Math. Oxford.

[1982] K. A. Brown, C. R. Hajarnavis, and A. B. MacEacharn, Noetherian rings of finite global dimension. Proc. London Math. Soc. (3) 44 (1982) 349–371.

[1983] K. A. Brown, C. R. Hajarnavis, and A. B. MacEacharn, Rings of finite global dimension integral over their centres. Communic. in Algebra 11 (1983) 67–93.

[1985] K. A. Brown and S. P. Smith, Bimodules over a solvable algebraic Lie algebra. Quart. J. Math. Oxford (2) 36 (1985) 129–139.

[1984] K. A. Brown and R. B. Warfield, Jr., Krull and global dimensions of fully bounded noetherian rings. Proc. Amer. Math. Soc. 92 (1984) 169–174.

[1988] K. A. Brown and R. B. Warfield, Jr., The influence of ideal structure on representation theory. J. Algebra 116 (1988) 294–315.

[1898] E. Cartan, Les groupes bilinéaires et les systèmes de nombres complexes. Ann. Fac. Sci. Toulouse 12 (1898) 1–99.

[1973] G. Cauchon, Les T–anneaux et la condition de Gabriel. C. R. Acad. Sci. Paris, Sér. A, 277 (1973) 1153–1156.

[1974] G. Cauchon, Sur l'intersection des puissances du radical d'un T–anneau noethérien. C. R. Acad. Sci. Paris, Sér. A, 279 (1974) 91–93.

[1976] G. Cauchon, Les T–anneaux, la condition (H) de Gabriel et ses consequences. Communic. in Algebra 4 (1976) 11–50.

[1960] S. U. Chase, Direct products of modules. Trans. Amer. Math. Soc. 97 (1960) 457–473.

[1980] A. W. Chatters and C. R. Hajarnavis, Rings with Chain Conditions. Research Notes in Math. No. 44, London (1980) Pitman.

[1974] P. M. Cohn, Algebra, Vol. 1. London (1974) Wiley.

[1977] P. M. Cohn, Algebra, Vol. 2. London (1977) Wiley.

[1988] B. Cortzen and L. W. Small, Finite extensions of rings. Proc. Amer. Math. Soc. 103 (1988) 1058–1062.

[1988] C. Dean, The middle annihilator conjecture for embeddable rings. Proc. Amer. Math. Soc. 103 (1988) 46–48.

[1942] J. Dieudonné, Sur le socle d'un anneau et les anneaux simples infinis. Bull. Soc. Math. France 70 (1942) 46–75.

[1926] P. A. M. Dirac, On quantum algebra. Proc. Cambridge Phil. Soc. 23 (1926) 412–418.

[1966] J. Dixmier, Représentations irréductibles des algèbres de Lie résolubles. J. Math. Pures Appl. (9) 45 (1966) 1–66.

[1968] J. Dixmier, Sur les algèbres de Weyl. Bull. Soc. Math. France 96 (1968) 209–242.

[1953] B. Eckmann and A. Schopf, Über injektive Moduln. Arch. der Math. 4 (1953) 75–78.

[1960] V. P. Elizarov, Rings of quotients for associative rings (Russian). Izvestija Akad. Nauk SSSR Ser. Mat. 24 (1960) 153–170. [Translation: Amer. Math. Soc. Transl. (2) 52 (1966) 151–170]

[1966] V. P. Elizarov, Modules of fractions (Russian). Sibirsk. Mat. Zh. 7 (1966) 221–226. [Translation: Siberian Math. J. 7 (1966) 178–181]

[1967] V. P. Elizarov, Flat extensions of rings (Russian). Doklady Akad. Nauk SSSR 175 (1967) 759–761. [Translation: Soviet Math. Doklady 8 (1967) 905–907]

[1969] V. P. Elizarov, Quotient rings (Russian). Algebra i Logika 8 (1969) 381–424. [Translation: Algebra and Logic 8 (1969) 219–243]

[1970] K. L. Fields, On the global dimension of residue rings. Pacific J. Math. 32 (1970) 345–349.

[1964] O. Forster, Über die Anzahl der Erzeugenden eines Ideals in einem Noetherschen Ring. Math. Zeitschrift 84 (1964) 80–87.

[1962] P. Gabriel, Des catégories abéliennes. Bull. Soc. Math. France 90 (1962) 323–448.

[1971] P. Gabriel, Représentations des algèbres de Lie résolubles (d'après J. Dixmier). In Séminaire Bourbaki 1968/69, pp. 1–22, Lecture Notes in Math., No. 179, Berlin (1971) Springer–Verlag.

[1966] I. M. Gelfand and A. A. Kirillov, Sur les corps liés aux algèbres enveloppantes des algèbres de Lie. Publ. Math. I.H.E.S. 31 (1966) 5–19.

[1939] I. Gelfand and A. Kolmogoroff, On rings of continuous functions on topological spaces. Doklady Akad. Nauk SSSR (n.s.) 22 (1939) 11–15.

[1960] E. R. Gentile, On rings with one–sided field of quotients. Proc. Amer. Math. Soc. 11 (1960) 380–384.

[1975] S. M. Ginn and P. B. Moss, Finitely embedded modules over noetherian rings. Bull. Amer. Math. Soc. 81 (1975) 709–710.

[1958a] A. W. Goldie, The structure of prime rings with maximum conditions. Proc. Nat. Acad. Sci. USA 44 (1958) 584–586.

[1958b] A. W. Goldie, The structure of prime rings under ascending chain conditions. Proc. London Math. Soc. (3) 8 (1958) 589–608.

[1960] A. W. Goldie, Semi–prime rings with maximum condition. Proc. London Math. Soc. (3) 10 (1960) 201–220.

[1964a] A. W. Goldie, Rings with Maximum Condition. Yale Univ., Dept. of Math. (1964).

[1964b] A. W. Goldie, Torsion–free modules and rings. J. Algebra 1 (1964) 268–287.

[1972a] A. W. Goldie, The structure of noetherian rings. In Lectures on Rings and Modules, pp. 213–321, Lecture Notes in Math., No. 246, Berlin (1972) Springer–Verlag.

[1972b] A. W. Goldie, Properties of the idealiser. In Ring Theory (R. Gordon, Ed.), pp. 161–169, New York (1972) Academic Press.

[1984] A. W. Goldie and G. Krause, Strongly regular elements of noetherian rings. J. Algebra 91 (1984) 410–429.

[1987] A. W. Goldie and G. Krause, Associated series and regular elements of noetherian rings. J. Algebra 105 (1987) 372–388.

[1975] K. R. Goodearl, Global dimension of differential operator rings. II. Trans. Amer. Math. Soc. 209 (1975) 65–85.

[1976] K. R. Goodearl, Ring Theory: Nonsingular Rings and Modules. New York (1976) Dekker.

[1986] K. R. Goodearl, Patch–continuity of normalized ranks of modules over one–sided noetherian rings. Pacific J. Math. 122 (1986) 83–94.

[1984] K. R. Goodearl and L. W. Small, Krull versus global dimension in noetherian P.I. rings. Proc. Amer. Math. Soc. 92 (1984) 175–178.

[1981] K. R. Goodearl and R. B. Warfield, Jr., State spaces of K_0 of noetherian rings. J. Algebra 71 (1981) 322–378.

[1974] R. Gordon, Gabriel and Krull dimension. In Ring Theory, Proceedings of the Oklahoma Conference (B. R. McDonald, A. R. Magid, and K. C. Smith, Eds.), pp. 241–295, New York (1974) Dekker.

[1973] R. Gordon and J. C. Robson, Krull dimension. Memoirs Amer. Math. Soc. No. 133 (1973).

[1974] R. Gordon and J. C. Robson, The Gabriel dimension of a module. J. Algebra 29 (1974) 459–473.

[1971] R. Hart, Krull dimension and global dimension of simple Ore–extensions. Math. Zeitschrift 121 (1971) 341–345.

[1981] A. G. Heinicke, On the Krull–symmetry of enveloping algebras. J. London Math. Soc. (2) 24 (1981) 109–112.

[1965] I. N. Herstein, A counterexample in Noetherian rings. Proc. Nat. Acad. Sci. U.S.A. 54 (1965) 1036–1037.

[1966] I. N. Herstein, A theorem on left Noetherian rings. J. Math. Anal. Appl. 15 (1966) 91–96.

[1986] I. N. Herstein, A nil–nilpotent type of theorem. In Aspects of Mathematics and its Applications (J. A. Barroso, Ed.), pp. 397–400, Amsterdam (1986) North–Holland.

[1890] D. Hilbert, Ueber die Theorie der algebraischen Formen. Math. Annalen 36 (1890) 473–534.

[1893] D. Hilbert, Ueber die vollen Invariantensysteme. Math. Annalen 42 (1893) 313–373.

[1903] D. A. Hilbert, Grundlagen der Geometrie. Leipzig (1903) Teubner.

[1937] K. A. Hirsch, A note on non–commutative polynomials. J. London Math. Soc. 12 (1937) 264–266.

[1969] M. Hochster, Prime ideal structure in commutative rings. Trans. Amer. Math. Soc. 142 (1969) 43–60.

[1889] O. Hölder, Zurückführung einer beliebigen algebraischen Gleichung auf eine Kette von Gleichungen. Math. Annalen 34 (1889) 26–56.

[1938] C. Hopkins, Nil–rings with minimal condition for admissible left ideals. Duke Math. J. 4 (1938) 664–667.

[1939] C. Hopkins, Rings with minimal conditions for left ideals. Annals of Math. (2) 40 (1939) 712–730.

[1974] T. W. Hungerford, Algebra. New York (1974) Holt, Rinehart and Winston.

[1945a] N. Jacobson, The radical and semi–simplicity for arbitrary rings. Amer. J. Math. 67 (1945) 300–320.

[1945b] N. Jacobson, A topology for the set of primitive ideals in an arbitrary ring. Proc. Nat. Acad. Sci. USA 31 (1945) 333–338.

[1956] N. Jacobson, Structure of Rings. Colloquium Publications, Vol. 37, Providence (1956) Amer. Math. Soc.

[1974] N. Jacobson, Basic Algebra I. San Francisco (1974) Freeman.

[1980] N. Jacobson, Basic Algebra II. San Francisco (1980) Freeman.

[1968] A. V. Jategaonkar, Left principal ideal domains. J. Algebra 8 (1968) 148–155.

[1969] A. V. Jategaonkar, A counter–example in ring theory and homological algebra. J. Algebra 12 (1969) 418–440.

[1973] A. V. Jategaonkar, Injective modules and classical localization in Noetherian rings. Bull. Amer. Math. Soc. 79 (1973) 152–157.

[1974a] A. V. Jategaonkar, Injective modules and localization in noncommutative Noetherian rings. Trans. Amer. Math. Soc. 190 (1974) 109–123.

[1974b] A. V. Jategaonkar, Jacobson's conjecture and modules over fully bounded Noetherian rings. J. Algebra 30 (1974) 103–121.

[1979] A. V. Jategaonkar, Noetherian bimodules. In Proc. Conf. on Noetherian Rings and Rings with Polynomial Identities (Durham Univ. 1979), pp. 158–169, Univ. of Leeds (1979) Mimeo.

[1981] A. V. Jategaonkar, Noetherian bimodules, primary decomposition, and Jacobson's conjecture. J. Algebra 71 (1981) 379–400.

[1982] A. V. Jategaonkar, Solvable Lie algebras, polycyclic–by–finite groups, and bimodule Krull dimension. Communic. in Algebra 10 (1982) 19–69.

[1986] A. V. Jategaonkar, Localization in Noetherian Rings. London Math. Soc. Lecture Note Series, No. 98, Cambridge (1986) Cambridge Univ. Press.

[1951] R. E. Johnson, The extended centralizer of a ring over a module. Proc. Amer. Math. Soc. 2 (1951) 891–895.

[1957] R. E. Johnson, Structure theory of faithful rings II. Restricted rings. Trans. Amer. Math. Soc. 84 (1957) 523–544.

[1869a] C. Jordan, Théorèmes sur les équations algébriques. J. Math. Pures Appl. (2) 14 (1869) 139–146.

[1869b] C. Jordan, Commentaire sur Galois. Math. Annalen 1 (1869) 141–160.

[1978] A. Joseph and L. W. Small, An additivity principle for Goldie rank. Israel J. Math. 31 (1978) 105–114.

[1970] I. Kaplansky, Commutative Algebra. Boston (1970) Allyn and Bacon.

[1970] G. Krause, On the Krull–dimension of left noetherian left Matlis–rings. Math. Zeitschrift 118 (1970) 207–214.

[1972] G. Krause, On fully left bounded left noetherian rings. J. Algebra 23 (1972) 88–99.

[1980] G. Krause, Middle annihilators in Noetherian rings. Communic. in Algebra 8 (1980) 781–791.

[1985] G. Krause and T. H. Lenagan, Growth of Algebras and Gelfand–Kirillov Dimension. Research Notes in Math., No. 116, London (1985) Pitman.

[1928a] W. Krull, Primidealketten in allgemeinen Ringbereichen. Sitzungsberichte Heidelberg. Akad. Wissenschaft (1928) 7. Abhandl., 3–14.

[1928b] W. Krull, Zur Theorie der zweiseitigen Ideale in nichtkommutativen Bereichen. Math. Zeitschrift 28 (1928) 481–503.

[1928c] W. Krull, Zur Theorie der allgemeinen Zahlringe. Math. Annalen 99 (1928) 51–70.

[1929] W. Krull, Idealtheorie in Ringen ohne Endlichkeitsbedingung. Math. Annalen 101 (1929) 729–744.

[1973] J. Lambek and G. Michler, The torsion theory at a prime ideal of a right noetherian ring. J. Algebra 25 (1973) 364–389.

[1974] J. Lambek and G. Michler, Localization of right noetherian rings at semiprime ideals. Canadian J. Math. 26 (1974) 1069–1085.

[1975] T. H. Lenagan, Artinian ideals in noetherian rings. Proc. Amer. Math. Soc. 51 (1975) 499–500.

[1977] T. H. Lenagan, Noetherian rings with Krull dimension one. J. London Math. Soc. (2) 15 (1977) 41–47.

[1959a] L. Lesieur and R. Croisot, Structure des anneaux premiers noethériens à gauche. C. R. Acad. Sci. Paris 248 (1959) 2545–2547.

[1959b] L. Lesieur and R Croisot, Sur les anneaux premiers noethériens à gauche. Ann. Sci. Ecole Norm. Sup. (Paris) (3) 76 (1959) 161–183.

[Pa] E. S. Letzter, Primitive ideals in finite extensions of noetherian rings. J. London Math. Soc.

[Pb] E. S. Letzter, Prime ideals in finite extensions of noetherian rings. J. Algebra

[1931] J. Levitzki, Über nilpotente Unterringe. Math. Annalen 105 (1931) 620–627.

[1939] J. Levitzki, On rings which satisfy the minimum condition for the right–hand ideals. Compositio Math. 7 (1939) 214–222.

[1945] J. Levitzki, Solution of a problem of G. Koethe. Amer. J. Math. 67 (1945) 437–442.

[1951] J. Levitzki, Prime ideals and the lower radical. Amer. J. Math. 73 (1951) 25–29.

[1963] L. Levy, Torsion–free and divisible modules over non–integral domains. Canad. J. Math. 15 (1963) 132–151.

[1893] S. Lie, Theorie der Transformationsgruppen, Band 3. Leipzig (1893) Teubner.

[1933] D. E. Littlewood, On the classification of algebras. Proc. London Math. Soc. (2) 35 (1933) 200–240.

[1981] M. Lorenz, Completely prime ideals in Ore extensions. Communic. in Algebra 9 (1981) 1227–1232.

[1972] A. T. Ludgate, A note on non–commutative noetherian rings. J. London Math. Soc. (2) 5 (1972) 406–408.

[1958] E. Matlis, Injective modules over noetherian rings. Pacific J. Math. 8 (1958) 511–528.

[1980] H. Matsumura, Commutative Algebra, Second Ed. Reading (1980) Benjamin / Cummings.

[1987] J. C. McConnell and J. C. Robson, Noncommutative Noetherian Rings. New York (1987) Wiley–Interscience.

[1949] N. H. McCoy, Prime ideals in general rings. Amer. J. Math. 71 (1949) 823–833.

[1969] A. C. Mewborn and C. N. Winton, Orders in self–injective semi–perfect rings. J. Algebra 13 (1969) 5–9.

[1893] T. Molien, Ueber Systeme höherer complexer Zahlen. Math. Annalen 41 (1893) 83–156.

[1974a] B. J. Müller, Localization in Non Commutative Rings. Monografias Inst. Mat., No. 1, Univ. Nacional Autónoma de México (1974).
[1974b] B. J. Müller, Localization of Non–Commutative Noetherian Rings at Semiprime Ideals. Algebra–Berichte, Seminar F. Kasch und B. Pareigis, München (1974) Verlag Uni–Druck.
[1976a] B. J. Müller, Localization in non–commutative Noetherian rings. Canad. J. Math. 28 (1976) 600–610.
[1976b] B. J. Müller, Localization in fully bounded Noetherian rings. Pacific J. Math. 67 (1976) 233–245.
[1979] B. J. Müller, Twosided localization in Noetherian PI–rings. In Ring Theory Antwerp 1978 (F. Van Oystaeyen, Ed.), pp. 169–190, New York (1979) Dekker.
[1980] B. J. Müller, Two–sided localization in Noetherian PI–rings. J. Algebra 63 (1980) 359–373.
[1985] B. J. Müller, Affine Noetherian PI–rings have enough clans. J. Algebra 97 (1985) 116–129.
[1950] M. Nagata, On the structure of complete local rings. Nagoya Math. J. 1 (1950) 63–70.
[1951] M. Nagata, On the theory of radicals in a ring. J. Math. Soc. Japan 3 (1951) 330–344.
[1962] M. Nagata, Local Rings. New York (1962) Interscience.
[1951] T. Nakayama, A remark on finitely generated modules. Nagoya Math. J. 3 (1951) 139–140.
[1921] E. Noether, Idealtheorie in Ringbereichen. Math. Annalen 83 (1921) 24–66.
[1923] E. Noether, Eliminationstheorie und allgemeine Idealtheorie. Math. Annalen 90 (1923) 229–261.
[1929] E. Noether, Hyperkomplexe Grössen und Darstellungstheorie. Math. Zeitschrift 30 (1929) 641–692.
[1983] E. Noether, Gesammelte Abhandlungen / Collected Papers, (N. Jacobson, Ed.). Berlin (1983) Springer–Verlag.
[1920] E. Noether and W. Schmeidler, Moduln in nichtkommutativen Bereichen, insbesondere aus Differential– und Differenzenausdrücken. Math. Zeitschrift 8 (1920) 1–35.
[1962] D. G. Northcott, An Introduction to Homological Algebra. Cambridge (1962) Cambridge Univ. Press.
[1967] Y. Nouazé and P. Gabriel, Idéaux premiers de l'algèbre enveloppante d'une algèbre de Lie nilpotente. J. Algebra 6 (1967) 77–99.
[1931] O. Ore, Linear equations in non–commutative fields. Annals of Math. 32 (1931) 463–477.
[1932a] O. Ore, Formale Theorie der linearen Differentialgleichungen (Erster Teil). J. reine angew. Math. 167 (1932) 221–234.
[1932b] O. Ore, Formale Theorie der linearen Differentialgleichungen (Zweiter Teil). J. reine angew. Math. 168 (1932) 233–252.
[1933] O. Ore, Theory of non–commutative polynomials. Annals of Math. 34 (1933) 480–508.
[1959] Z. Papp, On algebraically closed modules. Publ. Math. Debrecen 6 (1959) 311–327.
[1977] D. S. Passman, The Algebraic Structure of Group Rings. New York (1977) Wiley–Interscience.
[1942] S. Perlis, A characterization of the radical of an algebra. Bull. Amer. Math. Soc. 48 (1942) 128–132.
[1987] P. Polo, Bimodules sur une algèbre de Lie résoluble. J. Algebra 105 (1987) 271–283.
[1987] J. Rainwater, Global dimension of fully bounded noetherian rings. Communic. in Algebra 15 (1987) 2143–2156.
[1974] M. Ramras, Orders with finite global dimension. Pacific J. Math. 50 (1974) 583–587.

[1956] D. Rees, Two classical theorems of ideal theory. Proc. Cambridge Phil. Soc. 52 (1956) 155–157.

[1930] R. Remak, Über minimale invariante Untergruppen in der Theorie der endlichen Gruppen. J. reine angew. Math. 162 (1930) 1–16.

[1967] R. Rentschler and P. Gabriel, Sur la dimension des anneaux et ensembles ordonnés. C. R. Acad. Sci. Paris, Sér. A, 265 (1967) 712–715.

[1979] R. Resco, Transcendental division algebras and simple noetherian rings. Israel J. Math. 32 (1979) 236–256.

[1980] R. Resco, A dimension theorem for division rings. Israel J. Math. 35 (1980) 215–221.

[1982] R. Resco, L. W. Small, and J. T. Stafford, Krull and global dimensions of semiprime Noetherian PI–rings. Trans. Amer. Math. Soc. 274 (1982) 285–295.

[1979] R. Resco, L. W. Small, and A. R. Wadsworth, Tensor products of division rings and finite generation of subfields. Proc. Amer. Math. Soc. 77 (1979) 7–10.

[1986] R. Resco, J. T. Stafford, and R. B. Warfield, Jr., Fully bounded G–rings. Pacific J. Math. 124 (1986) 403–415.

[1967] J. C. Robson, Artinian quotient rings. Proc. London Math. Soc. (3) 17 (1967) 600–616.

[1959] A. Rosenberg and D. Zelinsky, Finiteness of the injective hull. Math. Zeitschrift 70 (1959) 372–380.

[1979] J. J. Rotman, An Introduction to Homological Algebra. New York (1979) Academic Press.

[1980] L. H. Rowen, Polynomial Identities in Ring Theory. New York (1980) Academic Press.

[1975] W. Schelter, Essential extensions and intersection theorems. Proc. Amer. Math. Soc. 53 (1975) 328–330.

[1928] O. Schreier, Über den Jordan–Hölderschen Satz. Abhandlungen Math. Sem. Hamburg. Univ. 6 (1928) 300–302.

[1968] I. Segal, Quantized differential forms. Topology 7 (1968) 147–172.

[1967] A. Seidenberg, Differential ideals in rings of finitely generated type. Amer. J. Math. 89 (1967) 22–42.

[1984] G. Sigurdsson, Differential operator rings whose prime factors have bounded Goldie dimension. Arch. der Math. 42 (1984) 348–353.

[1986] G. Sigurdsson, Links between prime ideals in differential operator rings. J. Algebra 102 (1986) 260–283.

[1966a] L. W. Small, Orders in artinian rings. J. Algebra 4 (1966) 13–41.

[1966b] L. W. Small, Correction and addendum: 'Orders in artinian rings'. J. Algebra 4 (1966) 505–507.

[1980] L. W. Small, Rings Satisfying a Polynomial Identity. Lecture Notes, Universität Essen (1980).

[1982] L. W. Small and J. T. Stafford, Regularity of zero divisors. Proc. London Math. Soc. (3) 44 (1982) 405–419.

[1988] R. L. Snider, Noncommutative regular local rings of dimension 3. Proc. Amer. Math. Soc. 104 (1988) 49-50.

[1976] J. T. Stafford, Completely faithful modules and ideals of simple Noetherian rings. Bull. London Math. Soc. 8 (1976) 168–173.

[1978] J. T. Stafford, Module structure of Weyl algebras. J. London Math. Soc. (2) 18 (1978) 429–442.

[1979] J. T. Stafford, On the regular elements of Noetherian rings. In Ring Theory, Proceedings of the 1978 Antwerp Conference, (F. Van Oystaeyen, Ed.), pp. 257–277, New York (1979) Dekker.

[1981] J. T. Stafford, Generating modules efficiently: algebraic K–theory for noncommutative noetherian rings. J. Algebra 69 (1981) 312–346. [Corrigendum: J. Algebra 82 (1983) 294–296]

[1982a] J. T. Stafford, Noetherian full quotient rings. Proc. London Math. Soc. (3) 44 (1982) 385–404.

[1982b] J. T. Stafford, Homological properties of the enveloping algebra $U(Sl_2)$. Math. Proc. Cambridge Phil. Soc. 91 (1982) 29–37.

[1985] J. T. Stafford, On the ideals of a noetherian ring. Trans. Amer. Math. Soc. 289 (1985) 381–392.

[1987] J. T. Stafford, The Goldie rank of a module. In Noetherian Rings and their Applications (L. W. Small, Ed.), pp. 1–20, Math. Surveys and Monographs, No. 24, Providence (1987) Amer. Math. Soc.

[1937] M. H. Stone, Applications of the theory of Boolean rings to general topology. Trans. Amer. Math. Soc. 41 (1937) 375–481.

[1967] R. G. Swan, The number of generators of a module. Math. Zeitschrift 102 (1967) 318–322.

[1963] T. D. Talintyre, Quotient rings of rings with maximum condition for right ideals. J. London Math. Soc. 38 (1963) 439–450.

[1966] T. D. Talintyre, Quotient rings with minimum condition on right ideals. J. London Math. Soc. 41 (1966) 141–144.

[1978] P. Vámos, On the minimal prime ideals of a tensor product of two fields. Math. Proc. Cambridge Phil. Soc. 84 (1978) 25–35.

[1972] R. G. Walker, Local rings and normalizing sets of elements. Proc. London Math. Soc. (3) 24 (1972) 27–45.

[1979a] R. B. Warfield, Jr., Bezout rings and serial rings. Communic. in Algebra 7 (1979) 533–545.

[1979b] R. B. Warfield, Jr., Modules over fully bounded Noetherian rings. In Ring Theory Waterloo 1978 (D. Handelman and J. Lawrence, Eds.), pp. 339–352, Lecture Notes in Math. No. 734, Berlin (1979) Springer–Verlag.

[1980] R. B. Warfield, Jr., The number of generators of a module over a fully bounded ring. J. Algebra 66 (1980) 425–447.

[1983] R. B. Warfield, Jr., Prime ideals in ring extensions. J. London Math. Soc. (2) 28 (1983) 453–460.

[1986] R. B. Warfield, Jr., Noncommutative localized rings. In Seminaire d'Algèbre P. Dubreil et M.–P. Malliavin, pp. 178–200, Lecture Notes in Math. No. 1220, Berlin (1986) Springer–Verlag.

[1970] D. B. Webber, Ideals and modules of simple Noetherian hereditary rings. J. Algebra 16 (1970) 239–242.

[1908] J. H. M. Wedderburn, On hypercomplex numbers. Proc. London Math. Soc. (2) 6 (1908) 77–118.

[1928] H. Weyl, Gruppentheorie und Quantenmechanik. Leipzig (1928) S. Hirzel.

[1944] O. Zariski, The compactness of the Riemann manifold of an abstract field of algebraic functions. Bull. Amer. Math. Soc. 50 (1944) 683–691.

[1934] H. Zassenhaus, Zum Satz von Jordan–Hölder–Schreier. Abhandlungen Math. Sem. Hamburg. Univ. 10 (1934) 106–108.

INDEX